FRESHWATER BIOMONITORING AND BENTHIC MACROINVERTEBRATES

FRESHWATER BIOMONITORING AND BENTHIC MACROINVERTEBRATES

Edited by

DAVID M. ROSENBERG

AND

VINCENT H. RESH

Chapman & Hall
New York London

First published in 1993 by
Chapman and Hall
an imprint of
Routledge, Chapman & Hall, Inc.
29 West 35th Street
New York, NY 10001-2291

Published in Great Britain by
Chapman and Hall
2-6 Boundary Row
London SE1 8HN

Printed in the United States of America on acid free paper.

Library of Congress Cataloging in Publication Data

Freshwater biomonitoring and benthic macroinvertebrates/edited by
David M. Rosenberg and Vincent H. Resh.
p. cm.
Includes bibliographical references and index.
ISBN 0-412-02251-6 (cloth)
1. Water quality bioassay. 2. Indicators (Biology) 3. Water
quality—Measurement. 4. Freshwater invertebrates—Ecology.
5. Environmental monitoring. I. Rosenberg, David M. II. Resh,
Vincent H.
QH96.8.B5F75 1992 92-13702
CIP

British Library Cataloguing in Publication Data also available.

Contributors

Leon A. Barmuta
Department of Zoology
University of Tasmania
GPO Box 252C
Hobart, Tasmania 7001
Australia

Ralph O. Brinkhurst
Aquatic Resources Center
P.O. Box 680818
Franklin, TN 37068-0818
USA

Arthur L. Buikema, Jr.
Department of Biology
Virginia Polytechnic Institute and
State University
Blacksburg, VA 24061
USA

John Cairns, Jr.
University Center for Environmental
and Hazardous Materials Studies
and Department of Biology
Virginia Polytechnic Institute and
State University
Blacksburg, VA 24061-0415
USA

Scott D. Cooper
Department of Biological Sciences
and Marine Sciences Institute
University of California
Santa Barbara, CA 93106
USA

Arthur Georges
Applied Ecology Research Group
University of Canberra
P.O. Box 1
Belconnen, A.C.T. 2616
Australia

John K. Jackson
Stroud Water Research Center
Philadelphia Academy of Natural
Sciences
R.D. 1, Box 512
Avondale, PA 19311
USA

Richard K. Johnson
Environmental Impact Assessment
Department
Swedish Environmental Protection
Agency
P.O. Box 7050
S-750 07, Uppsala
Sweden

K. Eric Marshall
Department of Fisheries and Oceans
Freshwater Institute Library
501 University Cresent
Winnipeg, MB R3T 2N6
Canada

Eric P. McElravy
Department of Entomological
Sciences
University of California
Berkeley, CA 94720
USA

Richard H. Norris
Water Research Centre
University of Canberra
P.O. Box 1
Belconnen, A.C.T. 2616
Australia

James R. Pratt
School of Forest Resources
The Pennsylvania State University
University Park, PA 16802
USA

Seth R. Reice
Department of Biology and
Curriculum in Ecology
University of North Carolina
Chapel Hill, NC 27599-3275
USA

Vincent H. Resh
Department of Entomological
Sciences
University of California
Berkeley, CA 94720
USA

David M. Rosenberg
Department of Fisheries and Oceans
Freshwater Institute
501 University Crescent
Winnipeg, MB R3T 2N6
Canada

J. Reese Voshell, Jr.
Department of Entomology
Virginia Polytechnic Institute and
State University
Blacksburg, VA 24061
USA

Ian R. Walker
Department of Biology
Okanagan College
1000 K.L.O. Road
Kelowna, BC V1Y 4X8
Canada

Torgny Wiederholm
Environmental Impact Assessment
Department
Swedish Environmental Protection
Agency
P.O. Box 7050
S-750 07, Uppsala
Sweden

Margaret Wohlenberg
Curriculum in Ecology
University of North Carolina
Chapel Hill, NC 27599-3275
USA

Contents

Preface

Biomonitoring may have come of age with the recent adoption, by North American and European governments, of national programs of environmental monitoring and assessment that include the use of aquatic biota. These programs will use a variety of "indicators" of environmental health; benthic macroinvertebrates are one of the most promising of them.

Benthic macroinvertebrates may be close to John Cairns' suggestion of the need for a "freeze-dried talking fish on a stick" for use in water quality monitoring. The chapters in this book deal with the many different approaches available for using benthic macroinvertebrates in biological monitoring programs.

Production of a book like this would not be possible without the cooperation of a number of colleagues who reviewed chapters in it. We thank J.D. Allan, G.T. Ankley, M.T. Barbour, R.O. Brinkhurst, J.L. Carter, S.D. Cooper, W.L. Fairchild, G.J. Griefer, H.B.N. Hynes, J.F. Klaverkamp, N.G. Kobzina, D.R. Lenat, P.R. Leavitt, D.F. Malley, J.L. Metcalfe-Smith, A. Morin, R.J. Naiman, R.H. Norris, J.A. Perry, J.R. Pratt, A.L. Sheldon, M. Stephenson, J.R. Voshell, Jr., and N.E. Williams. J.L. Carter, P.M. Grieef, D.L. Laroque, E.P. McElravy, and A.P. Wiens assisted in various ways. Our task was greatly simplified by the support of G.W. Payne, Science Editor at Chapman and Hall. Part of the book was completed during a Lady Davis Visiting Professorship to the Hebrew University of Jerusalem (V.H.R.).

We are pleased to dedicate our efforts on the volume to Trudy, Tina, Lee and Jonathan.

Enjoy!

David M. Rosenberg
Winnipeg, Manitoba

Vincent H. Resh
Berkeley, California
Rehovot, Israel

1

Introduction to Freshwater Biomonitoring and Benthic Macroinvertebrates

David M. Rosenberg and Vincent H. Resh

1.1. Setting the Stage

Since its beginnings in the early years of the twentieth century, the development of freshwater biomonitoring has been measurably slower than some of the other branches of science that emerged at about the same time. An extreme example, perhaps, is genetics; consider how that field has developed since T.H. Morgan's pioneering gene-linkage experiments using *Drosophila*. However, in recent years, biomonitoring has been marked by considerable progress. For example, the evolution of rapid assessment techniques, as a response to the time and cost constraints posed by quantitative biomonitoring approaches, has been dramatic, as has the use of multivariate statistical techniques for handling large data sets in contemporary biomonitoring programs. The application of experimental techniques in the laboratory and field, and use of paleolimnological approaches, also are rapidly developing adjuncts to biomonitoring programs.

The purpose of this book is to describe recent developments in the biomonitoring of fresh waters using benthic macroinvertebrates and to identify future directions in the science in order to provide an up-to-date source of information for those involved with water quality monitoring programs. The intended audience is managers, regulatory biologists, and researchers who are attempting to develop improved techniques for using benthic macroinvertebrates in successful biomonitoring activities.

Introductory chapters deal with historical aspects of the use of benthic macroinvertebrates in biomonitoring and access to the literature of biomonitoring. These are followed by the core of the book, a five-chapter unit dealing with population-, community-, and ecosystem-level approaches. The next group of three chapters deals with necessary, associated studies: paleolimnology, toxicity testing, and field experimentation. The book concludes with a chapter that speculates on what the future holds for freshwater biomonitoring using benthic macroinvertebrates.

Preparation of a book devoted to one group of organisms does not imply that only those organisms should be used in biomonitoring. Well-balanced monitoring programs involve physical, chemical, and biological measurements (Hynes, 1960; Hawkes, 1979; Reynoldson and Zarull, 1989). This book provides information that will enable benthic macroinvertebrates to be used to maximum effect within the context of a water quality monitoring program; other references provide details on physical and chemical support data required for such programs (e.g., Brown et al., 1970; Stainton et al., 1977; Suess, 1982a,b) and describe the use of other biological components (e.g., see James and Evison, 1979; Pridmore and Cooper, 1985; Hellawell, 1986).

This book can claim a number of progenitors. One group of publications on assessing water quality has been devoted to methods of collection and analysis of biota, including benthic macroinvertebrates (e.g., Slack et al., 1973; Weber, 1973; Hellawell, 1978; Suess, 1982c; American Public Health Association, 1985). Another group is composed of multiauthored, edited works on biomonitoring, which are usually the outcome of symposia or seminars, that include benthic macroinvertebrates as one of many topics addressed (e.g., Cairns and Dickson, 1973; Alabaster, 1977; James and Evison, 1979; Worf, 1980; Pridmore and Cooper, 1985; Richardson, 1987). Benthic macroinvertebrates are included in the texts by Phillips (1980, but this compendium is concerned mainly with estuarine and marine habitats), Hellawell (1986), and Abel (1989). Hart and Fuller (1974) have dealt specifically with the use of freshwater invertebrates in biomonitoring, but this work is now almost two decades old.

1.2. Definitions

It is important for our readers to understand the meaning of some general terms used throughout the book.

Fresh water: Refers to lentic (lakes, ponds) and lotic (rivers, streams) inland waters, as opposed to estuarine and marine habitats. Most of the book is concerned with freshwater habitats, although reference to estuaries and oceans may be made to illustrate a principle or to provide an interesting example.

Benthic macroinvertebrates: Refers to organisms that inhabit the bottom substrates (sediments, debris, logs, macrophytes, filamentous algae, etc.) of freshwater habitats, for at least part of their life cycle. *Macro*invertebrates are those retained by mesh sizes $\geq$200 to 500 μm (Slack et al., 1973; Weber, 1973; Wiederholm, 1980; Suess, 1982c) and are the main subject of the book, although the early life stages of some macroinvertebrate species are smaller than this size designation. Nectonic and surface-dwelling forms sometimes also are included.

Biomonitoring: Of the many definitions proposed, the following one best captures our intended meaning: "Biological monitoring can be defined as the systematic use of biological responses to evaluate changes in the environment with the intent to use this information in a quality control program. These changes often are due to anthropogenic sources. . . ." (Matthews, 1982, p. 129). Biomonitoring can be used for surveillance or to ensure compliance (McBride, 1985). Often, repetitive measurements are involved, and these are taken within the framework of a statistical design (McBride, 1985; Ward et al., 1986). The general terms "water quality monitoring" and "environmental surveillance" include physical and chemical as well as biological monitoring.

1.3. Uses of Benthic Macroinvertebrates for Biomonitoring

Perhaps the most common type of biomonitoring using benthic macroinvertebrates is *surveillance*. This approach includes surveys done before and after a project is completed (e.g., Wiens and Rosenberg, 1984; see also Abel, 1989) or before and after a toxicant is spilled (e.g., Sebastien et al., 1989). Surveillance also can be used to determine if water resource management techniques are working (McBride, 1985; Abel, 1989) or whether conservation measures are successful (Hellawell, 1986; Abel, 1989). The use of benthic invertebrates to predict environmental impacts prior to the start of a development is a specialized form of surveillance biomonitoring (e.g., Rosenberg and Snow, 1977). Historical biomonitoring, or long-term surveillance, ". . . can provide the evidence essential to the evaluation of apparent or emerging environmental problems" (Monitoring and Assessment Research Centre, 1985, p. viii). Historical biomonitoring can extend from several years (e.g., Novak et al., 1988) to several decades (e.g., Resh, 1976; Hall and Ide, 1987; Holopainen and Jónasson, 1989; Usseglio-Polatera and Bournaud, 1989) to millenia (e.g., Warwick, 1980).

The second major type of biomonitoring is done to ensure *compliance*—either to meet immediate statutory requirements (McBride, 1985) or to control long-term water quality (Wiederholm, 1980). Benthic macroinvertebrates can be used to test effluents and to ensure receiving water standards (Roper, 1985), or they can be used to ensure that standards are maintained during and after construction of a project.

Benthic macroinvertebrates are used to achieve the above objectives in a variety of ways, including monitoring changes in genetic composition, bioaccumulation of toxicants, toxicological testing in the laboratory and field, and measurements of changes in population numbers, community composition, or ecosystem functioning. These are some of the topics discussed in detail in the chapters that follow.

1.4. Advantages of Using Benthic Macroinvertebrates

The results of a literature survey reported by Hellawell (1986) revealed that algae and macroinvertebrates were the two groups of organisms most often recommended for use in assessing water quality. In practice, however, macroinvertebrates are by far the most commonly used group (Hawkes, 1979; Wiederholm, 1980; Suess, 1982c; Hellawell, 1986; Abel, 1989).

Benthic macroinvertebrates offer many advantages in biomonitoring, which explains their popularity. Some of these are intrinsic to the biology of the animals. First, they are ubiquitous (e.g., Lenat et al., 1980); therefore, they can be affected by environmental perturbations in many different types of aquatic systems and in habitats within those waters. Second, the large number of species involved offers a spectrum of responses to environmental stresses (Hellawell, 1986; Abel, 1989). Third, their basically sedentary nature allows effective spatial analyses of pollutant or disturbance effects (Slack et al., 1973; Hawkes, 1979; Penny, 1985; Hellawell, 1986; Abel, 1989). Fourth, they have long life cycles compared to other groups, which allows elucidation of temporal changes caused by perturbations (Gaufin, 1973; Slack et al., 1973; Weber, 1973; Lenat et al., 1980; Penny, 1985; Hellawell, 1986; Abel, 1989). As a result, benthic macroinvertebrates act as continuous monitors of the water they inhabit (Hawkes, 1979), enabling long-term analysis of both regular and intermittent discharges, variable concentrations of pollutants, single or multiple pollutants, and even synergistic or antagonistic effects (Gaufin, 1973; Hawkes, 1979; Lenat et al., 1980; Wiederholm, 1980).

Various technical developments have enabled benthic macroinvertebrates to be used advantageously in biomonitoring programs. First, qualitative sampling and sample analysis can be done using simple, inexpensive equipment (Hawkes, 1979; Wiederholm, 1980; Suess, 1982c; Penny, 1985; Hellawell, 1986). Second, the taxonomy of many groups is well-known, and keys to identification are available (Hawkes, 1979; Suess, 1982c; Hellawell, 1986; Abel, 1989). Third, many methods of data analysis, including biotic and diversity indices, have been developed and are widely used in community-level biomonitoring (Hellawell, 1986; but see below). Fourth, the responses of many common species to different types of pollution have been established (Hawkes, 1979; Suess, 1982c). Fifth, benthic macroinvertebrates are particularly well-suited to experimental approaches to biomonitoring (Rosenberg et al., 1986).

In the near future, other advantages probably will be added to the above list. Examples include the availability of biochemical and physiological measurements of stress in individual macroinvertebrates and the use of benthic macroinvertebrates as sentinel organisms, comparable to the "mussel watch" in estuarine and marine habitats (e.g., Bayne, 1989).

1.5. Difficulties of Using Benthic Macroinvertebrates

A balanced treatment requires that the difficulties of using benthic macroinvertebrates in biomonitoring also should be addressed. These include failure to indicate stress and problems of study design and analysis.

Benthic macroinvertebrates apparently do not respond to all impacts. For example, Hawkes (1979) reported only slight effects of low concentrations of a herbicide on the invertebrate fauna in a river, even though detrimental effects were indicated by angiosperms downstream of the effluent. This experience is a good example of the need for well-balanced monitoring programs (in this case, inclusion of more than one biological component) that was referred to above.

In terms of study design, quantitative sampling is difficult because the contagious distribution of benthic macroinvertebrates requires high numbers of samples to achieve desirable precision in estimating population abundance (Hawkes, 1979; Resh, 1979; Suess, 1982c; Hellawell, 1986; Abel, 1989). The resulting sample processing and identification requirements can be costly and time-consuming. However, alternative sampling designs and use of rapid assessment techniques may reduce this problem somewhat. Second, the distribution and abundance of benthic macroinvertebrates can be affected by factors other than water quality (e.g., natural conditions such as current velocity or nature of the substrate) (Hawkes, 1979; Suess, 1982c). This indicates the need for ecological knowledge of the species involved and carefully conceived experimental design in biomonitoring programs. Third, well-defined seasonal variations in abundance and distribution, especially of insects, may create sampling problems during specific periods or in specific habitats (Weber, 1973; Hawkes, 1979; Suess, 1982c), or may pose problems in comparing samples taken in different seasons. The potentially confounding effects of seasonal changes have to be accommodated in the design of biomonitoring programs; life-history knowledge of the species involved will help in this regard. Fourth, drift behavior in lotic waters can carry macroinvertebrates into areas in which they do not normally occur (Hellawell, 1977; Abel, 1989); therefore, sampling for localized effects of pollution sometimes may be misleading. Knowledge of habitat preferences and drifting behavior of certain species would be valuable in dealing with this difficulty.

In terms of analysis, two difficulties can be mentioned. First, certain groups are taxonomically difficult (e.g., larvae of Chironomidae, some Trichoptera, Oligochaeta), although progress is being made in developing adequate keys for them (Hellawell, 1986). Second, the multiplicity of biotic and diversity indices available for working with benthic macroinvertebrates may indicate that workers are not satisfied with the results that they provide (Hellawell, 1986). Experience eventually may provide a core group of measurements that will be favored by most workers, or nontraditional methods (e.g., Södergren, 1976; Fiance, 1978) may be adopted.

The chapters in this book will discuss ways of dealing with these and other challenges to the successful use of benthic macroinvertebrates in biomonitoring programs.

References

Abel, P.D. 1989. *Water Pollution Biology*. Ellis Horwood, Chichester, England.

Alabaster, J.S., ed. 1977. *Biological Monitoring of Inland Fisheries*. Applied Science Pubs., London.

American Public Health Association. 1985. *Standard Methods for the Examination of Water and Wastewater,* 16th ed. American Public Health Association, Washington, DC.

Bayne, B.L. 1989. Measuring the biological effects of pollution: the Mussel Watch approach. *Water Science and Technology* 21(10/11):1089–1100.

Brown, E., M.W. Skougstad, and M.J. Fishman. 1970. Methods for collection and analysis of water samples for dissolved minerals and gases. In *Techniques of Water-Resources Investigations of the United States Geological Survey,* Chapter A1, Book 5, pp. 1–160. U.S. Department of the Interior, Geological Survey, Washington, DC.

Cairns, J., Jr. and K.L. Dickson, eds. 1973. *Biological Methods for the Assessment of Water Quality*. American Society for Testing and Materials Special Technical Publication 528. American Society for Testing and Materials, Philadelphia, PA.

Fiance, S.B. 1978. Effects of pH on the biology and distribution of *Ephemerella funeralis* (Ephemeroptera). *Oikos* 31:332–9.

Gaufin, A.R. 1973. Use of aquatic invertebrates in the assessment of water quality. In *Biological Methods for the Assessment of Water Quality,* eds. J. Cairns, Jr. and K.L. Dickson, pp. 96–116. American Society for Testing and Materials Special Technical Publication 528. American Society for Testing and Materials, Philadelphia, PA.

Hall, R.J. and F.P. Ide. 1987. Evidence of acidification effects on stream insect communities in central Ontario between 1937 and 1985. *Canadian Journal of Fisheries and Aquatic Sciences* 44:1652–7.

Hart, C.W., Jr. and S.L.H. Fuller, eds. 1974. *Pollution Ecology of Freshwater Invertebrates*. Academic Press, New York.

Hawkes, H.A. 1979. Invertebrates as indicators of river water quality. In *Biological Indicators of Water Quality,* eds. A. James and L. Evison, Chap. 2. John Wiley, Chichester, England.

Hellawell, J. 1977. Biological surveillance and water quality monitoring. In *Biological Monitoring of Inland Fisheries,* ed. J.S. Alabaster, pp. 69–88. Applied Science Pubs., London.

Hellawell, J.M. 1978. *Biological Surveillance of Rivers: a Biological Monitoring Handbook*. Water Research Centre, Medmenham, England.

Hellawell, J.M. 1986. *Biological Indicators of Freshwater Pollution and Environmental Management*. Elsevier, London.

Holopainen, I.J. and P.M. Jónasson. 1989. Bathymetric distribution and abundance of *Pisidium* (Bivalvia: Sphaeriidae) in Lake Esrom, Denmark, from 1954 to 1988. *Oikos* 55:324–34.

Hynes, H.B.N. 1960. *The Biology of Polluted Waters*. Liverpool Univ. Press, Liverpool, England.

James, A. and L. Evison, eds. 1979. *Biological Indicators of Water Quality*. John Wiley, Chichester, England.

Lenat, D.R., L.A. Smock, and D.L. Penrose. 1980. Use of benthic macroinvertebrates as indicators of environmental quality. In *Biological Monitoring for Environmental Effects,* ed. D.L. Worf, pp. 97–112. D.C. Heath, Lexington, MA.

Matthews, R.A., A.L. Buikema, Jr., J. Cairns, Jr., and J.H. Rodgers, Jr. 1982. Biological monitoring. Part IIA. Receiving system functional methods, relationships and indices. *Water Research* 16:129–39.

McBride, G.B. 1985. The role of monitoring in the management of water resources. In *Biological Monitoring in Freshwaters: Proceedings of a Seminar, Hamilton, November 21–23, 1984. Part 1,* eds. R.D. Pridmore and A.B. Cooper, pp. 7–16. Water and Soil Miscellaneous Publication No. 82, National Water and Soil Conservation Authority, Wellington, NZ.

Monitoring and Assessment Research Centre. 1985. *Historical Monitoring*. MARC Report No. 31. Monitoring and Assessment Research Centre, London.

Novak, M.A., A.A. Reilly, and S.J. Jackling. 1988. Long-term monitoring of polychlorinated biphenyls in the Hudson River (New York) using caddisfly larvae and other macroinvertebrates. *Archives of Environmental Contamination and Toxicology* 17:699–710.

Penny, S.F. 1985. The use of macroinvertebrates in the assessment of point source pollution. In *Biological Monitoring in Freshwaters: Proceedings of a Seminar, Hamilton, November 21–23, 1984. Part 2,* eds. R.D. Pridmore and A.B. Cooper, pp. 205–15. Water and Soil Miscellaneous Publication No. 83, National Water and Soil Conservation Authority, Wellington, NZ.

Phillips, D.J.H. 1980. *Quantitative Aquatic Biological Indicators. Their Use to Monitor Trace Metal and Organochlorine Pollution*. Applied Science Pubs., London.

Pridmore, R.D. and A.B. Cooper, eds. 1985. *Biological Monitoring in Freshwaters: Proceedings of a Seminar, Hamilton, November 21–23, 1984*. Part 1. Water and Soil Miscellaneous Publication No. 82:1–188. Part 2. Water and Soil Miscellaneous Publication No. 83:190–379. National Water and Soil Conservation Authority, Wellington, NZ.

Resh, V.H. 1976. Changes in the caddis-fly fauna of Lake Erie, Ohio, and of the Rock River, Illinois over a fifty year period of environmental deterioration. In *Proceedings of the First International Symposium on Trichoptera, Lunz am See, Austria, September 16–20, 1974,* ed. H. Malicky, pp. 167–70. W. Junk Pubs., The Hague, The Netherlands.

Resh, V.H. 1979. Sampling variability and life history features: basic considerations in the design of aquatic insect studies. *Canadian Journal of Fisheries and Aquatic Sciences* 36:290–311.

Reynoldson, T.B. and M.A. Zarull. 1989. The biological assessment of contaminated sediments—the Detroit River example. *Hydrobiologia* 188/189:463–76.

Richardson, D.H.S., ed. 1987. *Biological Indicators of Pollution*. Royal Irish Academy, Dublin, Ireland.

Roper, D.S. 1985. The role of biological surveys and surveillance. In *Biological Monitoring in Freshwaters: Proceedings of a Seminar, Hamilton, November 21–23, 1984. Part 1,* eds. R.D. Pridmore and A.B. Cooper, pp. 17–20. Water and Soil Miscellaneous Publication No. 82, National Water and Soil Conservation Authority, Wellington, NZ.

Rosenberg, D.M., H.V. Danks, and D.M. Lehmkuhl. 1986. Importance of insects in environmental impact assessment. *Environmental Management* 10:773–83.

Rosenberg, D.M. and N.B. Snow. 1977. A design for environmental impact studies with special reference to sedimentation in aquatic systems of the Mackenzie and Porcupine River drainages. In *Proceedings of the Circumpolar Conference on Northern Ecology, September 15–18, 1975, Ottawa,* pp. 67–78, Section III. National Research Council of Canada, Ottawa, ON.

Sebastien, R.J., R.A. Brust, and D.M. Rosenberg. 1989. Impact of methoxychlor on selected nontarget organisms in a riffle of the Souris River, Manitoba. *Canadian Journal of Fisheries and Aquatic Sciences* 46:1047–61.

Slack, K.V., R.C. Averett, P.E. Greeson, and R.G. Lipscomb. 1973. Methods for collection and analysis of aquatic biological and microbiological samples. In *Techniques of Water-Resources Investigations of the United States Geological Survey,* Chapter 4A, Book 5, pp. 1–165. U.S. Department of the Interior, Geological Survey, Washington, DC.

Södergren, S. 1976. Ecological effects of heavy metal discharge in a salmon river. *Institute of Freshwater Research Drottningholm Report* 55:91–131.

Stainton, M.P., M.J. Capel, and F.A.J. Armstrong. 1977. *The Chemical Analysis of Fresh Water,* 2nd ed. Fisheries and Marine Service Miscellaneous Special Publication 25. Freshwater Institute, Winnipeg, MB.

Suess, M.J., ed. 1982a. *Examination of Water for Pollution Control. A Reference Handbook. Vol. 1. Sampling, Data Analysis and Laboratory Equipment*. Pergamon Press, Oxford, England.

Suess, M.J., ed. 1982b. *Examination of Water for Pollution Control. A Reference Handbook. Vol. 2. Physical, Chemical and Radiological Examination*. Pergamon Press, Oxford, England.

Suess, M.J., ed. 1982c. *Examination of Water for Pollution Control. A Reference Handbook. Vol. 3. Biological, Bacteriological and Virological Examination*. Pergamon Press, Oxford, England.

Usseglio-Polatera, P. and M. Bournaud. 1989. Trichoptera and Ephemeroptera as indicators of environmental changes of the Rhone River at Lyons over the last twenty-five years. *Regulated Rivers: Research and Management* 4:249–62.

Ward, R.C., J.C. Loftis, and G.B. McBride. 1986. The 'data-rich but information-poor' syndrome in water quality monitoring. *Environmental Management* 10: 291–7.

Warwick, W.F. 1980. Palaeolimnology of the Bay of Quinte, Lake Ontario: 2800 years of cultural influence. *Canadian Bulletin of Fisheries and Aquatic Sciences* 206:1–117.

Weber, C.I., ed. 1973. *Biological Field and Laboratory Methods for Measuring the*

Quality of Surface Waters and Effluents. EPA–670/4–73–001. U.S. Environmental Protection Agency, Cincinnati, OH.

Wiederholm, T. 1980. Use of benthos in lake monitoring. *Journal of the Water Pollution Control Federation* 52:537–47.

Wiens, A.P. and D.M. Rosenberg. 1984. Effect of impoundment and river diversion on profundal macrobenthos of Southern Indian Lake, Manitoba. *Canadian Journal of Fisheries and Aquatic Sciences* 41:638–48.

Worf, D.L., ed. 1980. *Biological Monitoring for Environmental Effects*. D.C. Heath, Lexington, MA.

2

A History of Biological Monitoring Using Benthic Macroinvertebrates

John Cairns, Jr. and James R. Pratt

2.1. Introduction

Biological monitoring in its most rudimentary form probably had its origin in the minds of fish wardens, river keepers, and minders of lakes and ponds. Anyone living near a water body has a sense of biomonitoring, although not necessarily a scientifically rigorous one. Sentinel human noses have smelled septic hydrogen sulfide at remarkably good analytical levels. Aristotle, who is credited with dabbling in nearly every known area of modern science, is known to have placed freshwater fish in salt water to observe their reactions.

We define biological monitoring as surveillance using the responses of living organisms to determine whether the environment is favorable to living material. Biological monitoring is not new, as the king's wine taster attests. In the early days of the industrial revolution, canaries were kept in underground coal mines. If a canary showed adverse reactions to conditions in the mine, the miners left. Biological monitoring also implies quality control in which some corrective action will be taken if expected conditions are not met; but the existence of a feedback mechanism that involves a response to indications of failed environmental health is problematic in most countries.

Not until the industrial revolution did the impact of human activities on water resources become clear to almost everyone. During the eighteenth century, the River Thames flowing through London produced a stench so nauseating that sheets soaked in vinegar were hung in Parliament to partly offset the noxious air wafting in from the river (Gameson and Wheeler, 1975). An uncharitable person might note that, even under these circumstances, Parliament was not quick to act to remedy the situation until human health concerns, such as typhoid fever and cholera (from the infamous Broad Street pump), surfaced as serious threats to the ebb and flow of "modern" life.

The protection of human health still drives much of the pollution control regulation and technology in the developed world. The realization that unmanaged ecosystems soon fail to provide free ecological services, such as drinking water, fish, and waste assimilation, has led to a considerably improved legislative and biological environment. However, much of what has been learned needs to be transferred to the management of ecosystems in developing nations.

The modern history of biomonitoring begins in Europe in the twentieth century at a time when human populations were sufficiently large to produce a cadre of well-trained scientists, the basic technology of science, and urban-industrial densities sufficient to affect severely a number of ecosystems. Studies of biological indicators took two paths: one identifying the species indicative of human degradation of lakes and streams, another the biological classification of lakes. During the first 20 years of the century, applied and theoretical limnology shared an unusual time of development in which each field fed the other. The development of pollution indicators (the Saprobien system) is described briefly here, while the origins of biological lake typology are described in Johnson et al. (Chapter 4).

Biological monitoring has not enjoyed the same degree of development as limnology and stream ecology. Stream ecology, especially, progressed from a descriptive spur of limnology to an experimental science focused on nutrient and energy dynamics. According to Minshall (1988), stream ecology had passed through the phase of description and simple quantification by the mid-twentieth century.

In many ways, biological monitoring still finds itself in this early phase. Because few applied studies have the freedom or resources to examine more than a few variables, community-based studies of macroinvertebrates form the basis of most biological studies of water quality for pragmatic reasons: macroinvertebrates are rather easy to collect and identify; macroinvertebrates are fish food and so are explainable to the general public; and analyses of macroinvertebrate communities allow inferences to be drawn about the food base (algae, leaves), habitat quality, and relative health of the community (many or few species).

Biomonitoring is just beginning to free itself from the grip of diversity indices (see also Norris and Georges, Chapter 7). New avenues of study are belatedly acknowledging the broad tolerances, biogeography, and regional nature of the macroinvertebrate biota. Comparable developments are occurring in both Europe and North America and provide good prospects for applying site-specific biological criteria to the management of ecosystems.

2.2. The European Tradition

The concept of biological indicators of environmental condition originated with the now-famous work of Kolkwitz and Marsson (1908, 1909) who developed the idea of saprobity (the degree of pollution) in rivers as a measure

Table 2.1. Pollution indicator organisms used in early classification schemes.*

	Classifications		
Indicators	Richardson (1929)	Richardson (1925)	Gaufin (1958)
Tubificids, some chironomid larvae, some gastropods	Pollutional	Pollutional	Tolerant
Sphaeriidae, Hirudinea, chironomid larvae	Sub-pollutional, tolerant		
Miscellaneous chironomid larvae	Sub-pollutional, doubtful		
Chironomidae, Sphaeriidae, a few Oligochaeta	Sub-pollutional, less tolerant		Facultative
Several species of gastropods, air-breathing insects	Pulmonate snails		
Two Pleurocercidae, one isopod, several Hydropsychidae species	Current-loving	Cleaner preference	
Some Crustacea, several intolerant Ephemeridae, Odonata, Trichoptera species	Cleaner water	Clean water	Intolerant

*Based on Wilhm (1975).

of the degree of contamination by organic matter (primarily sewage) and the resulting decrease in dissolved oxygen. Observations of the relative restriction in occurrence of certain taxa by environmental conditions led to the development of lists of indicator organisms (Table 2.1). The Saprobien system has been extended and revised repeatedly by European scientists since its inception (e.g., Kolkwitz, 1950; Liebmann, 1951, 1962; Fjerdingstad, 1965; Sládeček, 1965, 1973; Bick, 1971; Foissner, 1988).

The idea that certain species can be used to indicate certain types of environmental conditions is well-established. For example, trout are associated with a particular kind of habitat. Gardeners know that plants have certain preferences regarding soil, amount of sunlight, and temperature. The presence of a species indicates that the habitat is suitable and, because some of the environmental requirements are known for many species, their presence indicates something about the nature of the environment in which they are found. Thus, the concept of the presence of species indicating certain conditions is based on practical observations verifiable by almost anyone who has contact with the environment. However, determining the significance of the *absence* of a species is considerably more risky (see below).

Matched with the concept of indicator organisms was the assembling of lists of biological indicators. Rapidly reproducing populations of microorganisms, such as bacteria, fungi, Protozoa, and some algae, are among the best indicators of the continuous presence of organic wastes. Concurrent with the development of Kolkwitz and Marsson's Saprobien system, lake classification systems began to focus on oxygen content and biological components (Thienemann, 1925), especially macrobenthic invertebrates that were indicative of an oligotrophic to eutrophic/dystrophic gradient (Brinkhurst, 1974). Not surprisingly, the bottoms of eutrophic lakes had biotas reminiscent of polysaprobic rivers.

Biological zonation, based on organic pollution and dissolved oxygen concentration, is well-known on many continents; the concept is neither restricted to Europe (e.g., Forbes and Richardson, 1913) nor outdated (Sládeček, 1973). Considerable debate on the assignment of saprobic valences (a ranking of the indicator value of a species for levels of pollution) and the application of various biotic indices continues (Guhl, 1987; Foissner, 1988), and studies of impacts in European rivers and lakes regularly focus on saprobic characteristics. In contrast, physicochemical measures have predominated in the North American experience, despite the well-documented arguments of biologists that pollution assessment is primarily a biological problem (Wilhm, 1975).

Despite the acceptance of the Saprobien system in several central European countries (with caveats such as the expectation of good taxonomy; Guhl, 1987), this system has found little acceptance in North America for several reasons. First, species of organisms with limited geographical distributions were useless outside of Europe. Second, many of the most pressing North American problems concerned toxicity, rather than organic enrichment. (Interestingly, while "pollution control" in the form of municipal waste treatment has blossomed in North America in the past 20 years, some of Europe's largest cities still have little or even no sewage treatment, and they discharge wastes directly to the sea.) Third, North American benthic biologists have been skeptical of the indicator species concept, although the indicator value of certain taxa has been documented in North American streams (e.g., Lowe, 1974; Hilsenhoff, 1982). In short, the rejection of biological indicators served to perpetuate a belief in the importance of chemical measures of water quality.

Roback (1974, pp. 318–320) dismissed the indicator species concept as follows:

> The concept of indicator organisms is a beguiling one and has a long history in pollution biology. It promises a quick, easy way of making judgements. Unfortunately, there is a large gap between the promise and reality. As far as the insects are concerned, I am convinced that the concept has little validity, and that the presence or absence of any species in a stream indicates (as far as damage is concerned) no more or less than the bald fact of its presence or absence.
>
> The majority of insect species found in the average eutrophic body of water (the dominant type we have to deal with) can tolerate a broad enough range of water chemistry and physical conditions to render them useless, singly, as indicators of the degree of (or lack of) damage to any body of water.
>
> In addition to water chemistry, the presence or absence of a species can be determined by such factors as: (1) its presence or absence in the species pool available for colonizing the area studied; (2) the season in which the collection is made; (3) flow conditions at the time of the study; and (4) chance.
>
> It is probably more valid to speak of indicator assemblages of species (insects and other groups) or indicator communities, but at present our knowledge is not sufficient to define these units in the higher invertebrates.

Roback's comments serve as an adequate position statement contradicting the value of the indicator species concept and are perhaps reflective of misunderstandings about the Saprobien system, which has always been based on sampling a community. As noted by Roback, the most perplexing element of applying the indicator species concept is the problem of dealing with absent species. Species groups change continually and, sometimes, predictably. Species turnover, predicted by theories of dynamic community structure (MacArthur and Wilson, 1967; Schoener, 1983), provides an adequate explanation for the absence of some indicators at some times, especially when short-lived organisms are the targets of study. European ecologists have long advocated the use of the total biocenosis (community and associated environment) for developing assessments of ecosystem condition (Guhl, 1987). This position resulted from the realization that many of the best indicators of continuing damage were bacteria, algae, and Protozoa. Also, European workers were concerned about the subjectivity of the Saprobien system because, in many treatments, only the species judged to have acceptable indicator value (such as $G > 3$ on a scale of 5) were chosen (along with their relative abundances) in calculating the saprobity index (Pantle and Buck, 1955; Uhlmann, 1979). Contemporary workers, having assimilated new concepts of community ecology, now espouse some middle ground between saprobien-like indicator concepts and the total rejection of biological indicators.

The concept of saprobity has had a profound influence on the development

Table 2.2. Classification of Danish waters by indicator organisms. Conditions for Classes III, IV are not to be tolerated. Basic water quality limits are BOD5 = 6 mg/l, dissolved oxygen = 5 mg/l.*

Class	Saprobien System Classification	Indicator Organisms
I	Oligosaprobic	Clean water organisms: Trichoptera, Plecoptera
II	Beta-mesosaprobic	Abundant pollution-tolerant organisms; no dominance by *Chironomus* or *Tubifex*
III	Alpha-mesosaprobic	Tolerant species: *Chironomus, Tubifex, Asellus, Erpobdella*
IV	Polysaprobic	Exclusively *Eristalis, Tubifex, Chironomus*

*Based on Andersen (1977).

of laws and regulations for managing surface waters in Europe. For example, the Danish Environmental Protection Act of 1973 specified four biotic classifications based on classic Saprobien system oxygen, BOD, and biological indicators as references for biological quality (Table 2.2). Dischargers are enjoined from creating unacceptable biological conditions in waterways (Andersen, 1977). Similarly, in Great Britain, the Royal Commission on Sewage Disposal restricted effluent BOD and suspended solids as early as 1915. An excellent review of the history of water protection legislation, with special reference to the United Kingdom, is provided by Hellawell (1986).

Several indices (Table 2.3) have been developed based on general patterns of organism tolerance, distribution, and indicator value. These indices are quantitative or semi-quantitative means of evaluating biotic integrity and are philosophically close to much of the application of invertebrate biology in North America (see below and Chapter 6 by Resh and Jackson).

Early reliance on saprobity as the biological touchstone for monitoring ecosystem conditions gave way to diversity indices and then to "score systems" that combined indicator concepts of clean-water forms with elements of diversity, especially relative abundance of different indicator groups. A recent review of the development of macroinvertebrate indices in Europe is given by Metcalfe (1989). Score systems such as the Biological Monitoring Working Party (BMWP) score (Armitage et al., 1983) and the Belgian Biotic Index (BBI, De Pauw and Vanhooren, 1983) assign scores to biotic groups based on generally accepted organism sensitivities to pollution and habitat disturbance (i.e., stoneflies, caddisflies, and mayflies are given high scores based on their presence and abundance).

For the most part, score systems provide semi-quantitative estimates of ecosystem conditions. Sampling methods emphasize rapid information gen-

Table 2.3. Selected indices used in biological monitoring. Indicator species and diversity indices are continuous scores, whereas biotic indices are discrete scores. Detailed comparisons are given by Washington (1984), Hellawell (1986), Guhl (1987), and Resh and Jackson (Chapter 6).

Index	Input Data	Taxonomic Precision	Reference[1]
Indicator species			
Saprobity	Saprobity, abundance by species	High	(1)
Diversity			
Simpson diversity	Abundance by species or proportional abundance	Moderate-high	(2)
Shannon diversity	Abundance by species	Moderate-high	(3)
Brillouin	Abundance by species	Moderate-high	(4)
Margalef	Species number, total individuals	High	(5)
Sequential Comparison	Runs, total individuals, number of taxa	Low	(6)
Biotic			
BMWP[2] score BBI[3]	Macroinvertebrate family score	Low	(7,8,9)
Invertebrate Community Index	Abundance by species, expected values, categorical scores	Moderate-high	(10)
(see also Index of Biotic Integrity)			(11)

[1](1) Pantle and Buck (1955); (2) Simpson (1949); (3) Shannon and Weaver (1949); (4) Brillouin (1951); (5) Margalef (1958); (6) Cairns and Dickson (1971); (7) Woodiwiss (1964); (8) Armitage et al. (1983); (9) De Pauw and Vanhooren (1983); (10) Ohio EPA (1987); (11) Karr et al. (1986).

[2]Biological Monitoring Working Party.

[3]Belgian Biotic Index.

eration and require limited taxonomic precision; collected organisms are identified only to family. The score systems have not been universally accepted. Guhl (1987) has criticized the failure of scores to account for differences in tolerance within taxa as broad as families. Interestingly, developments in North America generally follow those in Europe, including the proliferation of acronyms.

Metcalfe (1989, pp. 136–137) summarizes one view of the application of macroinvertebrate community monitoring as follows:

> There have been suggestions that macroinvertebrate community structure is not sensitive enough to distinguish among various types and degrees of pollution. It is possible, however, that it is our method of detecting the response rather than the response itself which is insensitive. Theoretically,

> small changes in water quality should lead to alterations in the structure of a community which is already in a delicate balance.

Metcalfe predicts that changes should occur in natural communities exposed to anthropogenic influences; her conclusions contradict, to a degree, those of Roback (1974, see above). Streams are dynamic ecosystems, and it would be surprising if stream organisms responded quickly (or adversely) to *minor* changes in water quality. This does not detract in any way from the application of biological monitoring as an ecosystem management tool because the goal of biological monitoring should be to detect significant changes in ecosystems, not minor fluctuations that are quickly dampened.

Many biologists believe that most aquatic species have extremely broad tolerances (Cairns, 1982). This belief arises from the fact that sampling of healthy populations often coincides with the observation of somewhat degraded water quality during discrete sampling events. However, organisms integrate effects of altered environmental quality over a period of time. If degraded conditions are present for a period sufficient to produce population effects (mortality, altered growth, altered reproduction), then community structure will change. A simple example of this problem follows.

Many species of trout (*Salmo* spp.), adapted to cold water, cease to grow above some threshold temperature. Trout routinely may be found in water well above this threshold temperature in summer as long as a thermal refuge exists or elevated temperatures do not persist for long periods. However, if a stream is influenced by wastewater that elevates stream temperature above the threshold, the trout cease to grow and, if elevated temperatures persist, can be extirpated from the ecosystem (and perhaps replaced by bass, e.g., *Micropterus* spp.). Sampling the species under what appear to be adverse conditions is not necessarily predictive. In the first instance, only a short-term stress is operating; in the second instance, the stress is long-term and results in local extinction (or replacement). In the second instance, the long-term temperature stress may be quite small, but may result in greater consequences than a large, short-term, sublethal stress.

The point of this example is to show that organisms integrate effects over longer periods than biologists tend to sample or study. The concept of broad tolerances, true in the short term, is probably not true in the longer term. However, our knowledge of the long-term consequences of small environmental changes in temperature, nutrients, or substratum is modest, and our knowledge of the long-term effects of small changes in toxic chemicals in surface waters is considerably less. The observed consequences of changes in stream ecosystems are *not* modest. Large numbers of species can become locally extinct with uncertain consequences for ecosystem stability, stream function, and maintenance of food chains.

Evidence is accumulating that the responses of aquatic communities to stress involve the loss of some species and sometimes the replacement of species by functionally similar taxa. Biological surveillance of communities—with special emphasis on characterizing taxonomic richness and composition—is perhaps the most sensitive tool now available for quickly and accurately detecting alterations in aquatic ecosystems (Schindler, 1987; Schaeffer et al., 1988; Gray, 1989). This point is further supported by the current convergence of ideas about biological monitoring in both North America and Europe (see below). That is, although much disagreement exists among benthic biologists about indicator species and community indicators, a common consensus seems to be developing to the problem of community analysis in stressed ecosystems. Faunistic changes in streams are always very meaningful, although it is not always clear if altered water quality is the cause.

2.3. The North American Tradition

The indicator species concept was also developed in the United States by the classic studies of S.A. Forbes on the Illinois River. Forbes' limnological investigations began in the 1870s and demonstrated the indicator value of benthic fauna. However, the modern era of biological reconnaissance owes much of its strength to the work of Ruth Patrick and her co-workers who conducted a series of major river surveys in the eastern United States beginning in 1948.

Patrick's work was strongly influenced by her taxonomic studies on diatoms and her insistence on sampling several taxonomic groups. Her classic work on biological measures of stream conditions (Patrick, 1949) made direct use of the numbers and kinds of species in community indicator groups without particular regard to the identities of the taxa. Patrick was strongly influenced by the work of Preston (1948) on the canonical distribution of species and applied this concept directly to her work in streams (Fig. 2.1). While her own work dealt with diatoms colonizing both natural and artificial substrata, the stream surveys of the Academy of Natural Sciences (Philadelphia) set the tone for much of the later work on stream health in North America.

Foremost among the principles Patrick applied to investigating stream health was the developing concept of an island as applied to the colonization of natural habitats. Her work spawned not only an emergence of the study of diatoms, but also the thoughts of Roback, a collaborator and colleague, expressed above. Stream surveys produced large lists of taxa that were used as measures of biotic diversity using the now-classic upstream-downstream survey of point-source impacts.

Patrick's work influenced and was influenced by the work of MacArthur

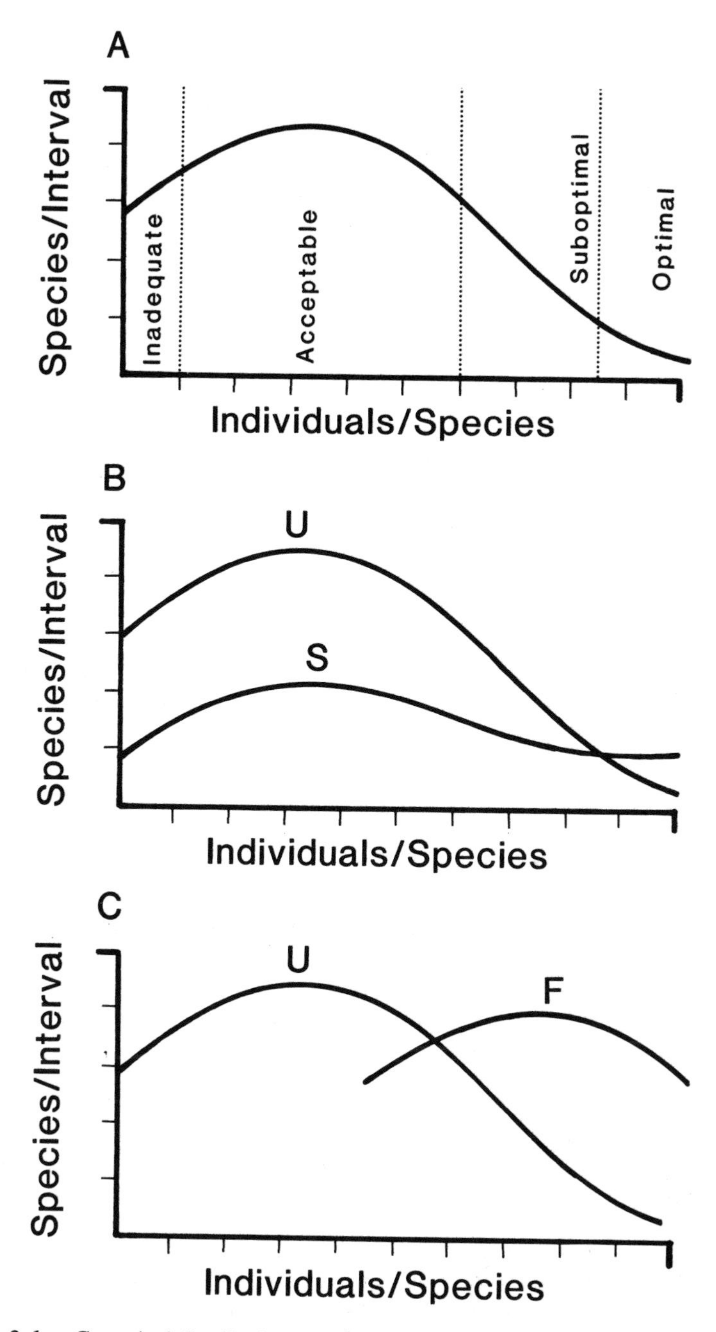

Figure 2.1. Canonical distribution of species abundances (based on Patrick, 1971). The scale for individuals/species is an octave (exponential) scale. A: Probable life conditions for species in a healthy community (based on Cairns, 1971). B: Effect of toxic or oxygen demanding stress on species distribution; U=unstressed, S=stressed. The number of species of moderate abundance is reduced. C: Effect of moderate fertilization on species distribution; U=unstressed, F=fertilized. Abundance of many species increases.

and Wilson (1967) whose community concept was based on dynamic processes that led to the continual immigration and extinction of species in local patches. The predicted continual turnover of species (Schoener, 1983) meant that communities were not static and that the indicator species concept was unlikely to be valid since perpetual residence of a particular taxon was probabalistic.

Although this brief chapter is not an appropriate place to discuss the merits of the MacArthur-Wilson hypothesis, the attention of both basic and applied ecologists turned rapidly to equilibrium theory and diversity measures (e.g., Shannon and Weaver, 1949) for evaluating impacts. An overwhelming array of indices coupling the number of taxa and their relative or absolute abundances blossomed (Hellawell, 1986), and they were adopted almost immediately in methods of environmental assessment (Wilhm and Dorris, 1968; Wilhm, 1970; Weber, 1973). Diversity indices proved practical, since the precise identities of the taxa were not needed (only the separation of taxa was required). Several indices of diversity have found wide application (Simpson, 1949; Margalef, 1958; Shannon and Weaver, 1949; Cairns and Dickson, 1971), and their properties have been repeatedly reviewed (Green, 1979; Washington, 1984; Hellawell, 1986). Additionally, J.L. Wilhm and other workers provided ranges for diversity indices that corresponded to habitat quality. That diversity calculations did not require taxonomy was demonstrated in the development of the Sequential Comparison Index (Cairns and Dickson, 1971). Diversity indices could be applied by even untrained field personnel, and this, of course, is precisely what happened. Untrained practitioners calculated diversity indices, in a cavalier fashion, regardless of their understanding of the sampled biota.

Reliance on diversity as a measure of biotic integrity was dealt a cruel blow by theoreticians who repeatedly questioned the relationship between diversity and system stability (Hurlbert, 1971) despite the simple, intuitive attractiveness of such a relationship. The death knell was pronounced by Green (1979, pp. 101–102):

> . . . the strongest argument against the use of diversity indices as derived criterion or predictor variables in environmental studies is that other statistical methods retain more of the information in the biological data while reducing them to a more useful and ecologically meaningful form.

Green quotes Poole (1974, pp. 99) that diversity indices are ". . . answers to which questions have not yet been found." Despite these arguments and the abandoning of diversity indices by European workers, diversity indices continue to find application in environmental management (see comments by Brinkhurst in Chapter 12), although the frequency of their use is decreasing.

More recently, the attention of benthic biologists has turned from analyses

Table 2.4. Environmental variables of value in analysis and classification of stream ecosystems.*

Geographic Variables	*Chemical Variables*
Latitude	pH
Longitude	Dissolved oxygen
Physiographic region	Total nitrogen
Altitude	Chloride
	Orthophosphate
	Alkalinity
Physical Variables	*Biological Variables*
Temperature	Riparian vegetation
Discharge	Macrophyte cover
Velocity	
Depth	
Width	
Sampling date	
Slope	
Distance from source	
Substratum type and composition	

*Based on Whittier et al. (1988) and Metcalfe (1989).

of community structure to evaluation of energetic relationships (Minshall, 1988). This change mirrors changing directions in ecology as a discipline and the development of general concepts of patterns of stream biota, as embodied in the River Continuum Concept (RCC; Vannote et al., 1980). This work, based primarily on patterns of functional groups of stream invertebrates (e.g., Cummins and Klug, 1979) and the relationship of biotic distribution to stream energy sources (allochthonous vs. autochthonous carbon), has had a profound influence on stream biology (Webster and Benfield, 1986). Virtually no study of stream benthos ignores the implications of the multiple stream gradients (Table 2.4) that affect both the kinds of organisms present (shredders, collector-gatherers, scrapers, filterers) and the energy balance of the system (net importing vs. net exporting systems).

Predictably, as soon as the elements of a river continuum hypothesis were stated, they were criticized (e.g., Winterbourn et al., 1981). The concepts required modification in biomes where no tree canopy existed or where canopies were coniferous and contributed quantities of highly refractory materials to streams (e.g., Minshall, 1988). Agricultural catchments also corrupted theoretical predictions, and it soon became apparent that the generality of the RCC was severely lacking despite its widespread acceptance. Nevertheless, streams studied by benthic biologists in the eastern and northeastern United States probably behave as predicted.

The lack of agreement among benthic biologists about theoretical concepts of community structure and energy flow notwithstanding, the river survey methodology of Patrick has been widely adopted in many countries for biological monitoring purposes. In the United States, state regulatory agencies place significant reliance on the results of benthic macroinvertebrate surveys for evaluating ecosystem health. Much of this activity could best be described as data "gathering" since the data are rarely rigorously analyzed. In many respects, the pattern has come full circle. Early benthologists could summarize the status of a system by qualitative analysis of a few samples. Then, more rigorous statistical analyses, including diversity indices and multivariate procedures for comparing community structure and water chemistry, were used to investigate community patterns. Now, many regulatory agencies, staffed with competent and experienced invertebrate biologists, have reverted to qualitative sampling (see Resh and Jackson, Chapter 6, for a detailed treatment of this topic).

Failure to apply rigorous methods routinely in benthic surveys has many sources. First, species-level taxonomy of immature insects and other taxa can be complex, and up-to-date works are often not available to nonspecialists (cf. Herricks and Cairns, 1982). This is clearly a result of the trend toward functional studies of systems at the expense of knowing which taxa are truly present. The situation probably will change, based on the current realization that we have names for only a small fraction of Earth's species (Wilson, 1984). Second, time constraints often require qualitative sampling. Third, retrievable data bases of species distributions are rarely established, also because of time (and funding) limitations. Fourth, benthic biologists continue to disagree on standards to ensure the comparability of collection methods. Fifth, human influence on ecosystems is so broad and diffuse (e.g., acid deposition) that identifying normal, healthy communities is sometimes problematic.

An important approach to standardizing collecting methods (and also separating natural and contaminant influences on the benthic fauna) has been the use of artificial substrata, such as multiplate samplers, rock baskets, and tiles (see the review by Rosenberg and Resh, 1982). Artificial substrata have found comparatively little use in Europe, but have been widely used in North America. Each type of artificial habitat is selective and may bear little resemblance to the extant or potential natural habitat. Colonization dynamics of artificial substrata for macroinvertebrates remain largely unknown, so colonization intervals of some length normally are used to ensure "complete" colonization. Nevertheless, artificial substrata can provide a degree of sampling replicability not otherwise available, especially when they are placed in comparable macrohabitats.

2.4. Current Developments

Considerable attention has focused on reducing the time required to obtain information about biological communities. Diversity indices attempted to provide a single value for the content of a community regardless of the need for taxonomic information or the ability to determine diversity nontaxonomically (e.g., Cairns and Dickson, 1971). However, benthic biologists (and perhaps biologists in general) have a long history of failing to agree on methods to evaluate environmental quality.

Two recent developments in the United States provide examples of new directions in applying sound environmental science to solving environmental impact problems. These developments are based on two concepts. First, physiographic regions delimited on the basis of bedrock geology, soils, and native vegetation have internal consistency (Hughes and Larsen, 1987; Omernik, 1987). That is, communities within delimited regions are likely to be more similar to each other than to communities from other regions. Second, within a rather homogeneous region, the natural structure and variability of communities can be determined (Karr et al., 1986). By carefully choosing reference sites for study, it should be possible to identify expected ranges of species numbers and composition (i.e., indicator assemblages or guilds) within the region. If the potential clean water flora and fauna can be identified, then a standard is available against which sites suspected of being adversely impacted can be compared.

These concepts have been applied to the identification of indicator species assemblages in the northwestern United States (Whittier et al., 1988), in Arkansas (Rohm et al., 1987), in Ohio (Ohio EPA, 1987), and in other areas. Unlike other, more general systems of impact assessment, these new developments acknowledge that different regions have different biotas. Expectations about the normal, healthy fauna can be used to ask questions about why a stream might not have an expected fauna. Answers may include obvious problems such as multiple discharges and cumulative impacts, interaction between point and nonpoint sources, and loss of appropriate habitat through siltation or channelization.

In Ohio, the Invertebrate Community Index (ICI) and the Index of Biotic Integrity (IBI, Karr et al., 1986) are used to evaluate stream reaches. Biological criteria are developed based on an understanding of the potential fauna and designated uses of the stream. When the resident biota is unhealthy (lacking species, containing pollution-tolerant forms), then action can be taken to manage not only point sources but nonpoint sources and structures that may have altered stream habitat. Benthic biology, therefore, may provide a forensic tool with which to evaluate expected biotic communities that support multiple uses of water.

2.5. Conclusions

Benthic macroinvertebrates have been attractive targets of biological monitoring efforts because they are a diverse group of long-lived, sedentary species that react strongly and, often, predictably to human influences on aquatic ecosystems. Early use of stream and lake benthos in biomonitoring focused on detection of organic pollution. Patterns of species change in streams were similar to community patterns in lakes of differing trophic status. As impacts became more severe, especially from toxic materials, studies focused on changing patterns of species richness and abundance, which led to the era of the diversity index. As concepts of stream ecosystems changed to include influences of terrestrial vegetation and stream gradients, biomonitoring changed to include evaluations of sensitive taxonomic groups, functional feeding groups, and regional patterns of species assemblages. Contemporary developments suggest that benthic biomonitoring is entering a more scientifically balanced era, one that includes faunistic and biogeographic analysis of natural and disturbed communities. The goal of these new efforts is to understand the native fauna and manage waters to maximize biological quality.

From the time of the first lake typology studies to now—one human lifetime—the population of the world has more than doubled, with populations in some countries still expanding at unacceptably high rates. Individual expectations of improved quality of life (i.e., more material goods and higher per capita energy use) suggest that pressures on the environment will increase at an even faster rate before the end of this century. It is imperative that biological monitoring systems be developed to follow environmental changes and, equally importantly, that corrective action be taken when this monitoring indicates trouble.

References

Andersen, T. 1977. Danish legislation on the use and protection of freshwater areas. *Folia Limnologica Scandinavica* 17:125–9.

Armitage, P.D., D. Moss, J.F. Wright, and M.T. Furse. 1983. The performance of a new biological water quality score system based on macroinvertebrates over a wide range of unpolluted running-water sites. *Water Research* 17:333–47.

Bick, H. 1971. The potentialities of ciliated Protozoa in the biological assessment of water pollution levels. In *Proceedings of the International Symposium on Identification and Measurement of Environmental Pollutants, Ottawa, ON, June 14–17, 1971,* Chmn. I. Hoffman, pp. 305–9. National Research Council of Canada, Ottawa, ON.

Brillouin, L. 1951. Maxwell's demon cannot operate: information and entropy I and II. *Journal of Applied Physics* 22:334–43.

Brinkhurst, R.O. 1974. *The Benthos of Lakes.* St. Martin's Press, New York.

Cairns, J., Jr. 1971. Factors affecting the number of species in fresh-water protozoan communities. In *The Structure and Function of Fresh-Water Microbial Communities,* ed. J. Cairns, Jr., pp. 219–48. Research Division Monograph 3, Virginia Polytechnic Institute and State University, Blacksburg, VA.

Cairns, J., Jr. 1982. Freshwater protozoan communities. In *Microbial Interactions and Communities, Vol. 1,* eds. A.T. Bull and A.R.K. Watkinson, pp. 249–85. Academic Press, London.

Cairns, J., Jr. and K.L. Dickson. 1971. A simple method for the biological assessment of the effects of waste discharges on aquatic bottom-dwelling organisms. *Journal of the Water Pollution Control Federation* 43:755–72.

Cummins, K.W. and M.J. Klug. 1979. Feeding ecology of stream invertebrates. *Annual Review of Ecology and Systematics* 10:147–72.

De Pauw, N. and G. Vanhooren. 1983. Method for biological quality assessment of watercourses in Belgium. *Hydrobiologia* 100:153–68.

Fjerdingstad, E. 1965. Taxonomy and saprobic valency of benthic phytomicro-organisms. *Internationale Revue der Gesamten Hydrobiologie* 50:475–604.

Foissner, W. 1988. Taxonomic and nomenclatural revision of Sládeček's list of ciliates (Protozoa: Ciliophora) as indicators of water quality. *Hydrobiologia* 166:1–64.

Forbes, S.A. and R.E. Richardson. 1913. Studies on the biology of the upper Illinois River. *Bulletin of the Illinois State Laboratory of Natural History* 9:481–574.

Gameson, A.L. and A. Wheeler. 1975. Restoration and recovery of the Thames estuary. In *Recovery and Restoration of Damaged Ecosystems,* eds. J. Cairns, Jr., K.L. Dickson, and E.E. Herricks, pp. 72–101. Univ. Press of Virginia, Charlottesville, VA.

Gaufin, A.R. 1958. The effects of pollution on a midwestern stream. *Ohio Journal of Science* 58:197–208.

Gray, J.S. 1989. Effects of environmental stress on species rich assemblages. *Biological Journal of the Linnean Society* 37:19–32.

Green, R.H. 1979. *Sampling Design and Statistical Methods for Environmental Biologists.* John Wiley, New York.

Guhl, W. 1987. Aquatic ecosystem characterizations by biotic indices. *Internationale Revue der Gesamten Hydrobiologie* 72:431–55.

Hellawell, J.M. 1986. *Biological Indicators of Freshwater Pollution and Environmental Management.* Elsevier, New York.

Herricks, E.E. and J. Cairns, Jr. 1982. Biological monitoring. Part III—Receiving system methodology based on community structure. *Water Research* 16:141–53.

Hilsenhoff, W. 1982. *Using a Biotic Index to Evaluate Water Quality in Streams.* Technical Bulletin No. 132, Wisconsin Department of Natural Resources, Madison, WI.

Hughes, R.M. and D.P. Larsen. 1987. Ecoregions: an approach to surface water protection. *Journal of the Water Pollution Control Federation* 60:486–93.

Hurlbert, S.H. 1971. The nonconcept of species diversity: a critique and alternative parameters. *Ecology* 52:577–86.

Karr, J.R., K.D. Fausch, P.L. Angermeier, P.R. Yant, and I.J. Schlosser. 1986. *Assessing Biological Integrity in Running Waters: a Method and Its Rationale.* Illinois Natural History Survey Special Publication 5, Champaign, IL.

Kolkwitz, R. 1950. Ökologie der Saprobien. *Schriftenreihe des Vereins für Wasser-, Boden- und Lufthygiene.* 4:1–64.

Kolkwitz, R. and M. Marsson. 1908. Ökologie der pflanzlichen Saprobien. *Berichte der Deutschen Botanischen Gesellschaft* 26A:505–19.

Kolkwitz, R. and M. Marsson. 1909. Ökologie der tierischen Saprobien. Beiträge zur Lehre von des biologischen Gewasserbeurteilung. *Internationale Revue der Gesamten Hydrobiologie und Hydrographie* 2:126–52.
Liebmann, H. 1951. The biological community of *Sphaerotilus* flocs and the physico-chemical basis of their formation. *Vom Wasser* 20:24.
Liebmann, H. 1962. *Handbuch der Frischwasser- und Abwasserbiologie,* Band I, 2 Auflag. Verlag R. Oldenbourg, München.
Lowe, R.L. 1974. *Environmental Requirements and Pollution Tolerance of Freshwater Diatoms.* EPA-670/4–74–005. U.S. Environmental Protection Agency, Cincinnati, OH.
MacArthur, R.H. and E.O. Wilson. 1967. *The Theory of Island Biogeography.* Princeton Univ. Press, Princeton, NJ.
Margalef, J.L. 1958. Information theory in ecology. *General Systems* 3:36–71.
Metcalfe, J.L. 1989. Biological water quality assessment of running waters based on macroinvertebrate communities: history and present status in Europe. *Environmental Pollution* 60:101–39.
Minshall, G.W. 1988. Stream ecosystem theory: a global perspective. *Journal of the North American Benthological Society* 7:263–88.
Ohio EPA (Environmental Protection Agency). 1987. *Biological Criteria for the Protection of Aquatic Life. Vol. III. Standardized Biological Field Sampling and Laboratory Methods for Assessing Fish and Macroinvertebrate Communities.* Division of Water Quality Monitoring and Assessment, Ohio Environmental Protection Agency, Columbus, OH.
Omernik, J.M. 1987. Ecoregions of the conterminous United States. *Annals of the Association of American Geography* 77:118–25.
Pantle, R. and H. Buck. 1955. Die biologische Überwachung der Gewässer und die Darstellung der Ergebnisse. *Gas- und Wasserfach* 96:604.
Patrick, R. 1949. A proposed biological measure of stream conditions, based on a survey of the Conestoga Basin, Lancaster County, Pennsylvania. *Proceedings of the Academy of Natural Sciences of Philadelphia* 101:277–341.
Patrick, R. 1971. Diatom communities. In *The Structure and Function of Fresh-Water Microbial Communities,* ed. J. Cairns, Jr., pp. 151–64. Research Division Monograph 3, Virginia Polytechnic Institute and State University, Blacksburg, VA.
Poole, R.W. 1974. *An Introduction to Quantitative Ecology.* McGraw-Hill, New York.
Preston, F.W. 1948. The commonness and rarity of species. *Ecology* 29:254–83.
Richardson, R.E. 1925. Illinois River bottom fauna in 1923. *Bulletin of the Illinois Natural History Survey* 15:391–422.
Richardson, R.E. 1929. The bottom fauna of the Middle Illinois River 1913–1925. Its distribution, abundance, valuation and index value in the study of stream pollution. *Bulletin of the Illinois Natural History Survey* 17:387–475.
Roback, S.S. 1974. Insects (Arthropoda: Insecta). In *Pollution Ecology of Freshwater Invertebrates,* eds. C.W. Hart, Jr. and S.L.H. Fuller, pp. 313–76. Academic Press, New York.
Rohm, C.M., J.W. Giese, and C.C. Bennett. 1987. Evaluation of an aquatic ecore-

gion classification of streams in Arkansas. *Journal of Freshwater Ecology* 4:127–40.

Rosenberg, D.M. and V.H. Resh. 1982. The use of artificial substrates in the study of freshwater benthic macroinvertebrates. In *Artificial Substrates,* ed. J. Cairns, Jr., pp. 175–235. Ann Arbor Science Pubs., Ann Arbor, MI.

Schaeffer, D.J., E.E. Herricks, and H.W. Kerster. 1988. Ecosystem health: I. Measuring ecosystem health. *Environmental Management* 12:445–55.

Schindler, D.W. 1987. Detecting ecosystem responses to anthropogenic stress. *Canadian Journal of Fisheries and Aquatic Sciences* 44 (Supplement 1):6–25.

Schoener, T. 1983. Rate of species turnover decreases from lower to higher organisms: a review of the data. *Oikos* 41:372–7.

Shannon, C.E. and W. Weaver. 1949. *The Mathematical Theory of Communication.* Univ. of Illinois Press, Urbana, IL.

Simpson, E.H. 1949. Measurement of diversity. *Nature* (London) 163:688.

Sládeček, V. 1965. The future of the saprobity system. *Hydrobiologia* 25:518–37.

Sládeček, V. 1973. System of water quality from the biological point of view. *Archiv für Hydrobiologie Ergebnisse der Limnologie* 7:1–218.

Thienemann, A. 1925. Die Binnengewasser Mitteleuropas. Eine Limnologische Einfuhrung. *Die Binnengewasser* 1:1–25.

Uhlmann, D. 1979. *Hydrobiology: a Text for Engineers and Scientists.* John Wiley, Chichester, England.

Vannote, R.L., G.W. Minshall, K.W. Cummins, J.R. Sedell, and C.E. Cushing. 1980. The river continuum concept. *Canadian Journal of Fisheries and Aquatic Sciences* 37:130–7.

Washington, H.G. 1984. Diversity, biotic and similarity indices. A review with special relevance to aquatic ecosystems. *Water Research* 18:653–94.

Weber, C.I., ed. 1973. *Biological Field and Laboratory Methods for Measuring the Quality of Surface Waters and Effluents.* EPA-670/4–73–001. U.S. Environmental Protection Agency, Cincinnati, OH.

Webster, J.R. and E.F. Benfield. 1986. Vascular plant breakdown in freshwater ecosystems. *Annual Review of Ecology and Systematics* 17:567–94.

Whittier, T., R.M. Hughes, and D.P. Larsen. 1988. Correspondence between ecoregions and spatial patterns in stream ecosystems in Oregon. *Canadian Journal of Fisheries and Aquatic Sciences* 45:1264–78.

Wilhm, J.L. 1970. Range of diversity index in benthic macroinvertebrate populations. *Journal of the Water Pollution Control Federation* 42:R221–4.

Wilhm, J.L. 1975. Biological indicators of pollution. In *River Ecology,* ed. B.A. Whitton, pp. 375–402. Univ. of California Press, Berkeley, CA.

Wilhm, J.L. and T.C. Dorris. 1968. Biological parameters for water quality criteria. *BioScience* 18:477–81.

Wilson, E.O. 1984. *Biophilia.* Harvard Univ. Press, Cambridge, MA.

Winterbourn, M.J., J.S. Rounick, and B. Cowie. 1981. Are New Zealand stream ecosystems really different? *New Zealand Journal of Marine and Freshwater Research* 15:321–8.

Woodiwiss, F.S. 1964. The biological system of stream classification used by the Trent River Board. *Chemistry and Industry* (London) 11:443–7.

3

The Literature of Biomonitoring

K. Eric Marshall

3.1. Introduction

The amount of scientific literature available today is enormous. The annual production of journal articles alone makes it impossible to keep abreast of what is being published, even in a highly specialized area, without help. Abstracting and indexing services can provide this help. The major service covering biological publications, *Biological Abstracts,* scans 9,379 currently active journals. In 1989, it indexed 530,000 items from 27,000 issues of journals and books (BIOSIS, 1989). Even a specialized abstracting service such as *Aquatic Sciences and Fisheries Abstracts* (ASFA), *Part 1 (Biological Sciences and Living Resources)* may include over 20,000 items in a year.

Most scientists collect references, reprints, and photocopies of papers pertinent to their interests. This can be done in several ways. First, scientists can visit a library regularly, scan the new issues of selected journals, and note or copy those papers of interest. If a record is kept of each journal issue scanned, this method can be quite effective. However, because no library carries every journal of potential relevance, and papers often appear in unexpected journals, this approach will not be totally effective. Second, scientists can use *Current Contents* (especially the section, "Agriculture, Biology and Environmental Sciences") as a convenient way of scanning the contents of many journal issues. Third, it is often useful to contact directly authors who publish regularly in a common area of interest. However, whatever method is used, it needs to be supplemented with the use of abstracting and indexing services, either on a regular basis or as the need arises. It is possible to arrange for regular notification of new publications that contain specific combinations of keywords, and this is available with each update of many online databases. The SDI feature on Dialog Information Services is one example. The CAN/SDI service, offered by the Canada Institute of

Scientific and Technical Information (CISTI) of the National Research Council of Canada, is another.

This chapter will describe sources of bibliographic information, discuss strategies to retrieve pertinent information, and suggest means of obtaining the documents required in biomonitoring. Examples based upon two of the chapters in this volume are presented to illustrate search requirements.

3.2. Bibliographic Sources

Printed abstracting and indexing services have been available for a long time. The *Zoological Record* was first published in 1864. *Biological Abstracts* started in 1926. Many specialized abstracting services, such as *Pollution Abstracts,* were begun in the 1960s and 1970s. In the 1970s, a few of these services became available in machine-readable form that could be searched interactively on a computer, and so the online search service was born. Many more abstracting services have become available for online searching since then. In the last two years, some of these services or databases have been issued on compact disc-read only memory (CD-ROM), which can be searched using a microcomputer linked to a CD reader. New databases are becoming available on CD-ROM at a rapid rate.

Several abstracting and indexing services contain information pertinent to biomonitoring. The most important ones are described in Table 3.1.

3.3. Case Histories

The principles of searching are the same, whether printed indexes, electronic media of an online search service, or CD-ROMs are used. First, significant key words or phrases for the desired search must be chosen. A review of significant words in the titles of papers that the researcher knows to be most pertinent should provide basic search terms. Alternative spellings (e.g., ALUMINUM and ALUMINIUM), and the use of truncation to retrieve singulars, plurals, and other variations of a root word (e.g., PLANKT* will retrieve PLANKTON, PLANKTER, PLANKTERS, PLANKTONIC, etc.) should be noted, because in an electronic search, only the exact form of the search terms entered will be selected. Printed and online indexes will help identify these alternative search terms. Second, perusal of the initial results of a search may indicate a need to narrow or broaden it. If broadening is required, then material found in the original effort may suggest additional search terms. Third, a comprehensive search will require that many different databases be used. Fourth, reference lists in the papers identified may provide additional titles.

Searches were made to find material pertinent to Chapters 4 (Johnson et

Table 3.1. Abstracting and indexing services relevant to biomonitoring and their availability online and as CD-ROMs in 1990. (Addresses of issuing agencies and vendors are given in Appendix 3.1.)

Abstracting/ Indexing Service	Issuing Agency[1]	Initial Date of Publication	Material Covered	Frequency of Publication	Availability[2]		Comments
					Online	CD-ROM	
Aquatic Sciences and Fisheries Abstracts (ASFA)	FAO & Cambridge Scientific Abstracts	1958	Literature of marine, freshwater, and brackish habitats	3 parts; 1 vol each with 12 issues/yr	1978 +	1982 +	Part 1: biological sciences and living resources; Part 2: ocean technology, policy, nonliving resources; Part 3: aquatic pollution and environmental quality; started as FAO Current Bibliography for Aquatic Sciences and Fisheries which merged with Aquatic Biology Abstracts to become ASFA; Part 3 started in 1990
Biological Abstracts	BIOSIS	1926	Primary biological literature	Biweekly (2 vol/yr)	1969 + (BIOSIS Previews)	1990 + (BA on CD)	The premier biological literature source; BIOSIS Connection is an online source for recent literature
Biological Abstracts/ RRM (Reports, Reviews, Meetings)	BIOSIS	1965	Biological government reports, reviews, reports on meetings, popular journal articles	Biweekly (2 vol/yr)	1969 + (BIOSIS Previews)	No	Complementary publication to Biological Abstracts proper; does not include abstracts; previously called BioResearch Titles (1965–67) and BioResearch Index (1968–80)
Current Contents	ISI	1958	Contents of major scientific journals	Weekly	1988 +	No	Printed version issued in many sections; Agriculture, Biology & Environmental Sciences most relevant; also available on diskette (Current Contents on Diskette) for searching on a PC

Environment Abstracts & Environment Index	Bowker	1971	Environmental publications, including government reports	Monthly	1971 + (Enviroline)	1971 +	Started as Environment Information Access (1971), then became Environment Abstracts (1974); annual cumulative index is Environment Index; CD called Acid Rain, Energy, Environment
Environment Library	OCLC	1988	Environmental publications (no journal articles)	Annual	See Comments	1988 +	Exclusively a CD covering material selected from the OCLC online cataloging database; EPA reports partially covered
Environmental Periodicals Bibliography	Environmental Studies Institute	1972	Contents of environmental journals	Monthly	1973 +	1973 +	Printed version started in 1972; NISC issued CD in 1990
Excerpta Medica	Elsevier	1947	Includes literature on environmental contamination	Monthly	1974 + (EMBase)	1984 +	Printed version has many subject sections, including Environmental Health & Pollution Control
Freshwater and Aquaculture Contents Tables	FAO	1978	Contents of main freshwater journals	Monthly	No	No	Each issue covers ~24 journal issues; full index in ASFA
Freshwater Biological Association Current Awareness Service	FBA	1955	Includes journal articles received in FBA library	Monthly	No	No	Printed version only; useful listings arranged under broad subject headings
Life Sciences Collection	Cambridge Scientific Abstracts	1966	Primary biological literature	Monthly or quarterly	1978 +	1982 +	Cambridge Scientific Abstracts publishes a number of abstracting journals, including Ecology Abstracts, which form this database

(Continued)

Table 3.1. (Continued)

Abstracting/ Indexing Service	Issuing Agency[1]	Initial Date of Publication	Material Covered	Frequency of Publication	Availability[2]		Comments
					Online	CD-ROM	
Micrometalog Canadian Research Index	Micromedia	1973	Canadian federal & provincial government reports	Monthly	1979 + (Micrometalog)	1979 +	Printed version originally called Profile Index, became Micrometalog Index, now Micrometalog Canadian Research Index; online only on CAN/OLE (Canadian OnLine Enquiry, National Research Council of Canada)
National Technical Information Service	NTIS	1965	US government research reports	Biweekly	1964 +	1980 +	Printed version called Government Reports Announcements & Index; includes coverage of EPA reports; CD available from Silver Platter, Dialog, and OCLC
Pollution Abstracts	Cambridge Scientific Abstracts	1970	Literature of pollution	Bimonthly	1970 +	1981 + (PolTox)	CD version includes related materials from other Cambridge Scientific Abstracts and other sources, including Medline
Science Citation Index	ISI	1945	3,000 major scientific journals	Quarterly (annual & multiyear cumulations)	1974 +	1986 +	Relatively small number of ecological journals covered; indexing is based on references cited allowing a unique approach to literature searching
Selected Water Resources Abstracts	NTIS	1968	Water resources literature, especially from USA	Monthly	1968 +	1968 +	Strong coverage of US government reports; CD available from NISC and OCLC
University of Guelph Library Catalogue	University of Guelph	1988	Books and reports in U of G library, including many Canadian government publications	Annual?	No	1988 +	Only issued as CD; limited coverage of biological material

Wildlife and Fish Worldwide	NISC	1990	Aquatic material, especially fisheries	Twice yearly	No	1971 +	CD version of Wildlife Review and Fisheries Review (formerly Sport Fishery Abstracts)
Zoological Record	Zoological Society of London & BIOSIS	1864	Zoological literature	Annual volume of 20 parts	1978 +	No	The 20 parts of the printed version each cover a phylum or other group of animals

[1]Explanation of acronyms: BIOSIS, registered name of the company publishing *Biological Abstracts;* FAO, Food and Agriculture Organization of the United Nations; FBA, Freshwater Biological Association; ISI, Institute for Scientific Information; NISC, registered name of company (derived from National Information Services Corporation); NTIS, National Technical Information Service; OCLC, registered name of company (formerly known as Online Computer Library Center).

[2]Parentheses under "Online" and "CD-ROM" indicate names of databases or CD products. All of the online databases are available on Dialog unless otherwise indicated.

al.: Section 4.2.6, "Sentinel Organisms") and 6 (Resh and Jackson) in this volume in order to illustrate practical aspects of online and CD searching. Sample searches were made on the CD products available in the Freshwater Institute (Winnipeg, Manitoba, Canada) library; the practical aspects of online and CD searching are the same, so the principles illustrated here also can be applied to online searches. Comparisons were made between the references obtained by the authors of these chapters and those obtained by searching CD databases. The material on sentinel organisms for metals in Chapter 4 concentrated on readily available journal articles, whereas the material on rapid assessment approaches in Chapter 6 frequently referenced technical reports and other items of "grey" literature. Therefore, it was anticipated that searches for references in Chapter 6 would produce relatively few successes, whereas more references would be retrieved for Chapter 4.

3.3.1. Searching for Material for Chapter 4 (Section 4.2.6, "Sentinel Organisms")

A list of 72 references on sentinel organisms for metals was provided to start the search. Twenty-nine of the references were published either before 1981 or in 1988 or 1989 and therefore would not appear in the ASFA-CD database used (Table 3.1). The remaining references were published in readily available journals or came from other sources that ASFA covers.

Searches were started with a simple strategy, which then was modified according to the results obtained. The following, initial search strategy was used (* indicates truncation):

FRESHWATER AND BENTH AND (BIOMONITOR* OR BIOACCUMULAT* OR BIOCONCENTRAT* OR SENTINEL)*

The 1987–88 CD was examined, and 14 references were found, only one of which was on the original list (Table 3.2, A).

The next search was run against the 1982–86 CD using:

((HEAVY OR TRACE) AND (METALS OR ELEMENTS)) OR COPPER OR MOLYBDENUM OR CADMIUM OR CHROMIUM OR LEAD OR ZINC

combined with the first search using AND logic. A total of 25 references resulted, only one of which was on the original list (Table 3.2, B). However, 10 references not on the original list were deemed to be "possibly relevant" and in need of further examination before a decision on inclusion could be made.

Clearly, the search was too narrow. A close look at the references on the original list revealed that, although many papers dealt with benthic organisms, often only the name of the organism was given in the title with no

Table 3.2. Number of references retrieved searching the Aquatic Sciences and Fisheries Abstracts (ASFA)-CD database for benthic invertebrates as sentinel organisms for metals.

Search Strategy[1]	Total	On Original List	Initially Perused but not Included on Original List	Possibly Relevant[2]	Obscure Journals[3]	Irrelevant
A	14	1	8	2	3	0
B	25	1*	0	10	11	2
C	90	8	5	61	2	14
D	261	14	4	90	7	146

[1]See text for description of search strategies.
[2]Need further examination.
[3]References are relevant but come from poorly known/limited circulation sources; therefore, not used.
*Two citations to the same reference were retrieved.

indication that it was benthic. Therefore, the BENTH* part of the search strategy was expanded as follows:

> *CRAYFISH OR AMPHIPOD* OR AQUATIC ORGANISMS OR INVERTEBRAT* OR CLAM OR EPHEMEROPT* OR TRICHOPT* OR MUSSEL* OR BIVALV* OR SNAIL* OR ANODONTA* OR CHIRONOMID* OR BENTH* OR ZOOBENTH*.*

It was also apparent that the BIOMONITOR part of the strategy needed to be expanded, so the following terms were added:

> *ACCUMULAT* OR UPTAKE OR CONCENTRAT* OR MEASUR* OR INDICATOR* OR REGULATION OR ADAPTATION OR DISTRIBUTION*

The expanded search was run on the 1987–88 (Table 3.2, C) and 1982–86 (Table 3.2, D) CDs, respectively, and 90 and 261 references were retrieved. Searches C and D produced a total of 22 references out of the original 72 on the list. Strangely, a check showed that at least five of the original references were not in the database. Of the total of 351 references retrieved from searches C and D together, 160 clearly were not relevant (45.5%) and 151 references (43.0%) were "possibly relevant." Nine of the references retrieved had been examined before the original list was compiled, and excluded for various reasons.

The results obtained by quickly searching a database easily available in the Freshwater Institute library indicated that no single database can provide complete success. In fact, many of the references on the original list were obtained by a combination of regular subscription (CAN/SDI) searches made on BIOSIS Previews (see Biological Abstracts, Table 3.1), by browsing through new journal issues in the library, and by perusing the bibliographies of relevant papers. In a similar fashion, searches on different databases contributed to a bibliography by Marshall (1980) on the subject of mayflies and pollution. Because each database has its own unique references, a variety of databases should be searched, using the same basic strategies.

3.3.2. Searching for Material for Chapter 6

The first task was to decide what terms to use in a search strategy for rapid assessment procedures. A list of 20 references provided on this subject indicated that a series of different searches was needed to cover all aspects. The following search strategies were used:

A) *(ASSESS* OR BIOASSESS*) AND PROTOCOL**
B) *BIOTIC INDEX*)*
C) *MONITOR* AND (BIOTA OR BIOTIC) AND RESPONSE**
D) *TROPHIC CLASSIFICATION*
E) *QUALITY AND (ASSURANCE OR ASSESSMENT OR STANDARD* OR SURVEILLANCE)*
F) *IMPACT AND ASSESSMENT*

Each of the above searches was combined using AND logic with INVERTEBRAT* OR MACROINVERTEBRAT* OR BENTH*. In the event that many references resulted (the system gives numbers), the final set could be combined using AND logic with FRESHWATER to eliminate marine studies. In this way, the search would be built up interactively, depending on how many references were retrieved.

Table 3.3 indicates how many references were found for each strategy. The search for rapid assessment procedures had to be made quite broad in order to cover the known material. Attempts to narrow the searches by limiting the output to material on lakes, streams, or rivers eliminated the known pertinent references. Only two of the 78 references retrieved appeared on the original list of 20. Two of the remaining 76 references were pertinent to Chapter 6, but were not on the list. Of the 20 original references, 18 were not covered by the databases used (i.e., one was published in 1972, before the database began; three were published too recently to be included; one was in press; and 13 were references to government reports, which are not well-covered by the databases used). Therefore, the level of precision attained in the search was two out of five, or if the two new references are added, four out of seven (57%).

This search was made knowing that most of the original references were not likely to be found in the CD products used, because it is the practice in the Freshwater Institute library to use CDs in order to fine-tune strategies

Table 3.3. Number of references retrieved using CD searches for benthic assessment protocols.

CD Used[2]	Search Strategies[1]						Total No. of References
	A	B	C	D	E	F	
ASFA[3] 82/86	1	11	0	6	7	24	49
ASFA 87/88	0	2	4	3	1	10	20
University of Guelph Library Catalogue	0	2	0	0	2	0	4
Environment Library	0	4	0	1	0	0	5
					Total:		78

[1]See text for description of A–F.
[2]See Table 3.1 for description of databases.
[3]Aquatic Sciences and Fisheries Abstracts.

for online searches. However, in this case, it is even unlikely that the currently available online databases would contain the relatively obscure references to government reports on the original list. Nonetheless, the CD searches produced two new and pertinent references.

3.4. Obtaining Copies of References Retrieved from Bibliographic Databases

This concluding section recommends ways of obtaining hard copies of references retrieved during the course of literature searches. Five main types of references are involved: journal literature, books, conference publications, grey literature, and theses.

3.4.1. Journal Literature

Most of the references found in any literature search probably will be journal articles. Access to a good library will allow the pertinent articles to be photocopied. Most libraries will arrange interlibrary loans for material not held by them. It is also possible to obtain a reprint of a pertinent paper by contacting the author directly.

3.4.2. Book Material

If the book title you are seeking is not listed in the library catalog of your institution or in other libraries that can be accessed, then an interlibrary loan can be arranged. Authors of chapters in books sometimes are able to provide reprints.

3.4.3. Conference Material

It may be difficult to recognize from the title of a volume that it represents the proceedings of a conference. Moreover, papers read at meetings often are difficult to obtain; sometimes preprints of papers are given only to those in attendance. Many conference papers are never published in full; however, much of the information may appear in journal articles, and literature searches already done may include the information presented. A letter to the author may result in a copy of the conference paper or reprints of journal articles stemming from the work reported.

3.4.4. Reports and Grey Literature

Reports and grey literature may be difficult to obtain. In the United States, the National Technical Information Service (NTIS) collects U.S. federal government research reports and sells them either as paper copies or on microfiche. NTIS also issues indexes *(Government Reports Announcements and Index)* to available material (Table 3.1). In Canada, Micromedia, a private company, offers the same type of service for Canadian federal, provincial, and municipal reports. An index, *Microlog Canadian Research Index,* is issued listing the reports available (Table 3.1). Many libraries will have accounts with either or both of these organizations. Reports from the United Kingdom may be available from the British Library Document Supply Centre (Boston Spa, Wetherby, West Yorkshire, LS23 7BQ, UK). For reports not available from these sources, either the organization issuing the report or the organization for which the work was done should be contacted. Because of the many frustrations caused when readers try to obtain this type of material, a few journals are now asking authors to indicate in their reference lists where these documents may be borrowed or bought.

3.4.5. Theses

University Microfilms International (300 North Zeeb Road, Ann Arbor, MI, 48106–1346) makes available a variety of dissertation information sources and supplies doctoral theses from U.S. and some European universities in microform or as photocopies. Canadian theses, both doctoral and masters, can be obtained on microform from the National Library of Canada (395 Wellington Street, Ottawa, ON, K1A ON4). Theses from some European universities are published in primary journals. Theses not available from these sources may be available from the university concerned.

References

BIOSIS. 1989. *Serial Sources for the BIOSIS Previews® Database,* Vol. 1989. BIOSIS, Philadelphia, PA.

Marshall, K.E. 1980. Online computer retrieval of information on Ephemeroptera: a comparison of different sources. In *Advances in Ephemeroptera Biology*, eds. J.F. Flannagan and K.E. Marshall, pp. 467–89. Plenum Press, New York.

Appendix 3.1. Addresses of Issuing Agencies and Vendors Listed in Table 3.1

BIOSIS, 2100 Arch Street, Philadelphia, PA 19103–1399, USA
R.R. Bowker, 245 W. 17th Street, New York, NY 10011, USA
Cambridge Scientific Abstracts, 7200 Wisconsin Avenue, Bethesda, MD 20814, USA
CAN/OLE, Canada Institute for Scientific Information, National Research Council of Canada, Ottawa, ON, Canada K1A OS2
Dialog Information Services, Inc., 3460 Hillview Avenue, Palo Alto, CA 94304, USA
Elsevier Science Publishers, P.O. Box 211, 1000 AE Amsterdam, The Netherlands
Environmental Studies Institute, 2074 Alameda Padre Serra, Santa Barbara, CA 93103, USA
Food and Agriculture Organization of the United Nations, Via delle Terme di Caracalla, 00100 Rome, Italy
Freshwater Biological Association, The Ferry House, Ambleside, Cumbria LA22 OLP, UK
Institute for Scientific Information, 3501 Market Street, Philadelphia, PA 19104, USA
Micromedia Limited, 158 Pearl Street, Toronto, ON, Canada M5H 1L3
National Technical Information Service, Springfield, VA 22161, USA
NISC, Suite 6, Wyman Towers, 3100 St. Paul Street, Baltimore, MD 21218, USA
OCLC, 6565 Franz Road, Dublin, OH 43017–0702, USA
University of Guelph, Guelph, ON, Canada N1G 2W1
Zoological Society of London, Regents Park, London NW1 4RY, UK

4

Freshwater Biomonitoring Using Individual Organisms, Populations, and Species Assemblages of Benthic Macroinvertebrates

Richard K. Johnson, Torgny Wiederholm, and David M. Rosenberg

4.1. Introduction

The concept of indicator species is of central importance in the use of benthic macroinvertebrates in biological monitoring. "Indicator species" is defined here as a species (or species assemblage) that has particular requirements with regard to a known set of physical or chemical variables such that changes in presence/absence, numbers, morphology, physiology, or behavior of that species indicate that the given physical or chemical variables are outside its preferred limits. The factor or factors that regulate population abundance or presence/absence may act at any stage of the life cycle, and may be of abiotic (e.g., chemical variables: O_2, H^+, or trace metal concentration; physical variables: sedimentation) or biotic (e.g., competition, predation, parasitism) origin.

Ideally, indicator organisms are those species that have narrow and specific environmental tolerances. The principal underlying assumption in using indicator organisms (and, in fact, in using species assemblages or communities) for water quality assessment is that the presence of the indicator is a reflection of its environment. Thus, its presence in abundance signifies that its physical, chemical, and nutritional requirements are being met. Therefore, if the environmental factors that are most commonly limiting to the species concerned are known, the presence of the organism will indicate specific environmental conditions. Conversely, organisms that have wide tolerances for different environmental conditions, and whose patterns of distribution or abundance are only slightly affected by substantial variations in environmental quality, are poor indicators. Whereas the presence of a species assures us that certain minimal conditions have been met, absence of a species does not, of course, tell us that the critical environmental factors are not being met. The absence of a taxon also might result from geographical barriers (i.e., the animal has not been introduced to an area but might well survive if it was), occupation of its functional niche (e.g., competitive exclusion by an ecological analog), or normal life-cycle events (e.g., population abundance may be below detection limits as the result of emergence, or pressure due to intense levels of predation or high rates of parasitism).

The "ideal" indicator should have the following characteristics (Rosenberg and Wiens, 1976; Hellawell, 1986):

1. *Taxonomic soundness and easy recognition by the nonspecialist.* Taxonomic uncertainties will complicate long-term monitoring and between-site interpretation.
2. *Cosmopolitan distribution* (or distribution involving an ecological analog). Choice of a cosmopolitan species would allow for comparative studies on regional, national, and international scales.
3. *Numerical abundance.* Numerical predominance of an indicator species allows for ease of sampling and for conclusions regarding quantitative distribution patterns.
4. *Low genetic and ecological variability.* Again, indicators should have relatively narrow ecological demands.
5. *Large body size.* This would facilitate sampling and sorting.
6. *Limited mobility and relatively long life history.* These would allow for ease of integration on spatial and temporal scales.
7. *Ecological characteristics are well known.* Background physiological and autecological information should be available widely.
8. *Suitable for use in laboratory studies.* These may allow for determination of causality.

Bioaccumulative indicators are a special kind of indicator organism (Hellawell, 1986). Commonly called "sentinel" organisms, these benthic macroinvertebrates accumulate pollutants from their surroundings and/or food so that an analysis of their tissues provides an estimate of environmental concentrations of the pollutants.

In this chapter, we will examine the properties and responses of *individual organisms, populations,* and *species assemblages* that may be used to characterize the quality of the environment. These properties and responses include: (1) biochemical to life-history changes and bioaccumulation at the organism level, and (2) presence/absence or numerical predominance of indicator organism populations and species assemblages. Changes in *community* composition are considered in Resh and McElravy (Chapter 5), Resh and Jackson (Chapter 6), and Norris and Georges (Chapter 7).

4.2. Organism-Level Indicators

4.2.1. Biochemical Indicators

Responses to environmental stress originate at the biochemical and physiological levels of the organism. Therefore, changes at these levels should be sensitive indicators of the health of benthic macroinvertebrates, and should provide the earliest possible warning of adverse effects (Kingett, 1985; Mitin, 1985; Graney and Giesy, 1986, 1987, 1988; Giesy, 1988; Khoruzhaya, 1989). The goal is to use suborganismal indicators as sensitive, integrative measures of sublethal stress, which could be monitored in organisms in their natural environment (Giesy et al., 1983).

Use of biochemical and physiological indicators in freshwater benthic macroinvertebrates presently is limited because of a lack of basic knowledge of these processes in most organisms (Graney and Giesy, 1987; Giesy, 1988). For example, it is essential to know the normal background variability of a particular indicator so that changes associated with a stress can be identified (e.g., Graney and Giesy, 1986, 1987; McKee and Knowles, 1989; Khoruzhaya, 1989). Also, it is important that suborganismal indicators be related to ecologically relevant endpoints for individuals and populations (Giesy et al., 1983; Graney and Giesy, 1986, 1988; Farris et al., 1988). A comprehensive biomonitoring approach would include suborganismal indicators along with techniques used to assess pollutant impacts at population, community, and ecosystem levels (Di Giulio et al., 1989).

Biochemical indicators of environmental stress in freshwater benthic macroinvertebrates fall into the following general categories: (1) energy metabolism; (2) enzyme activities; (3) RNA, DNA, amino acid, and protein content; and (4) ion regulation. Most effort has been applied to studies of ion regulation in macroinvertebrates.

4.2.1.1. Energy Metabolism

"In general the concentrations of adenylates and cations of all the adenylates . . . seem to serve as an overall regulator of many catabolic and anabolic processes. The advantage of an energy based nonspecific measure of stress is that it should act as an integrative measure across several independent or synergistic stressors and should be proportional to the physiological state of the organism" (Giesy et al., 1983, p. 283). The adenylate energy charge (AEC) of both species of freshwater clams listed in Table 4.1 was reduced after exposure to Cd. However, Giesy et al. (1983) pointed out that although the technique offers potential for monitoring stress, natural seasonal fluctuations must be known, a greater number of species and variety of stressors must be tested, and changes in AEC have to be related to more traditional measures of chronic stress such as alterations in growth and reproduction.

4.2.1.2. Enzyme Activities

Examples of studies concerned with the effects of various stresses on enzyme activities in freshwater benthic macroinvertebrates are summarized in

Table 4.1. Biochemical indicators of stress in freshwater benthic macroinvertebrates: energy metabolism.

Organism(s)	Phosphoadenylate Involved	Stress	Effects	Reference
Corbicula fluminea (freshwater clam)	ATP, ADP, AMP concentrations; adenylate energy charge [AEC = (ATP + $\frac{1}{2}$ADP)/(ATP + ADP + AMP)]	$CdCl_2$	AECs of foot muscle tissue were depressed in exposures to Cd below the 96 h LC_{50} (32 mg Cd/l); responses of adenylate concentrations were more variable	Giesy et al. (1983)
Anodonta cygnea (freshwater clam)	AEC	Cd	AEC significantly lowered in all tissues after 12 wk exposure to 50 μg Cd/l; viability of *A. cygnea* closely related to AEC level	Hemelraad et al. (1990a)

Table 4.2. Biochemical indicators of stress in freshwater benthic macroinvertebrates: enzyme activities.

Organism(s)	Enzymes Involved	Stress	Effects	Reference
Argia vivida nymphs (damselfly)	EST, G6PDH, LAP, LDH, MDH, TO[1]	Thermal gradient	Qualitative (LAP, LDH, TO) and quantitative (G6PDH) increases in isozymes in response to warm temperatures; no change in EST and MDH	Schott and Brusven (1980)
Epeorus latifolium, *Rhithrogena lepneva* nymphs (mayflies)	LDH, NE[2]	Mine tailings water and effluent from an ore concentrating plant	Frequency of occurrence of zones of enzyme activity increased below pollution sources; nymphs collected below mining effluents and having more variable isozymes were more resistant than those collected from a control stream	Mitin (1985)
Barytelphusa guerini (freshwater crab)	GPa, GPab, LDH, SDH[3]	$CdCl_2$	GPa and GPab concentrations increased; SDH decreased; LDH decreased (except at 4 d in heart and thoracic ganglion tissues where it increased)	Reddy et al. (1989)
Acroneuria lycorias, *A. abnormis* nymphs (stoneflies)	AChE[4]	Fenitrothion	AChE activity significantly reduced over time at all concentrations of fenitrothion (1, 2, 40 μg/l), although the two lower concentrations were similar; degree of inhibition was related to fenitrothion concentration and duration of exposure	Flannagan et al. (1978)
Several species of aquatic macroinvertebrates; field-collected: *Claasenia* sp. nymphs (stonefly), *Ephemerella* sp. nymphs (mayfly), *Hydropsyche slossonae/betteni* larvae (caddisfly); lab-cultured: *Hyalella azteca* (freshwater shrimp)	AChE	Organophosphate (OP) insecticides: azinphosmethyl (*Ephemerella*, *H. slossonae/betteni*, *H. azteca*); chlorpyrifos (*Claasenia*); fenitrothion (*Claasenia*)	Mixture in responses of AChE activities in aquatic macroinvertebrates at sublethal concentrations possible during field spraying activities; only exposure of *Claasenia* to concentrations of chlorpyrifos approaching lethality (20, 40, 60, 80 μg/l) for 24, 48, and 72 h significantly depressed AChE activity	Day and Scott (1990)

Barytelphusa guerini (freshwater crab)	AChE	NaF	AChE activity increased in all tissues after 4 d of exposure to 30 mg/l, but activity inhibited after 15 d of exposure	Reddy and Venugopal (1990a)
Procambarus clarkii (crayfish)	ATPase	Cd, Pb, Hg	Sublethal concentrations of Cd, Pb, and Hg did not cause notable effects on gill ATPase activity in exposures lasting up to 96 h	Torreblanca et al. (1989)
Corbicula sp. (freshwater clam)	Cellulolytic activity (exo- and endocellulases)	Cu and Zn in power plant effluents	Cellulolytic activity significantly reduced in: (1) field-located artificial streams, after 10 and 20 d exposure to 16 and 56 μg Cu and Zn/l respectively; (2) power plant outfall, after 10 d exposure to 80–345 μg Cu/l and 47–78 μg Zn/l in 1985; after 30 d exposure to 23–47 μg Cu/l and 15–47 μg Zn/l in 1986	Farris et al. (1988)
		$ZnSO_4$	Total enzyme activity declined with increasing Zn exposure concentration (nominal concentrations of 0, 50, and 1,000 μg Zn/l) and time (0, 5, 10, 20, and 30 d) in laboratory and field-laboratory systems; avoidance behavior of clams exposed to 1,000 μg/l resulted in delayed effects on cellulolytic activity	Farris et al. (1989)

[1]EST = esterase; G6PDH = glucose-6-phosphate dehydrogenase; LAP = leucine aminopeptidase; LDH = lactate dehydrogenase; MDH = malate dehydrogenase; TO = tetrazolium oxidase.
[2]LDH = as above; NE = nonspecific esterases.
[3]GPa = glycogen phosphorylase "a"; GPab = glycogen phosphorylase "ab"; LDH = as above; SDH = succinate dehydrogenase.
[4]AChE = acetylcholinesterase.

Table 4.2. Most of the enzymes used are metabolic or they deal with nerve transmission [i.e., acetylcholinesterase (AChE) activity]. The two electrophoretic studies (Schott and Brusven, 1980; Mitin, 1985) reported increased kinds and quantities of metabolic enzymes in response to stress; however, the quantitative responses reported by Reddy et al. (1989) were mixed. AChE activity should be inhibited by exposure to most stressors; for organophosphate (OP) insecticides, ≥20% depression in AChE activity indicates exposure, whereas ≥50% depression generally indicates a threat to life (Day and Scott, 1990). However, the responses described in Table 4.2 were variable. Day and Scott (1990) noted the variability in levels of AChE activity in other groups of animals exposed to low concentrations of OPs and advised that it was ". . . important to establish the 'normal' range of values of AChE levels within a species before attempting to document exposure" (Day and Scott, 1990, p. 111). These authors concluded that measurement of AChE activity was useful only for detecting acute toxicity following exposure of OPs in the field and that the choice of species may be important. Cellulase activity in the freshwater clam *Corbicula* sp. was depressed by exposure to Cu and Zn in both studies by Farris et al. (1988, 1989); Farris et al. (1988) claimed greater sensitivity (i.e., earlier detection of effects) by measuring cellulolytic enzyme activity than was obtained from bioassays or measurements of macroinvertebrate diversity. Future development of this method shows promise.

Except for Farris et al. (1988, 1989) and Day and Scott (1990), ecological implications of the changes documented in Table 4.2 were not fully known or considered and, in fact, four of the studies (Flannagan et al., 1978; Reddy et al., 1989; Torreblanca et al., 1989; Reddy and Venugopal, 1990a) were limited to the laboratory. Moreover, the gill ATPase study of Torreblanca et al. (1989) yielded no observable effects. In general, additional research involving different species, different kinds and concentrations of stressors, and studies of natural variability is required before enzyme activities of freshwater benthic macroinvertebrates can be applied routinely as a biochemical monitoring tool.

4.2.1.3. RNA, DNA, Amino Acid, and Protein Content

Selected studies of the effects of various stresses on RNA, DNA, amino acid, and protein content of freshwater benthic macroinvertebrates are summarized in Table 4.3. It is evident, as for the previous biochemical indicators, that only a limited range of species and environmental disturbances has been examined. Also evident is the variability in responses. For example, the studies that used total and individual free amino acid (FAA) concentrations as a bioindicator (Gardner et al., 1981; Bhagyalakshmi et al., 1983; Graney and Giesy, 1986, 1987, 1988; Day et al., 1990; Saber Hussain

Table 4.3. Biochemical indicators of stress in freshwater benthic macroinvertebrates: RNA, DNA, amino acids, and protein.

Organism	Protein/ Constituent(s) Involved	Stress	Effects	Reference
Austropotamobius pallipes (crayfish)	Total protein (hemolymph)	Mining (suspended sediments)	Concentrations lower in crayfish from polluted area than controls	Kabré and Chaisemartin (1986–1987)
Neochetina eichhorniae adults (water hyacinth weevil)	RNA, DNA, protein	Cd, Mn, Zn (at 10, 25, and 50 ppm)	Cd decreased RNA, DNA, and protein concentrations at all three heavy metal doses; Mn was similar but had greatest effect at 50 ppm; Zn increased 3 variables over range of concentration used	Saber Hussain and Jamil (1989)
	Total protein, total free amino acids (FAA)	Hg	Total protein decreased; decrease greatest in weevils fed on leaves contaminated at lowest concentration (25 ppm cf. 50 and 75 ppm) because of feeding avoidance at higher concentrations; total FAA increased, possibly indicating use of supplementary energy source to meet requirements of Hg stress	Saber Hussain and Jamil (1990)
Barytelphusa guerini (freshwater crab)	Total protein, total FAA	NaF	Effects of exposure to sublethal concentrations (30 mg/l) of NaF measured in gills, muscle, hepatopancreas, heart, and thoracic ganglion at 4 and 15 d; total protein concentrations decreased significantly after 15 d in all tissues except thoracic ganglion; after 4 d, total FAA concentrations decreased in gills and muscle and increased in the three other tissues; after 15 d, total FAA decreased in all tissues	Reddy and Venugopal (1990b)
Gammarus pseudolimnaeus (freshwater shrimp)	Total and individual FAA	Pentachlorophenol (PCP) and osmotic stress	Exposure to acutely toxic concentrations of PCP significantly decreased total FAA pool at higher concentration (1.68 mg PCP/l) and significantly changed FAA profile at lower concentration (1.13 mg PCP/l); hyperosmotic conditions had no effect, whereas hypoosmotic conditions significantly decreased FAA pool	Graney and Giesy (1987)

(*Continued*)

Table 4.3. (Continued)

Organism	Protein/ Constituent(s) Involved	Stress	Effects	Reference
		PCP	Long-term (45 d) exposure to sublethal concentrations of PCP significantly decreased concentrations of total FAA after 5 d at 0.77 and 1.25 mg PCP/l (but not at 1.06 mg/l); individual FAA not affected at any PCP concentration	Graney and Giesy (1986)
Oziotelphusa senex senex (freshwater crab)	Total FAA	Sumithion	Concentration of total FAA increased during acute exposure and decreased during chronic exposure	Bhagyalakshmi et al. (1983) (as cited in Graney and Giesy, 1987)
Amblema plicata (freshwater clam)	Total and individual FAA	Coal acid mine drainage and lead mine tailings (Pb and Cd)	Significantly higher total FAA concentrations in polluted than control streams, but relative compositions of FAA were similar	Gardner et al. (1981)
Corbicula fluminea (freshwater clam)	Total and individual FAA	Sodium dodecyl sulfate (SDS)	Total FAA concentrations in two tissue types increased significantly in short-term (96 h) exposures to 25 mg SDS/l and chronic (60 d) exposures to 0.65 and 3.0 mg SDS/l; alterations of individual FAA were similar in both exposures: alanine concentrations and % contribution to total FAA pool were higher, whereas glutamic acid concentrations either were unaffected or decreased	Graney and Giesy (1988)
Elliptio complanata (freshwater mussel)	Total and individual FAA	Yamaska River, Quebec, catchment waters	Mussels caged and placed in various locations in the Yamaska River catchment for exposure periods of 27–29 d and 77–79 d; total FAA concentrations in two of three tissue types were elevated at a number of sites affected by various sources of contamination; however, total FAA concentrations also decreased at some contaminated sites; decreases in % contribution to total FAA pool of glycine, serine, threonine, and valine at several sites, and % increases in glutamine and glutamic acid at same sites	Day et al. (1990)

Table 4.4. Summary of responses to stress as measured by changes in RNA, DNA, amino acids, and protein for studies listed in Table 4.3 (stresses and abbreviations as given in Table 4.3).

Protein/ Constituent(s)	Response to Stress			
	Quantitative Changes (Concentration)			
	Decrease	Increase	None	Qualitative Changes
Total protein	Mining (suspended sediment)			
	Cd, Mn	Zn		
	Hg			
	NaF			
RNA, DNA	Cd, Mn	Zn		
Total FAA pool	NaF			
	PCP (acute)[1], hypoosmotic conditions	Hg	Hyperosmotic conditions	
	PCP (chronic)			

(*Continued*)

Table 4.4. (Continued)

Protein/ Constituent(s)	Response to Stress			
	Quantitative Changes (Concentration)			Qualitative Changes
	Decrease	Increase	None	
	Sumithion (chronic)	Sumithion (acute)		
		Acid mine drainage and lead tailings		
	Yamaska River water	Yamaska River water	Yamaska River water (gill tissue)	
		SDS		
Individual FAA			PCP (chronic)	PCP (acute)[2]
			Acid mine drainage and lead tailings	
	Yamaska River water (glycine, serine, threonine, valine)	Yamaska River water (glutamine, glutamic acid)	Yamaska River water	
	SDS (glutamic acid)	SDS (alanine)	SDS (glutamic acid)	

[1]At higher concentration.
[2]At lower concentration; concentration profile changed.

and Jamil, 1990) revealed somewhat inconsistent responses (Table 4.4). In discussing the difference between the Gardner et al. (1981) study and theirs, Graney and Giesy (1987, p. 170) commented: "The reason for the differences is unknown; however, given the variety of mechanisms by which the total FAA concentration can be altered, it is not surprising that the response of the total FAA concentration varies in different organisms stressed under different conditions." Fuller understanding of this variability is needed before changes in FAA concentrations can be used effectively as an *in situ* indicator of stress (Graney and Griesy, 1988; Day et al., 1990).

Graney and Giesy (1987, p. 171) noted that ". . . when developing a general biochemical indicator of stress, understanding the mechanism is not always essential as long as the changes observed are consistent, can be quantitatively or qualitatively related to exposure, can be separated from the "background noise" or variability normally associated with field conditions and are related to ecologically relevant end points." Although Graney and Giesy (1986, 1988) attempted to link stress-related changes in total and individual FAA to "ecologically relevant end points," many of the criteria described by Graney and Giesy (1987) are not fulfilled by the studies summarized here. It is obvious that more work is required before this group of biochemical indicators can be included in routine biomonitoring programs using freshwater benthic macroinvertebrates.

4.2.1.4. Ion Regulation

Effects of environmental stresses on ion regulation have received more attention than the previous categories of biochemical indicators; most of these studies concern acid-stress effects on crayfish (Table 4.5). Emphasis on many different species of crayfish under a variety of experimental conditions has revealed certain trends in ion regulation. Typically, hemolymph concentrations of Na^+ and Cl^- decrease in response to acid stress, whereas Ca^{++} concentrations increase, a similar response to that observed in bivalves (Table 4.6). Apparently, crayfish and clams can mobilize Ca^{++} from their exoskeleton and shell and mantle, respectively, to buffer acid effects (Wood and Rogano, 1986; Malley et al., 1988; Pynnönen, 1990). The crayfish *Cambarus robustus* appears to be an exception to the trend, perhaps because this species is acid-tolerant (Hollett et al., 1986). Instances of decreased Ca^{++} concentrations in crayfish are unexplained.

In contrast, concentrations of most major ions in aquatic insects that have been exposed to acid stress appear to decrease or remain unchanged (Table 4.5), and little consistency in response among ions is evident (Table 4.6). Obvious physiological differences exist between insects and crustaceans in ion regulation responses to acid stress (e.g., ability of crayfish to mobilize Ca^{++} from their exoskeleton) but far more work on ion regulation in insects is required before trends in this group can be identified with certainty.

Table 4.5. Biochemical indicators of stress in freshwater benthic macroinvertebrates: ion regulation. (Designation of ions follows source publication.)

Organism(s)	Ions Involved	Effects	Reference
A: Naphthalene			
Chironomus attenuatus 4th instar larvae (midge)	Na^+, K^+, Cl^- (hemolymph)	Exposure for 1, 2, and 4 h to concentrations of 0, 1, 5, and 10 mg/l significantly elevated hemolymph Na^+, K^+, and Cl^- concentrations	Darville et al. (1983)
B: Cadmium			
Anodonta cygnea zellensis (freshwater clam)	Na^+, K^+, Ca^{++}, Mg^{++}, Fe^{++}, Zn^{++}, other elements (various tissues, hemolymph)	Exposure to 50 μg Cd/l for 12 wk; most obvious effect: Na^+ concentrations in total body, separate tissues, and hemolymph declined dramatically (wk 2–8 of exposure) and then stabilized; K^+ concentration in total body declined after 8 wk exposure; minor changes for other ions (Ca^{++}, Fe^{++}, Zn^{++}); no change for many others; hemolymph concentrations of some elements (Zn, S) showed opposite response to that of tissues	Hemelraad et al. (1990b)
C: Acid stress (often involving Al and other trace metals)			
1. Mollusca *Anodonta grandis grandis* (freshwater clam)	Na^+, K^+, Ca^{++}, Mg^{++}, Cl^-, $SO_4^=$ (blood); Ca, Na, Mn, Al, Cd (tissues)	Clams from a neutral-pH lake introduced into an acidic one (pH 5.9); changes in blood: marked elevation of Ca^{++}, temporary increase in Cl^-, decline in Mg^{++}, and no change in Na^+, K^+, or $SO_4^=$; after experimental addition of alum that caused short-term extremes of pH 4.5 and Al concentration of 2,237 μg/l, Na^+ and Cl^- concentrations declined and Ca^{++} concentrations increased even more; after alum addition, Al, Ca, and Mn in tissues increased, Na decreased, and Cd did not change	Malley et al. (1988)
Anodonta anatina, A. cygnea, Unio pictorum, U. tumidus (freshwater clams)	Na^+, K^+, Ca^{++}, Cl^- (hemolymph)	Exposure to acidified (pH 4.0–4.5) soft water (4.6 mg Ca/l) for 2 wk and hard water (18.5 mg Ca/l) for 4 wk; in general, exposures decreased concentrations of Na^+, Cl^-, and K^+ (although K^+ not much affected) and increased concentrations of Ca^{++} in hemolymph; elevation of hemolymph Ca^{++} positively correlated with decreased hemolymph pH; soft-water conditions accelerated hemolymph pH decrease and Ca^{++} increase, whereas Na^+ and Cl^- concentrations changed less rapidly	Pynnönen (1990)

2. Insecta			
Corixa punctata adults (water boatman)	Na^{+} (hemolymph)	Although Na-influx declined with increasing Al concentrations, and was lower overall at pH 3 than pH 4, hemolymph Na^{+} concentrations increased significantly; no mortality over 20 h of test	Witters et al. (1984)
Limnephilus pallens 4th and 5th instar larvae (caddisfly); *Chironomus riparius*, *Orthocladius consobrinus* 3rd and early 4th instar larvae (midges)	Na, Cl, Ca	Acid-sensitive *O. consobrinus* suffered net loss of total-body Na and Cl at low pH and showed a significant negative correlation between mortality and Na and Cl content; its Ca concentrations decreased at low pHs in initially acidic water; acid-tolerant *L. pallens* and *Ch. riparius* did not suffer Na loss when exposed to acid water (except at pH 2.8 for *L. pallens*); Cl concentrations were lower at all pHs in initially acidic water than reference water for *L. pallens* but did not change for *Ch. riparius* regardless of pH treatment and source of water; *L. pallens* lost Ca at pH 3.0 in reference water and at all pHs in initially acid water; *Ch. riparius* showed a similar pattern; K was not lost from either midge species	Havas and Hutchinson (1983)
Pteronarcys proteus nymphs (stonefly)	Na, K, Ca, Mg	Exposure to pH 3.0 for 120 h caused no significant differences in body burdens of Ca, Mg, or K compared to controls; however, Na body burden progressively decreased with time and became significantly lower than control values from 72 h onward	Lechleitner et al. (1985)
Leptophlebia cupida ≥26th instar nymphs (mayfly); *Prosimulium fuscum/mixtum* 5th–7th instar larvae (black fly)	Na, Ca	Both species transplanted from pH 6.2 to 4.2 streamwater showed significant decreases in whole-body concentrations of Ca; whole-body Na decreased in black flies	Hall et al. (1988)
3. Crustacea: crayfish			
Astacus astacus, *Pacifastacus*	Na^{+}, K^{+}, Ca^{++}, Cl^{-} (hemolymph)	*Experiment 1:* exposure of *A. astacus* to pH 3.7 for 3 d significantly reduced the hemocyte fraction of K^{+} but Na^{+} was	Appelberg (1985)

(Continued)

Table 4.5. (Continued)

Organism(s)	Ions Involved	Effects	Reference
leniusculus intermolt stage (C4)		unaffected; plasma Na^+ and Cl^- were significantly reduced, Ca^{++} was increased, and K^+ was unaffected; *Experiment 2:* total hemolymph concentrations in *A. astacus* of Na^+, K^+, Ca^{++}, and Cl^- were significantly decreased at pH 4.0 after 11 d; significant losses of Na^+ and Ca^{++} occurred at pH 5.0 after 11 d, but these were smaller than at pH 4.0; *Experiment 3:* concentrations of Na^+ in hemolymph of both species were significantly reduced after 14 d exposure at pH 5.0 to different concentrations of Al^{+++}	
Orconectes propinquus intermolt adults, *O. rusticus* intermolts	Na^+, K^+, Ca^{++}, Cl^- (hemolymph)	After 5 d at pH 4.0, concentrations of Na^+ and Cl^- in *O. propinquus* decreased significantly, Na^+ more so than Cl^-; K^+ also decreased but a similar change occurred in controls; Ca^{++} concentrations increased significantly; in *O. rusticus* Na^+ and Cl^- decreased progressively, by similar amounts, over time but not to the same extent as in *O. propinquus;* K^+ concentrations almost doubled; Ca^{++} increased significantly, the overall increase being twice that in *O. propinquus*	Wood and Rogano (1986)
Cambarus robustus intermolt adults and stage III juveniles, *Orconectes rusticus* stage III juveniles	Na^+, Ca^{++} (adult hemolymph), Na^+ (juvenile whole body)	After 4 d at pH 3.8, no change in concentrations of Na^+ and Ca^{++} in hemolymph of adult *C. robustus* (cf. adult *O. rusticus* in Wood and Rogano, 1986); total-body Na^+ of juveniles remained constant in all treatments; in contrast, juvenile *O. rusticus* exposed to pH 3.8 for 4 d lost ~50% of their total-body Na^+ after 48 h; no effects at pH 5.0 in juveniles	Hollett et al. (1986)
Austropotamobius pallipes (4–7 cm in length)	Na^+ (hemolymph)	Na^+ concentrations in caged *A. pallipes* were significantly lower in the three acid zones (A: untreated, pH 5.3, 0.32 mg Al/l; B: acidified to pH 4.5; C: acidified to pH 4.5, Al concentration raised by 0.3–0.4 mg/l) (overall mean hemolymph concentration: 167 meq Na/l) than in the limed zone (D: overall mean: 295 meq Na/l) of an experimentally manipulated stream	Weatherley et al. (1989)
Astacus astacus intermolt adults	Na^+, K^+, Ca^{++}, Mg^{++}, Cl^- (hemolymph)	pH tended to increase, decrease, and decrease strongly under hypoxic, acidic, and hypoxic-acidic conditions, respectively; most obvious ion concentration changes: Ca^{++} and Mg^{++} increased and Cl^- decreased under hypoxia; Na^+ decreased under hypoxic-acidic conditions	Nikinmaa et al. (1983)

Table 4.6. Summary of responses to acid stress as measured by changes in major ion concentrations for studies listed in Table 4.5. (Designation of ions follows source publication.)

Organism(s)	Experimental Conditions	Effects on Ion Concentrations			Reference
		Increase	Decrease	No change	
A: Insects					
Corixa punctata	Hemolymph	Na^{+}			Witters et al. (1984)
Limnephilus pallens	Total body		Ca, Cl	Na	Havas and Hutchinson (1983)
Chironomus riparius			Ca	Na, K, Cl	
Orthocladius consobrinus			Na, Ca, Cl	K	
Pteronarcys proteus	Total body		Na	K, Ca, Mg	Lechleitner et al. (1985)
Leptophlebia cupida	Total body		Ca	Na	Hall et al. (1988)
Prosimulium fuscum/ mixtum			Na, Ca		
B: Clams					
Anodonta grandis grandis	Blood levels: transplantation	Ca^{++}, Cl^{-}	Mg^{++}	Na^{+}, K^{+}, $SO_4^{=}$	Malley et al. (1988)
	: alum addition	Ca^{++}	Na^{+}, Cl^{-}		
	Tissues: alum addition	Ca, Mn, Al	Na	Cd	
Anodonta anatina, A. cygnea, Unio pictorum, U. tumidus	Hemolymph	Ca^{++}	Na^{+}, K^{+}, Cl^{-}		Pynnönen (1990)
C: Crayfishes					
Astacus astacus	Experiment #1: hemocytes		K^{+}	Na^{+}	Appelberg (1985)
	: plasma	Ca^{++}	Na^{+}, Cl^{-}	K^{+}	

(continued)

Table 4.6. (Continued)

Organism(s)	Experimental Conditions	Effects on Ion Concentrations			Reference
		Increase	Decrease	No change	
	Experiment #2: total hemolymph		Na^+, K^+, Ca^{++}, Cl^-		
	Experiment #3: hemolymph		Na^+		
Pacifastacus leniusculus	Experiment #3: hemolymph		Na^+		
Orconectes propinquus	Hemolymph	Ca^{++}	Na^+, Cl^-, K^+(?)		Wood and Rogano (1986)
Orconectes rusticus		K^+, Ca^{++}	Na^+, Cl^-		
Cambarus robustus	Hemolymph: adults			Na^+, Ca^{++}	Hollett et al. (1986)
	Total body: juveniles			Na^+	
Orconectes rusticus	Total body: juveniles		Na^+		
Austropotamobius pallipes	Hemolymph		Na^+		Weatherley et al. (1989)
Astacus astacus	Hemolymph: acidic conditions		Ca^{++}		Nikinmaa et al. (1983)
	: hypoxic-acidic conditions		Na^{++}, Cl^- (♀♀)		

Given the information available for crayfish, routine monitoring programs could be established using field populations presumed to be exposed to acid stress. However, changes in major ions described above should be interpreted in the light of acid sensitivity or tolerance of the species; comparison with a known reference population is essential. Subjective measurements of carapace rigidity in the crayfish *Orconectes virilis* (France, 1987a; Davies, 1989) have been used to indicate disturbance of Ca^{++} metabolism by acid stress; however, in reality, this method is a physiological indicator of ion regulation difficulty.

4.2.1.5. Other Biochemical Measures

The incipient application to freshwater benthic macroinvertebrates of biochemical indicators that currently are being used in other biotic groups should be mentioned. For example, the metal-binding protein metallothionein (MT) has been used to indicate metal exposure in mammals and fish (e.g., Tohyama et al., 1981; Hamilton and Mehrle, 1986). Although MTs or MT-like proteins are known to occur in freshwater macroinvertebrates (e.g., mayflies: Aoki et al., 1989; chironomids: Yamamura et al., 1983; clams: Doherty et al., 1987b), use of these proteins as indicators has not yet been reported.

A number of other biochemical indicators have potential for adaptation to freshwater macroinvertebrates. Perhaps the most promising of these is the induction of cytochrome P450-dependent monooxygenases, which has been proposed as an indicator of exposure to polychlorinated biphenyls and polycyclic aromatic hydrocarbons in fish and marine invertebrates (G.T. Ankley, U.S. Environmental Protection Agency, Duluth, MN, personal communication). Eventually, biochemical markers such as these also may be used in freshwater macroinvertebrates.

4.2.2. Physiological Indicators

Physiological indicators include measurements of changes in activities such as heartbeat (e.g., Costa, 1970; Trueman et al., 1973), bioelectric action potential (e.g., Idoniboye-Obu, 1977), or respiration rate, and the recent adaptation from marine to freshwater organisms of "scope for growth" measurements (Naylor et al., 1989), which use respiration in their calculation. Respiratory metabolism, which is a broad integrator of all physiological functions (Maki et al., 1973), has received most of the attention in work with freshwater benthic macroinvertebrates.

Many studies have measured the effects of a toxicant on respiration in the laboratory, but did not attempt to relate their methods or findings to field conditions. The studies listed in Table 4.7 showed some degree of application to the field. For example, the response of aquatic organisms to naph-

Table 4.7. Effects of environmental stresses on respiratory metabolism in freshwater benthic macroinvertebrates.

Stress	Organism	Measurements Involved	Effects	Reference
Naphthalene	*Somatochlora cingulata* nymphs (dragonfly)	O_2 consumption rate (mg O_2/h/g wet wgt) at naphthalene concentrations of 0.0 (control), 0.01, and 0.1 mg/l at 2, 6, 12, and 24 h	O_2 consumption increased with concentration of naphthalene (0.1 mg/l = 2× that of controls); differences after 24 h were significant (AOV: $P < 0.001$)	Correa and Coler (1983)
	Chironomus attenuatus, Tanytarsus dissimilis 4th instar larvae (midge)	O_2 consumption rate (μl/h/mg dry wgt) at naphthalene concentrations of 0, 1, 5, 10, and 12 mg/l for 1-h exposure	O_2 consumption rates of both species depressed (at 12 mg/l: *Ch. attenuatus* by 34%, *T. dissimilis* by 49%)	Darville and Wilhm (1984)
Trichloroacetic acid (TCAA)	*Somatochlora cingulata* 0.1–0.4 g nymphs (dragonfly)	O_2 consumption rate (μg O_2/h/g wet wgt) at TCAA concentrations of 0.0 (control), 0.01, 0.1, and 1.0 mg/l for 8-h exposure	O_2 consumption increased with concentration of TCAA (1 mg/l ~2× that of controls); differences between controls and experimental animals significant ($P < 0.01$)	Correa et al. (1985a)
	Aeshna, Basiaeshna nymphs (dragonfly)	O_2 consumption rate (mg O_2/h/g wet wgt) at TCAA concentrations of 0, 1, 10, 100, and 1,000 ppb at 2, 4, 8, 12, 24, 48, and 96 h	O_2 consumption increased with concentrations of TCAA (1,000 ppb ~2.5× that of controls); 100 and 1,000 ppb significantly different from controls (AOV, Duncan's New Multiple Range Test)	Calabrese et al. (1987)
TCAA and dichloroacetic acid (DCAA)	*Aeshna umbrosa* near final instar nymphs (dragonfly)	O_2 consumption rate (mg O_2/h/g dry wgt) at concentrations of 0 (control), 1, 10, 100, and 1,000 μg/l of TCAA, DCAA, and TCAA + DCAA for 8-h exposure; the TCAA + DCAA treatment doubled the nominal concentrations	Dose-dependent increase in O_2 consumption for TCAA, DCAA, and TCAA + DCAA >10 μg/l (AOV: $P < 0.001$); 100 and 1,000 μg/l differed significantly from controls for the 3 chemical treatments (Dunnett's Multiple Comparison Test: $P < 0.05$); effect of TCAA and DCAA on O_2 consumption rate not significantly different (AOV: $P > 0.05$); no evidence of interaction between TCAA and DCAA	Dominguez et al. (1988)

$Al + H^+$	*Somatochlora cingulata* nymphs of 4 different wgt classes: 0.01–0.5 g (dragonfly)	O_2 consumption rates ($\mu g\ O_2/h/g$ wet wgt) at pHs of 6.75 (control), 4.20, and 3.59, and concentrations of 0 (control), 10, 20, and 30 mg Al/l at pH 4.20	O_2 consumption decreased with increasing concentrations of Al and H^+; greatest reduction in smaller nymphs; significant differences between control animals and those exposed to low pH and low pH + Al (AOV: $P < 0.01$); Al concentrations at low pH as toxic as low pH alone	Correa et al. (1985b)
	Limnephilus sp. larvae of 3 different wgt classes: 0.04–0.14 g (caddisfly)	O_2 consumption rates (mg O_2/h/g wet wgt) at pH 6.75 (control), pH 4.0, and pH 4.0 + 0.3 mg Al/l; measurements done every 24 h for 4 d; larvae removed from cases for experiments	No significant differences between controls and experimental animals exposed to low pH and low pH + Al (AOV: $P > 0.05$)	Correa et al. (1986)
	Heptagenia fuscogrisea 1.5–4 mg nymphs, *H. sulphurea* 2–5 mg nymphs, and *Ephemera danica* 12–22 mg nymphs (mayflies)	O_2 consumption rates (mg O_2/24 h/g dry wgt) at 0 (control), 0.5, and 2.0 mg Al/l, and pHs 4.0 and 4.8 for 10 d	O_2 consumption rates increased significantly at both Al concentrations (AOV); effect of pH less important; base-level respiration rate differed among species used	Herrmann and Andersson (1986)
	Libellula julia last instar nymphs (dragonfly)	O_2 consumption rates ($\mu g\ O_2/h/g$ wet wgt) at pH 6.1 + 0.01 mg Al/l (control), and pH 4.0 at 0.02 (acidified water control), 0.3, 3.0, and 30 mg Al/l; O_2 uptake determined at 24, 40, 48, 72, 88, and 96 h	O_2 consumption rates significantly depressed over controls at all treatment levels of Al (AOV: $P < 0.01$) and declined at all Al concentrations, although only 3.0 and 30.0 mg/l significantly different from controls (Duncan's New Multiple Range Test: $P < 0.05$); 0.02 and 0.3 mg Al/l significantly different from 30 mg Al/l; low pH (4.0) alone did not appear to inhibit O_2 uptake as much as low pH and Al concentrations ≥ 3 mg/l; no significant differences in O_2 consumption among times ($P > 0.05$)	Rockwood et al. (1990)

(continued)

Table 4.7. (Continued)

Stress	Organism	Measurements Involved	Effects	Reference
Cd, Hg, NaPCP[1]	*Limnodrilus hoffmeisteri*, *Stylodrilus heringianus*, *Monopylephorus cuticulatus*[2] (oligochaete worms)	Respiration rate for each species (μl O_2/h/mg dry wgt) at 2 different concentrations each of Cd, Hg, and NaPCP and 2 different levels of modified environmental factors (pH, salinity, and temperature); another set of experiments measured respiration rate for each species using paired, modified environmental factors and toxicants	Single stress factors: results variable (see table 3, p. 689), but exposures caused detectable changes in respiration rate; paired stress factors: exposures caused additional changes in respiration rate in 41/54 comparisons (see table 4, p. 695)	Brinkhurst et al. (1983)
Dibrom (pesticide)	*Hydroperla crosbyi* nymphs of various sizes (stonefly), *Corydalus cornutus* nymphs of various sizes (hellgrammite); stonefly chosen as "intolerant" organism, whereas hellgrammite chosen as "very tolerant" organism	O_2 consumption rate (μl 0_2/h/mg dry wgt) at predetermined sublethal exposures to Dibrom (stonefly: 8 ppb; hellgrammite: 5 ppm) for 20 min to 4 h (depending on time of death)	Significant difference between O_2 consumption rates in presence and absence of Dibrom (test of homogeneity of coefficient of regression); large stonefly nymphs consumed significantly more O_2 in presence of pesticide than did comparable-sized nymphs in controls, whereas smaller nymphs showed reverse response; O_2 consumption of hellgrammites significantly greater than controls in presence of pesticide over all sizes	Maki et al. (1973)

[1]Sodium pentachlorophenol.
[2]Estuarine species.

thalene, as measured by oxygen consumption, in general appears to be highly variable (Darville and Wilhm, 1984). However, the two studies reviewed in Table 4.7 are not directly comparable because of different experimental conditions (e.g., naphthalene concentrations, measurement periods of oxygen consumption) and different species of aquatic insects (see comments under low pH and Al below). Darville and Wilhm (1984) claimed that *Tanytarsus* was more sensitive to naphthalene than *Chironomus* because the former was a "clean-water" genus, whereas the latter was a "tolerant" genus. However, assigning pollution tolerances at the genus level is questioned by Resh and Unzicker (1975) (see also discussion in Resh and McElravy, Chapter 5).

Fairly uniform results were obtained by the three studies that examined the effects of environmentally relevant concentrations of tri- and dichloroacetic acids, which are by-products of water chlorination, on odonate nymphs collected from various aquatic habitats in Massachusetts (Table 4.7). Oxygen consumption rates were dose-dependent; concentrations of 100 and 1,000 μg/l of these haloform chemicals caused significant increases in respiration rate over controls. Differences among the studies in respiratory response observed at the highest concentration tested were attributed to differences in experimental design (different taxa of odonates used, observations based on wet weight vs. dry weight, etc.; see Dominguez et al., 1988, p. 267).

Despite relative uniformity in experimental conditions of the studies examining low pH and Al effects on respiration (i.e., similar H^+ and Al concentrations, similar durations of exposure), responses of the aquatic insects used differed somewhat (Table 4.7). Al seemed important in the responses of the mayflies and the dragonfly *Libellula julia,* but not in the dragonfly *Somatochlora cingulata*. Oxygen consumption rates were depressed in both dragonfly species, unaffected in the caddisfly, and increased in the three mayfly species. Contrasting responses between the mayflies of Herrmann and Andersson's (1986) study, and the dragonflies of Correa et al. (1985b) and Rockwood et al. (1990) may result from morphological differences (Rockwood et al., 1990). It is possible that mayflies are capable of cleaning their gill membranes of mucus accumulations and/or Al hydroxide precipitates that were caused by the treatment, whereas dragonflies cannot do so because their gills are confined in a rectal gill chamber.

Only Herrmann and Andersson (1986) tried to relate their findings to the natural distribution of the animals in their study. *Ephemera danica* is restricted to Swedish streams having high pH and low Al concentration; *Heptagenia fuscogrisea* occurs in streams of low pH and high Al concentration; and *H. sulphurea* occurs in conditions intermediate between the two other species. Accordingly, increased Al concentrations in the laboratory studies caused highly significant respiration increases and the highest overall increase ($\simeq$100%) in *E. danica*. The respiratory response was less pronounced for the two species of *Heptagenia*, and no clear order of response could be discerned between them.

Brinkhurst et al. (1983) (Table 4.7) examined the sublethal effects of environmental variables and toxicants, alone and in pairs, on respiration rates, critical oxygen levels, and degree of respiratory regulation in three species of aquatic oligochaetes. The species were selected on the basis of their pollution tolerance/intolerance as determined from field studies. Results indicated that the three respiratory measures were not correlated with field-determined tolerance to organic pollution (cf. Maki et al., 1973 in Table 4.7). The authors concluded that, although laboratory studies may be useful in comparing harmful toxicants, they are less reliable in determining responses of natural fauna to these toxicants under field conditions; laboratory tests simply cannot duplicate important biotic factors that occur in the field.

Overall, it appears that respiratory responses to sublethal stresses may be species and stress specific (Brinkhurst et al., 1983). Although respiration of oligochaetes has been used in marine studies as an indicator of sediment quality (Long and Chapman, 1985; Chapman, 1986), experiments in fresh waters will have to be done with a wider range of organisms and toxicants, and under more environmentally realistic conditions, before generalizations can be made and respiration becomes a routinely applied measurement in biomonitoring.

The "scope for growth" (SfG) technique of Naylor et al. (1989) involves monitoring how various components of the energy budget of individual *Gammarus pulex* are affected by stress, that is:

$$C - F - R = SfG$$

where C = energy consumed as food, F = energy lost as feces, and R = energy metabolized (measured as respiration). Positive values of SfG indicate that energy is available for growth and reproduction, whereas negative values indicate that animals are under stress and are using reserves for essential metabolism. If it can be assumed that observed changes in the energy budget are related to subsequent changes in growth and fecundity, then the approach provides rapid measures of "potential" growth, and it provides insight into which components of the energy budget are most affected by stress. For example, Naylor et al. (1989) reported that energy absorbed (= C−F) was more sensitive than energy lost (R) to the effects of Zn and pH. The preliminary work of Naylor et al. (1989) showed that SfG yielded order of magnitude more sensitive measures of Zn and pH stress than those obtained by acute toxicity testing.

4.2.3. Morphological Deformities

The occurrence of morphological deformities in benthic macroinvertebrates living in polluted environments has been known for over two decades

(Brinkhurst et al., 1968); the idea that such abnormalities could be used to indicate contaminant stress in field situations has persisted since then (Table 4.8). The organisms involved belong to only two groups of invertebrates: the Insecta and the Oligochaeta. However, most reports concern the insect family Chironomidae (see review in Warwick, 1988), and particularly the genus *Chironomus*. Most commonly reported are deformities of the head capsule, especially the mouthparts. Causative agents in field studies of highly contaminated sites are vaguely referred to as "industrial and agricultural pollutants," whereas experimental studies deal with specific toxicants (e.g., dichlobenil herbicide, DDE, coal liquid, copper) (Table 4.8).

The frequency of occurrence of deformities—natural and otherwise—varies widely. For example, Warwick (1988) reported that 19% of *Cryptochironomus* larvae from uncontaminated Maskwa Lake, Saskatchewan, showed some deformity. Many studies are poorly controlled, so attribution of deformities to contaminants is uncertain, although observations of high percentages in locations known to be severely contaminated generally are not in doubt. Some authors have reported increased incidences with increasing contamination in field-collected samples (e.g., Milbrink, 1983; Wiederholm, 1984a; Warwick et al., 1987; Dickman et al., 1990); however, the best dose-response relationships obviously came from experimental laboratory studies (e.g., Kosalwat and Knight, 1987).

Currently, morphological deformities in benthic macroinvertebrates remain only a qualitative measurement of the presence of contaminants in an ecosystem and, unfortunately, a measurement that is restricted to only a few taxonomic groups. The search for morphological deformities should be expanded to a wider group of benthic macroinvertebrate taxa. Moreover, to make the use of deformities more quantitative, the following aspects should be pursued:

1. Deformities and their frequencies of occurrence in response to environmental contaminants must be firmly established using adequate controls. In the absence of suitable control sites, paleolimnological techniques may have to be used (e.g., Warwick, 1980a; Wiederholm, 1984a; see Walker, Chapter 9).
2. The full extent of deformities in various body parts must be documented and indices to their ranges of severity developed. Indices to deformities in the antennae of *Chironomus* and the ligulae and antennae of *Procladius* are available (Warwick, 1985, 1991), but these indices are highly technical, which may inhibit their widespread, general use.
3. Cause-effect relationships need to be established by experimentation. The toxicants or groups of toxicants (and possible synergisms) that produce particular morphological deformities should be iden-

Table 4.8. Morphological deformities in benthic macroinvertebrates.

Organism	Location	Morphological Deformity	Causal Agent(s)	Comments	Reference
Hemiptera: Corixidae *Corixa punctata* *Sigara dorsalis*	Experimental ponds in the Lincolnshire Fens, England	Lack of pigmentation in nymphs and adults	Dichlobenil herbicide	—	Tooby and Macey (1977)
Plecoptera: Capniidae *Isocapnia integra* *Utacapnia columbiana*	Bow River, Alberta	Enlarged or unusually shaped segments of the antenna, maxillary palpus, labial palpus, and cercus; an extra antenna; unusually shaped femur; labial or maxillary palpus missing segments or present as a short bud	Domestic and industrial pollutants	11–14% of *U. columbiana* and 15–80% of *I. integra* deformed at affected stations (cf. 0–3% at control stations)	Donald (1980)
Plecoptera: Perlidae *Phasganophora capitata*	Gooseberry Creek, New York	Tracheal gills of nymphs atrophied	Chlorine	100% of specimens deformed at affected site; specimens at control sites were normal	Simpson (1980)
Trichoptera: Hydropsychidae *Cheumatopsyche* sp.	Gooseberry Creek, New York	Tracheal gills of larvae atrophied	Chlorine	83% of specimens deformed at affected site; specimens at control sites were normal	Simpson (1980)
Hydropsyche pellucidula	Río Manzanares, central Spain	Darkened anal papillae and damaged tracheal gills of larvae	Chlorine	62% of specimens deformed at affected site; specimens at control site were normal; similar deformities produced in the laboratory	Camargo (1991)
Diptera: Chironomidae *Chironomus* *Procladius, Protanypus, Chironomus, Stictochironomus,* unidentified Tanytarsini and Chironomini	W end of L. Erie Skaha and Okanagan lakes, British Columbia	Menta and mandibles of larvae; head capsules heavily pigmented and thickened; thickened body walls of thorax and abdomen	Industrial and agricultural pollutants	~1% of specimens collected from Skaha and Okanagan lakes deformed; less in Lake Erie (see Warwick, 1980b, p. 262, for data)	Hamilton and Sæther (1971)

Chironomus (s.s.)? *cucini* (*salinarius* grp.); single deformed specimens of *Procladius* sp. and *Micropsectra* sp. *praecox* grp. also collected	Parry Sound, Georgian Bay, Ontario	Menta and mandibles of larvae	Industrial effluents	0–78% of *C. cucini* affected; station 9 in Parry Sound Harbor worst	Hare and Carter (1976)
Chironomus thummi and *Ch. plumosus*	Teltowkanal, West Berlin, GDR	Menta of larvae	Industrial effluents (heavy metals?)	25–38% of *Ch. thummi* at stations 1–5 had deformed menta	Koehn and Frank (1980)
Chironomus, Procladius	Bay of Quinte, Lake Ontario	Menta or ligulae of larvae	Industrial and/or agricultural pollutants (?)	Incidence of deformities increased from 0.1% in pre-European sediments to 1.0% in 1951 and 2.0% in 1972	Warwick (1980a)
Chironomus spp.; single deformed specimen of *Procladius* sp. also collected	Pasqua Lake, Saskatchewan	Larval mouthparts (mentum, mandible, premandible, epipharyngeal pecten)	Unidentified contaminants (possibly industrial and/or agricultural in origin)	2.3% of specimens deformed; (see Warwick, 1988, p. 297, for revised data)	Warwick (1980b)
Chironomus decorus	Experimental ponds located at Oak Ridge National Laboratory, Roane City, Tennessee	Menta of larvae	Coal liquid (synthetic crude oil)	4.6, 2.7, and 3.0% of specimens deformed at 375, 75, and 15 ml oil/m^3, respectively, although control had incidence of 3.5%; aberrations of medial portion of mentum appeared dose-related	Cushman (1984)
Chironomus, Micropsectra, Tanytarsus	12 Swedish lakes	Mouthparts (mentum, mandible, premandible) of larvae	Industrial effluents (heavy metals?)	Occurrence of mentum deformities tended to increase with increasing contamination, i.e., subfossil material from unpolluted sites: 0–0.8%; recent material from unpolluted sites: 0–0.7%; slightly polluted sites: 1.4–	Wiederholm (1984a)

(Continued)

Table 4.8. (Continued)

Organism	Location	Morphological Deformity	Causal Agent(s)	Comments	Reference
				1.6%; polluted sites: 4.1–10.7%; strongly polluted sites: 3.8–25.0%	
Chironomus spp.	Tobin Lake, Saskatchewan	Antennae of larvae	Industrial and agricultural pollutants	7.0 and 8.4% of specimens at two sites in Tobin Lake were affected (cf. 1.2% in comparatively uncontaminated Last Mountain Lake)	Warwick (1985)
Chironomus tentans	Reexamined material from experiments of Hamilton and Sæther (1971) at the Freshwater Institute, Winnipeg	Antennae of larvae	DDE	Incidence of antennal deformities decreased as the concentration of DDE increased (1–20 μg/l: 25.0–4.6%)	Warwick (1985)
Chironomus decorus	Laboratory experiment, University of California, Davis	Epipharyngeal plates of larvae	Food-substrate-bound copper	Incidence of deformities increased significantly with increasing concentration of copper ($Y = 0.024X - 1.14$, where Y = % deformities, X = substrate copper, mg/kg); incidence ranged from ~2% at 0 mg/kg substrate Cu to ~62% at 2,660 mg/kg substrate Cu	Kosalwat and Knight (1987)
Chironomus spp. (*Ch. thummi*, *Ch. anthracinus*, and *Ch. plumosus* larval types represented)	Port Hope Harbour, Lake Ontario	Menta of larvae	Radiation; heavy metals and elevated water temperatures may also be involved	Incidence of deformities greater in more polluted inner harbor (83%) than in outer harbor (14%); radiation dose rate in chironomids estimated to be 1 mGy/day	Warwick et al. (1987)

Chironomus, *Cryptochironomus*, *Procladius* [*Ch. plumosus*, *Ch. salinarius* type larvae; *Procladius (Holotanypus)* sp., *P. (Psilotanypus) bellus*, and *Procladius* spp.; *Cryptochironomus* sp.]	Tobin Lake, Saskatchewan	Antennae, menta, mandibles, premandibles, epipharyngeal pectens, labral lamellae, and labral setae of larvae	Industrial and agricultural pollutants	Incidence of deformities: Genus / Site (%)*: 2, 17, I *Chironomus*: 41.3, 31.8, 16.2 *Procladius*: 4.4, 2.1, 1.5 *Cryptochironomus*: 4.9, 6.7, 0.8 *sites 2 and 17: Tobin Lake; site I: mildly contaminated Last Mountain Lake, Saskatchewan	Warwick and Tisdale (1988)
Procladius paludicola	Murray and Darling rivers, Australia	Ligulae of larvae	Possibly pesticides or heavy metals	Incidence of deformities at six locations varied from 0–24% in December 1985 survey; none found in July 1986 survey; coincident measurements of contaminants not done	Pettigrove (1989)
Procladius spp.	A number of sites across Canada	Antennae, ligulae, paraligulae, mandibles, dorsomenta, and hypopharyngeal pectens of larvae	Organic chemicals, heavy metals, sewage, radionuclides	Incidence of deformities at 21 sites varied from 0–9.6%; *Procladius* appeared to be less sensitive than *Chironomus*	Warwick (1989)
Chironomidae genera (excluding Tanypodinae)	Niagara River catchment (Welland, Niagara, and Buffalo rivers; pond in St. Catherines, Ontario)	Labial plate of larvae	Agricultural, industrial, and domestic pollutants	Incidence of gross deformities ranged from 9% at a "control" site (downstream of an agricultural area) to 47% downstream of a tire manufacturing company on the Welland River; vinyl chloride suspected	Dickman et al. (1990)

(*continued*)

Table 4.8. (Continued)

Organism	Location	Morphological Deformity	Causal Agent(s)	Comments	Reference
Oligochaeta: Tubificida: Tubificidae *Potamothrix hammoniensis*	Lake Vänern, Sweden	Chaetae	Industrial and municipal wastes, especially effluents from pulp mills and chloralkali plants containing heavy metals; mercury the foremost suspect	Incidence of deformities varied from 16–78% in areas of heavily contaminated sediments to $<10\%$ in peripheral areas; specimens with "gravely deformed chaetae" were most often associated with the former areas, whereas specimens with "less deformed chaetae" were associated with the latter	Milbrink (1983)

tified, and relationships between doses, frequencies of deformities, and their severities established. More studies like Kosalwat and Knight (1987) are needed. Effects of exposure time should be an integral part of such considerations.

Larval *Chironomus* seem especially susceptible to morphological deformities caused by environmental contaminants (Table 4.8). Species in this genus have a number of advantages that favor their use as a key organism in the development of quantitative approaches to using deformities as indicators of stress: they are widespread in distribution, usually numerically abundant, taxonomically and biologically well known, and easily cultured. However, their use in intensive research programs should not preclude a wider search for other species that could act similarly as indicators.

4.2.4. *Behavioral Responses*

Marcucella and Abramson (1978) defined "behavioral toxicity" as a behavioral change, which is induced by stress, that exceeds the normal range of variability. Behavioral impairment is more sensitive than disability (*sensu* Depledge, 1989) to pollutant effects (Depledge, 1989). Because deviations from normal behavior in response to a specific pollutant are likely the result of physiological disorder, they should provide a sensitive, early warning measure of sublethal toxicity (Rand, 1985). The use of aquatic macroinvertebrates in behavior/toxicity assessment is at an early stage compared to similar studies on fish (Little et al., 1985).

Benthic macroinvertebrates may be considered ideal organisms for behavior/toxicity assessment because of their close affiliation with sediments, which are often a sink for toxic compounds. Furthermore, ecologically relevant results can be obtained in short-term laboratory experiments. For example, oligochaete respiratory behavior may be a rapid means of obtaining sublethal responses to environmentally limiting factors (Coler et al., 1988). Kapu and Schaeffer (1991) used alterations in movements and body shape of the planarian *Dugesia dorotocephala* to do rapid (≤1 h), preliminary evaluations of the toxicity of several metals. Heinis et al. (1990) showed that impaired feeding behavior of the chironomid *Glyptotendipes pallens* was a sensitive indicator of cadmium-induced stress. By quantifying behavioral responses using an impedance conversion technique, these authors showed that feeding behavior was affected at concentrations 100 times lower than the LC50 value. These studies indicate that sublethal effects can be obtained easily, often in the time necessary for an acute toxicity test, and that the limits of detection may be far below standard LC50 measures.

Behavioral bioassays can become quite complex. For example, Detra and Collins (1991) showed the existence of a statistically significant relationship between altered swimming behaviors of larval *Chironomus riparius*, parathion concentration, duration of exposure, and acetylcholinesterase inhibition. Swimming behaviors were categorized and assigned numerical values according to the degree of poisoning observed. These authors developed predictive models between altered swimming behavior and the three other variables measured.

Prior to the 1980s, few studies were done on macroinvertebrate behavioral responses to environmental contamination. However, in the last decade, the number of field-oriented studies related to pollutant-behavioral responses has increased substantially. Most such studies have involved anomalies in feeding or both active and passive avoidance (drift) behavior. Moreover, these studies have dealt with a broad range of taxonomic groups, for example, net-spinning behavior of trichopterans (Petersen and Petersen, 1984); activity and molting of ephemeropterans (Henry et al., 1986); preference-avoidance behavior of plecopterans (Scherer and McNicol, 1986); crawling and valve closure behavior of molluscs (Kitching et al., 1987; Doherty et al., 1987a); respiratory behavior of oligochaetes (Coler et al., 1988); and chironomid feeding behavior (Heinis et al., 1990). Although the use of macroinvertebrate behavior as a measure of environmental stress shows promise, more research is needed into normal behavioral responses under field conditions. For example, before causative relationships with contaminants can be inferred, a better understanding of the importance of intra- and interspecific interactions on behavioral responses is necessary.

4.2.5. Life-History Responses

Butler (1984, p. 25) defined *life cycle* as ". . . the sequence of morphological stages and physiological processes that link one generation to the next" and *life history* as ". . . the qualitative and quantitative details of the variable events that are associated with the life cycle . . . ." The life history of a benthic macroinvertebrate ultimately is defined by factors that govern the survival and subsequent reproduction of a species or population. Thus, important proximate life-history descriptors include: survival and mortality, fecundity, growth rate, development stage, size, and longevity. Life-history indicators, or endpoints, of environmental stress in freshwater benthic macroinvertebrates generally fall into three categories: (1) survival, (2) growth, and (3) reproduction. Sample studies of the effects of various contaminants on these life-history endpoints are summarized in Table 4.9. These studies involved many different taxonomic groups and a variety of stressors, of which metals and low pH were the most common. Most studies used more than one life-history endpoint, although no clear preference existed for the combinations chosen.

Life-history endpoints are now commonly used as measures of stress in pollutant assessment studies, especially those that deal with the toxicological

Table 4.9. Life-history indicators of stress in freshwater benthic macroinvertebrates.

Organism	Life-History Variable[1]			Stress	Reference
	S or M	G or D	R or E		
Porifera					
Ephydatia fluviatilis			x	Cd, Hg	Mysing-Gubala and Poirrier (1981)
Coelenterata					
Hydra littoralis		x		Ni, Cd	Santiago-Fandino (1983)
H. oligactis	x		x	PCBs	Adams and Haileselassie (1984)
Mollusca					
Ancylus fluviatilis			x	Heated effluent	Hadderingh et al. (1987)
Ferrissia wautieri			x	Heated effluent	Hadderingh et al. (1987)
Acroloxus lacustris			x	Heated effluent	Hadderingh et al. (1987)
Planorbella trivolvis	x	x		Low pH	Hunter (1988)
Helisoma trivolvis	x	x	x	Trisodium nitrilotriacetate	Flannagan (1974)
Lymnaea palustris	x	x		Pb	Borgmann et al. (1978)
Crustacea					
Astacus astacus		x		$[H^+]$	Appelberg (1984)
Orconectes virilis			x	Low pH	France (1987b), Davies (1989)
Hyalella azteca	x	x		Low pH	France and Stokes (1987)
Gammarus pulex	x	x		Low pH, humus	Hargeby and Petersen (1988)
Gammarus sp.	x		x	PCBs	Nebeker and Puglisi (1974)
Pontogammarus robustoides	x	x		Thermal pollution	Kititsyna and Pidgayko (1974)
Insecta					
Brachycentrus numerosus	x			Fenitrothion	Symons and Metcalfe (1978)
Clistoronia magnifica	x	x		Low pH, Ni	van Frankenhuyzen and Geen (1987)
Rhithrogena semicolorata		x	x	Tannery pollution	Hamilton and Timmons (1980)
Ephemerella ignita		x	x	Tannery pollution	Hamilton and Timmons (1980)
Aedes aegypti		x		Hg, Zn, Cr	Abbasi et al. (1985)
Polypedilum nubifer			x	Cd	Hatakeyama (1987)

(continued)

Table 4.9. (Continued)

Organism	Life-History Variable[1]			Stress	Reference
	S or M	G or D	R or E		
Chironomus attenuatus	x			Low pH, phenol, NaCl	Thornton and Wilhm (1975)
Ch. maturus		x		Low pH, $[H^+]$	Bates and Stahl (1985)
Tanytarsus dissimilis	x	x		Cu, Cd, Zn, Pb	Anderson et al. (1980)
T. dissimilis	x		x	PCBs	Nebeker and Puglisi (1974)
Oligochaeta					
Tubifex tubifex	x	x	x	Cu, NH_4	Bonacina et al. (1987)
T. tubifex	x	x	x	Cu, Zn, Hg, Cd	Wiederholm et al. (1987)
T. tubifex			x	Contaminated sediments	Reynoldson et al. (1991)
Limnodrilus hoffmeisteri	x	x	x	Cu, Zn, Hg, Cd	Wiederholm et al. (1987)
L. udekemianus	x	x	x	Cu, Zn, Hg, Cd	Wiederholm et al. (1987)
L. claparedeanus	x	x	x	Cu, Zn, Hg, Cd	Wiederholm et al. (1987)
Potamothrix hammoniensis	x	x	x	Cu, Zn, Hg, Cd	Wiederholm et al. (1987)

[1]S = survival; M = mortality; G = growth; D = development; R = reproduction; E = emergence.

assessment of H^+ concentrations and trace metal contaminants (Wiederholm, 1984b; Table 4.9). Recent work with high H^+ concentrations has shown that invertebrate susceptibility often depends on developmental stage (Økland and Økland, 1986). For example, France and Stokes (1987) showed that the resistance and tolerance of the amphipod *Hyalella azteca* to low pH was related strongly to size and developmental stage.

A microcosm study of the influence of trace metal pollutants on the survival, growth, and reproduction of five oligochaete species (Wiederholm et al., 1987) revealed that reproduction was the most sensitive measure of sediment toxicity, followed by growth; oligochaete survival was a poor predictor of copper stress, except at high concentrations. Field studies of oligochaete distribution supported the microscosm experiment results (Wiederholm et al., 1987; Wiederholm and Dave, 1989). A more recent standardized bioassay, which uses reproduction of *Tubifex tubifex* as an ecologically relevant endpoint, has been described by Reynoldson et al. (1991). Four variables were examined in this sediment bioassay: (1) survivorship of breeding adults (a measure of acute toxicity); (2) total number of cocoons produced (an indicator of the effects of sediment quality on gametogenesis); (3) hatching rate of cocoons (a measure of the effects of sediment toxicity on embryogenesis); and (4) number of young (an estimate of embryogenesis and toxicity to newly hatched individuals). The method requires 28 days and provides repeatable results and good discriminatory power.

Recent approaches have become more integrated; they have combined field and laboratory studies to evaluate the responses of macroinvertebrates to contamination. For example, using a combination of field and laboratory data, France and LaZerte (1987) derived a simple model to explain the restricted distribution of *Hyalella azteca* within acidified lakes in Ontario, Canada. Their model suggested that the population declines of *Hyalella* within acidified lakes could be explained by direct mortality resulting from short-term pH fluctuations. Engblom and Lingdell (1984) also found good agreement between distribution patterns of ephemeropterans in Sweden and laboratory determined pH-tolerance limits. Along these lines, Chapman et al. (1987) suggested a composite index, "the sediment quality triad," which consists of chemical, bioassay, and field data as a means of integrating measures of contamination assessment. The integration of field studies and traditional laboratory measures of chronic stress to determine causative factors is a promising approach in biomonitoring studies.

Before life-history information can be used fully in biomonitoring, a sound understanding of within-population variability normally associated with field conditions is necessary; simply put, inference of stress only can be made if deviations from normal life-history patterns are known. Food quality and quantity and habitat conditions (e.g., substratum, temperature, oxygen) are factors that vary markedly within and among systems; they can have profound effects on features of macroinvertebrate life history, such as growth and voltinism (Anderson and Cummins, 1979). For example, Jónasson (1972) showed that the midge *Chironomus anthracinus* had a one-year life cycle at depths <17 m in Lake Esrom, Denmark, because of high food availability and optimal temperatures. However, at depths >17 m, where the effects of eutrophication were more pronounced, the larvae required two years because of unfavorable temperature and oxygen conditions. Resh and Rosenberg (1989) have discussed the spatial and temporal variability of macroinvertebrates in both lakes and streams. When using life-history variables as ecologically relevant endpoints for environmental stress, it is desirable although not necessary to understand fully the mechanisms involved. However, it *is* necessary that changes be attributable to exposure. A lack of basic knowledge complicates data interpretation and often precludes or limits the use of life-history variables in field-oriented biomonitoring studies.

4.2.6. *Sentinel Organisms*

The ability of organisms to concentrate pollutants from ambient water into their bodies has been known since the turn of this century (Phillips, 1980). However, the use of sentinel organisms to monitor aquatic pollution did not begin until ≈ 25 years ago, when coastal and oceanic organisms were involved in the measurement of radionuclides in marine ecosystems: the low concentrations of radionuclides in seawater prevented their accurate measurement; use of bioaccumulators facilitated the analysis (Phillips and Segar, 1986). Since then, the use of macroinvertebrates as sentinels has gained far more support in marine and estuarine than freshwater ecosystems, as shown in the review by Phillips (1980; see also, e.g., Lauenstein and O'Connor, 1988, for progress made in the U.S. "Mussel Watch Project"); however, this situation has begun to change in recent years.

The use of sentinel organisms provides a number of advantages over the direct analysis of pollutants in water or sediment (Phillips, 1980, 1991; Metcalfe et al., 1988). For example, measurement of a pollutant in an organism signifies that the pollutant is bioavailable and may be a threat not only to the organism itself but also to other parts of the food web and the ecosystem (V.-Balogh, 1988a; Lower and Kendall, 1990). In addition, the ideal sentinel organism provides a time-integrated measure of pollutant availability, in contrast to the instantaneous nature of pollutant concentrations measured in water or surficial sediment (e.g., Nehring et al., 1979). In fact, bioaccumulation may facilitate the analysis of contaminants when ambient concentrations in water and sediment are too low for direct analysis (e.g., Nehring, 1976; Nehring et al., 1979; Graney et al., 1983).

4.2.6.1. *The "Ideal" Sentinel Organism*

The "ideal" sentinel organism should have the following characteristics (Phillips, 1980; Hellawell, 1986):

1. Individuals of a species should show the same simple correlation between their pollutant content and the average pollutant concentration in the environment, at all locations and under all conditions.
2. Individuals of a species should not be killed or rendered incapable of long-term reproduction by maximum levels of the pollutant in the environment.
3. The species should be sedentary, so findings relate to the area being studied.
4. The species should be large enough or present in high enough numbers to provide sufficient tissue (either whole-body or organ) for analysis.
5. Individuals of a species should be widespread enough to facilitate comparisons among different areas.
6. The species should be sufficiently long-lived to enable sampling of several year classes and to provide evidence of long-term effects.
7. Individuals of a species should be easy to collect.
8. Individuals of a species should be hardy enough to survive laboratory or field handling (e.g., caging), if required.

In reality, these desirable features rarely are found all in the same organism (Hellawell, 1986).

4.2.6.2. Examples of the Use of Freshwater Benthic Macroinvertebrates as Sentinel Organisms

Freshwater benthic macroinvertebrates have been used most commonly to monitor metals and organic contaminants such as pesticides and polychlorinated biphenyls (PCBs). Table 4.10 lists examples of their use as sentinel organisms for metal pollution; it is important to be able to monitor metals because they have a pervasive global distribution, and freshwater ecosystems are particularly sensitive to them (Nriagu, 1990).

In general, Table 4.10 indicates that studies have been done on a variety of metals in both lotic and lentic habitats. Bivalve molluscs and crayfish appear to be the most frequently used organisms, perhaps because they have characteristics that come closest to the "ideal" sentinel organism (see above). The most obvious advantage is their relatively large size, which allows for ease of collection (point 7) and provides sufficient tissue biomass for effective analysis (point 4).

The literature reveals that two basic approaches—survey and experimental—have been applied in using freshwater benthic macroinvertebrates as sentinel organisms for metals. Because a continuum exists between the two approaches, the distinction between them in Table 4.10 sometimes is arbi-

Table 4.10. Examples of the use of freshwater benthic macroinvertebrates as sentinel organisms for metal pollution. (See text for explanation of "Approach": S = survey; E = experimental.)

Taxon	Metal	Source	Location	Approach S	Approach E	Reference
Oligochaeta (freshwater worms)						
Mainly *Limnodrilus hoffmeisteri* or *Tubifex tubifex*	Co, Cr, Cu, Fe, Ni, Pb, Zn	Industrial	Port Hope Harbour, Lake Ontario, ON	x		Hart et al. (1986)
Crustacea: Amphipoda (freshwater shrimp)						
Gammarus pulex	Cu, Pb, Zn	Urban	A river in England		x	Bascombe et al. (1990)
Gammarus tigrinus	Cd	Unidentified	Rivers Werra and Wesser, Germany		x	Zauke (1981)
Crustacea: Decapoda (crayfish)						
Cambarus bartoni	Cd, Cu, Fe, Mg, Mn, Ni, Zn	Smelter	3 lakes near Sudbury, ON		x	Alikhan et al. (1990)
Cambarus bartoni, C. robustus, Orconectes obscurus, O. propinquus, O. virilis	Hg	Natural	13 lakes in S-central Ontario		x	Allard and Stokes (1989)
Orconectes rusticus	Cu	Experimental	Laboratory study; specimens from Indian Creek, OH		x	Evans (1980)
Orconectes virilis	Hg	Unidentified	Clay Lake, ON; Saskatchewan River at Grand Rapids, MB; Lake Winnipegosis at Denbeigh Point, MB		x	Vermeer (1972)
Orconectes virilis	Cd, Cu, Pb, Zn	Urban, industrial	Fox River, IL		x	Anderson and Brower (1978)
Orconectes virilis, O. rusticus	Hg	Industrial	Upper Wisconsin River, WI	x		Rada et al. (1986)
Orconectes virilis (mostly), *O. propinquus, O. rusticus, Cambarus diogenes*	Hg	Industrial (S zone)	Wisconsin River, WI	x		Sheffey (1978)
Pacifastacus leniusculus	Cd, Cu, Hg, Pb	Urban	Lake Washington, WA		x	Stinson and Eaton (1983)
Procambarus clarkii	Cd, Pb	Mining, urban	Marshes of the Guadalquivir River, Spain		x	Rincon-Leon et al. (1988)

Procambarus clarkii	Al, Cd, Pb	Roadside influences, agricultural runoff	Central and NW Louisiana	x		Madigosky et al. (1991)
Insecta: Ephemeroptera (mayflies), Plecoptera (stoneflies), Trichoptera (caddisflies)						
Trichoptera (*Plectrocnemia conspersa*)	Cu	Mining	Darley Brook, England		x	Darlington et al. (1987)
Ephemeroptera (*Ephemerella grandis*), Plecoptera (*Pteronarcys californica*)	Ag, Cu, Pb, Zn	Natural, experimental	Willow Creek, CO, and laboratory study		x	Nehring (1976)
Plecoptera (*Claasenia sabulosa, Hesperoperla pacifica*), Trichoptera (*Arctopsyche grandis, Brachycentrus* sp., *Hydropsyche* sp., *Limnephilus* sp.)	Al, As, Cd, Cu, Fe, Pb, Zn	Mining	Blackfoot River, MT	x		Moore et al. (1991)
Ephemeroptera (*Cinygmula* sp., *Drunella doddsi, D. grandis*), Plecoptera (*Megarcys signata, Pteronarcys badia*), Trichoptera (*Hydropsyche*)	Mo	Mining	East-Gunnison River catchment, CO	x		Colborn (1982)
Insecta: mixture of species						
Ephemeroptera (*Hexagenia*), Diptera (Chironomidae: *Chironomus, Clinotanypus, Procladius*: midges), Megaloptera (*Sialis*: alderflies)	As, Cd, Cu, Pb, Zn	Mines, smelters, long-range atmospheric deposition	17 lakes in S Ontario and Quebec		x	Hare et al. (1991)
Ephemeroptera (Heptageniidae), Diptera (Tipulidae: craneflies), Trichoptera (*Macronemum*), Odonata (Anisoptera: dragonflies), Plecoptera (*Paragnetina, Nemoura*), (a single genus or species of each)	Pb	Mining, experimental	Chalus River, Iran, and laboratory study		x	Nehring et al. (1979)

(*continued*)

Table 4.10. (Continued)

Taxon	Metal	Source	Location	Approach S	Approach E	Reference
Ephemeroptera, Plecoptera, Coleoptera (beetles) (many species)	Cd, Pb, Zn	Mining	River Derwent, England	x		Burrows and Whitton (1983)
Mollusca: Gastropoda (freshwater snails)						
Lymnaea stagnalis	Cd, Cr, Cu, Fe, Hg, Mn, Pb, Zn	Agricultural, reservoir, highway, unidentified	Various locations around Lake Balaton, Hungary		x	V.-Balogh et al. (1988)
Physa integra, Pseudosuccinea columella, Helisoma trivolvis, Campeloma decisum, and the pelecypod *Sphaerium partumeium*	Pb	Highways, experimental	Weston's Mill Pond and Farrington Lake, NJ, and laboratory studies		x	Newman and McIntosh (1982; 1983a,b)
Semisulcospira bensoni	Cd, Cu, Pb, Zn	Mining	Sasu and Se rivers, Japan	x		Ishizaki and Hamada (1987)
Mollusca: Pelecypoda (freshwater clams and mussels)						
Amblema perplicata	Cd, Zn	Electroplating	Williamson Ditch and Trimble Creek, IN		x	Adams et al. (1981)
Anodonta anatina	Cd, Cu, Hg, Ni, Pb, Zn	Urban, agricultural	Urban and rural locations along Thames River, England		x	Manly and George (1977)
Anodonta grandis grandis, Fusconaia flava, Lampsilis radiata luteola	Mn	Unidentified	Various locations in Ohio		x	Tevesz et al. (1989)
Corbicula fluminea	Cd, Cu, Zn	Experimental	Field artificial stream in Virginia and laboratory study		x	Graney et al. (1983, 1984)
Corbicula fluminea	Cd, Co, Cu, Fe, Mn, Ni, Pb, V, Zn	Industrial, agricultural, urban, experimental	Shatt al-Arab River, Iraq, and laboratory study		x	Abaychi and Mustafa (1988)
Corbicula manilensis	As, Cd, Cr, Cu, Hg, Mn, Pb, Zn	Possibly upstream industrial and agricultural areas, barge traffic on river itself	Apalachicola River, FL	x		Elder and Mattraw (1984)

Elliptio complanata	Cu, Fe, Mn, Pb, Zn	Mining and smelting	3 lakes in Rouyn-Noranda, Quebec, region		x	Tessier et al. (1984)
Lampsilis radiata siliquoidea	Cd, Pb	Industrial	Lake St. Clair and Canadian side of Detroit and St. Clair rivers (Huron-Erie corridor)	x		Pugsley et al. (1988)
Lampsilis ventricosa	Cd, Pb	Mining	Big River, MO	x		Czarnezki (1987)
Quadrula quadrula	Cu	Electroplating, experimental	Muskingum River, OH, and laboratory study		x	Foster and Bates (1978)
Unio cf. *elongatulus*	Hg	Mining	Paglia River, Italy, and laboratory study		x	Renzoni and Bacci (1976)
Unio mancus var. *elongatulus*	Cd, Cu, Fe, Mn, Zn	Electroplating	River Bardello, Italy	x		Merlini et al. (1978)
Unio pictorum	Cd, Cr, Cu, Hg, Pb, Zn	Urban, agricultural	Lake Balaton, Hungary, and 3 tributaries to it	x		V.-Balogh and Salánki (1987)
Unio pictorum	Cd, Cu, Pb, Zn	Anticorrosion/ antifouling paints	Lake Balaton, Hungary, and a boat harbor on it	x		V.-Balogh (1988b)
Velesunio ambiguus	Cd, Fe, Mn, Zn	Mining (Molonglo River), experimental	Murray and Molonglo rivers, Australia, and laboratory studies		x	Jones and Walker (1979), Millington and Walker (1983)
Many different taxa						
Zoobenthos: mainly Oligochaeta (*Tubifex tubifex*) and Chironomidae (*Stictochironomus* gr. *psammophilus*, *Procladius skuse* larvae)	Cd, Co, Cu, Fe, Mn, Ni, Pb, Zn	Precipitation, agricultural runoff (?)	Lake Piaseczno, Poland	x		Radwan et al. (1990)
Oligochaeta, Diptera (*Chaoborus*, *Chironomus*) (profundal zone); *Helisoma* sp., *Anodonta implicata*, *Elliptio complanata* (littoral zone)	As, Hg	Mining	Headwater lakes of Shubenacadie River, NS		x	Metcalfe and Mudroch (1987)
Crustacea (Gammaridae); Insecta (Odonata); mostly Mollusca (Gastropoda and Pelecypoda)	Cd, Co, Cu, Fe, Mn, Ni, Pb, Zn	Industrial and agricultural	Lower Danube River, including Kiliya Delta, U.S.S.R.	x		Yevtushenko et al. (1990)
Insecta: Hemiptera (*Iliocoris cimicoides*, *Notonecta glauca*:	Cu, Mn, Pb, Zn	Urban, industrial	Usman' and Ivnitsa rivers, U.S.S.R.	x		Zhulidov et al. (1980)

(continued)

Table 4.10. (Continued)

Taxon	Metal	Source	Location	Approach S	Approach E	Reference
true bugs), Coleoptera (*Acilius canaliculatus, Dytiscus marginalis*); Gastropoda: *Lymnaea stagnalis, Planorbareus corneus*						
Insecta: chironomid larvae; Pelecypoda: *Anodonta cygnea, Unio pictorum*	Cd, Cu, Fe, Hg, Mn, Pb, Zn	Unidentified	Lake Balaton, Hungary	x		Salánki et al. (1982)
Insecta: Coleoptera (*Acilius canaliculatus, Dytiscus marginalis*); Mollusca: Gastropoda (*Lymnaea stagnalis*), Pelecypoda (*Anodonta cygnea*); and other species of freshwater macroinvertebrates	Hg	Unidentified	Various locations in U.S.S.R.		x	Nikanorov et al. (1988)

trary. In the survey approach, the intent is to yield information on the current status of contaminants in the ecosystem being examined. The concentrations of selected contaminants in the tissues of macroinvertebrates that are collected periodically over wide geographic areas are analyzed; the macroinvertebrates may be treated as a batch or separated into recognizable taxa. In contrast, the intent of the experimental approach is to calibrate the use of organisms (see below) so that they can provide ongoing assessment of contaminant levels in the ecosystem being examined. Experimental studies usually are site-specific (although more than one site may be involved), sampling may be repeated on a regular basis, and the species level of identification is used. Characteristics of the ideal sentinel organism outlined above apply to both survey and experimental approaches, although points 2, 6, and 8 are of less concern to surveys than to the experimental approach.

4.2.6.3. Calibrating the Use of Sentinel Organisms

Although all of the characteristics of an ideal sentinel organism listed above are pertinent to effective use, point 1—the need for a simple correlation between pollutant content of the organism and average pollutant concentration in the environment—generally has been the one most difficult to satisfy. The frequent violation of this first characteristic requires that calibration studies be done. This is the topic that has received most of the attention in contemporary research on sentinel organisms.

The calibration process identifies the effects of abiotic (e.g., temperature, water chemistry, sediment geochemistry, contaminant form) and biotic (e.g., physiological influences such as lipid content, uptake and depuration rates, tolerance; life-history characteristics such as age/size/weight, growth, diet, sex, or reproductive condition) variables on the uptake of contaminants by sentinel organisms and relates the concentration of a contaminant in a sentinel organism to that of its environment (e.g., Jones and Walker, 1979; Millington and Walker, 1983; Russell and Gobas, 1989; Muncaster, et al., 1990).

The key lies in *understanding* the factors involved in determining the final contaminant level observed in a sentinel organism; considerable current research is directed at obtaining this understanding (e.g., Campbell and Stokes, 1985; Stephenson and Mackie, 1988; Timmermans, 1991). It is also important that contaminant levels in an organism eventually be related to life-history and population effects (Reynoldson and Metcalfe-Smith, 1992; e.g., see Borgmann et al. 1991).

Optimal use of a sentinel organism is achieved when the amount of unexplained variance between the concentration in the organism and the environment is minimized. Expressed another way, the higher the "signal" to "noise" ratio, the greater the probability that the sentinel organism will indicate accurately changes of contaminants in the environment.

4.3. Populations and Species Assemblages as Indicators

This section considers population-level and species assemblage approaches to the assessment of water quality based on macroinvertebrate indicators. Not all types of pollution are considered, and only selected references rather than exhaustive listings are used to illustrate points.

4.3.1. Biotic Indices

The use of indices to summarize information and to assess pollution effects on aquatic communities has appealed to applied scientists since the early part of the twentieth century. Three basic types of indices exist: diversity, comparison (similarity or dissimilarity), and biotic. These indices also are discussed in detail by Resh and Jackson (Chapter 6) and Norris and Georges (Chapter 7).

Biotic indices differ from both diversity and comparison indices in that they are often specific to the type of pollution or the geographical area involved. They are used to classify the degree of pollution in an aquatic system by determining the tolerance or sensitivity of an animal to a given pollutant (the indicator organism concept). Indicator organisms in a sample are assigned scores according to their tolerance or intolerance to the pollutant in question; some indices also score abundance. The sum of individual scores provides a value by which the pollution status of the site can be identified.

Historically, biotic indices were devised to deal with organic pollution; their application to other forms of pollution may be problematical. However, even their use for monitoring organic pollution may be complicated because contemporary sewage effluents often contain metals and other contaminants. Despite this and other problems in their use (see below), biotic indices remain popular because they incorporate biological responses in a numerical expression that can be understood easily.

Empirically derived biotic indices assume that polluted sites or systems generally contain fewer species than unimpacted ones and that species will tend to be removed selectively along a pollution gradient according to their relative susceptibility to the pollutant. Generally, this may be true, but more research that integrates laboratory and field-derived tolerance data into biotic indices is needed.

4.3.2. Univariate Studies

The first models constructed to explain macroinvertebrate distribution in response to pollution involved a single variable. For example, at the turn of the century in Europe, water quality affected by sewage effluents was the major concern facing most applied aquatic ecologists. Kolkwitz and Marsson (1909) tried to quantify changes occurring in the biological community along sewage pollution gradients and thus developed their now classic "Saprobien system" (see Cairns and Pratt, Chapter 2). Later, indirect effects of inadequate domestic sewage and waste control caused eutrophication of temperate lentic and lotic systems. Again, applied ecologists attempted to quantify species assemblage and community changes in nutrient-poor (oligotrophic) to nutrient-rich (eutrophic) systems by constructing biotic indices (*sensu* Wiederholm, 1976; Sæther, 1979). More recently, acidification has been recognized as an environmental threat, and researchers have developed indices that use species assemblages (Raddum and Sæther, 1981) or entire communities (e.g., Raddum et al., 1988).

4.3.2.1. Organic Enrichment/Eutrophication

Disposal of organic matter such as sewage and domestic waste into aquatic systems is, undoubtedly, one of the oldest and most fully documented forms of pollution. Non-point source discharges of fertilizers from agricultural areas can occur via surface water runoff or groundwater infiltration; although less studied, these discharges similarly can supply large quantities of nutrients to lakes and rivers, resulting in cultural eutrophication. Although intermediate levels of organic enrichment may favor certain suspension- or deposit-feeding macroinvertebrate groups (such as blackfly and chironomid larvae), changes in substratum (through increased sedimentation of organic matter) and low dissolved oxygen concentrations that often occur at high levels of organic enrichment usually result in the disappearance of intolerant taxa (Hynes, 1960; Hellawell, 1986). Interpretation of population changes may be complicated, however, by the relative importance of site-specific local factors such as morphometry and flow in modifying macroinvertebrate responses (e.g., Corkum and Ciborowski, 1988).

4.3.2.1.1. Organic Enrichment/Eutrophication: Lotic Systems

1. The Saprobien system and modifications

The Saprobien system of Kolkwitz and Marsson (1909) is one of the oldest classification schemes using benthic macroinvertebrates in pollution analysis. Kolkwitz and Marsson (1909) recognized four zones: polysaprobic (a zone grossly polluted by organic matter); alpha- and beta-mesosaprobic (zones where dissolved oxygen concentration increases); and oligosaprobic (a zone of high oxygen content and diverse fauna and flora). Palearctic and nearctic macroinvertebrates likely to be found in each of these zones have been summarized from several publications and are presented in Appendix 4.1, which also lists responses to acidification (see below) for numerous taxa. Caution should be exercised in using Appendix 4.1. First, it is not a comprehensive listing of indicator taxa for organic pollution and acidification. Second, the absence of a taxon does not imply that it is intolerant; other factors may be responsible for nonoccurrence. However, presence of the organism is proof that it tolerates the conditions under which it has been listed. Regarding organic pollution, Appendix 4.1 shows that most of the polysaprobic species are to be found among the Oligochaeta and Chironomidae.

The Saprobien system has been criticized for being limited to evaluating sewage effects (Chutter, 1972), for not working in turbulent streams and systems receiving toxic or nonbiodegradable wastes (Chandler, 1970), for requiring a relatively high degree of taxonomic expertise (USEPA, 1973), and for taking no account of local factors (such as substratum type) (Hynes, 1960; see review by Washington, 1984). However, most of these criticisms also apply to many of the more recently derived biotic index or score classification schemes. Moreover, many of these criticisms do not apply to the conditions under which the Saprobien system was intended for use (i.e., lotic systems of continental Europe where sewage pollution has been and continues to be of major concern). The Saprobien system still is used widely in the original or a modified form; for example, Germany recently accepted a substantially improved and simplified version of the Saprobien system as a standardized method for the biological-ecological examination of lotic systems (Friedrich, 1990). The sabrobity index of a site is calculated as:

$$S = \frac{\sum_{i=1}^{n} s_i \cdot A_i \cdot G_i}{\sum_{i=1}^{n} A_i \cdot G_i}$$

where S = saprobity index of the site, s_i = saprobic value of the ith taxon, A_i = abundance index of the ith taxon, and G_i = indication weight of the ith taxon. The terms are scored according to the explanation given in Friedrich (1990).

2. The Trent Biotic Index (TBI) and modifications

Biotic indices or scores that consist of the numerical ranking of sites or systems according to the relative abundance (or presence/absence) of macroinvertebrate indicator taxa often are preferred in the United Kingdom and North America (Metcalfe, 1989). In the TBI (Woodiwiss, 1964), originally developed for use in the Trent River Authority area, England, rankings are made according to the presence or absence of six groups of key organisms: (1) Plecoptera nymphs; (2) Ephemeroptera nymphs; (3) Trichoptera larvae; (4) *Gammarus*; (5) *Asellus*; and (6) tubificids and/or red chironomid larvae. The rankings presumably reflect the relative oxygen requirements of these indicator organisms: Plecoptera and Ephemeroptera nymphs need higher dissolved oxygen concentrations than do tubificid worms and red chironomid larvae. Each sample is given a score ranging from 0 (= polluted site where none of the above animal groups are present) to 10 (= clean water site where more than one species of Plecoptera and several other taxonomic groups are present). The TBI often has been criticized for being insensitive (Washington, 1984, and papers cited therein) and providing erroneous results (e.g., Abel, 1989).

Chandler's Biotic Score (CBS), a frequently used variation of the TBI that was developed in Scotland (Chandler, 1970), also divides the fauna into key indicator groups, but it requires quantitative data and a higher degree of taxonomic expertise. Each taxon in a sample is scored according to its sensitivity to pollution and its relative abundance; the sum of individual scores yields the site score. CBSs generally range from 0 (= polluted site where no taxa are present) to >3,000 (= unpolluted site).

Because CBSs were low in unpolluted, headwater sites, Balloch et al. (1976) modified the CBS by normalizing it for the number of groups present in the sample. The resulting score, the average CBS, has values ranging between 0 and 100 and is regarded by many as providing an adequate measure of water quality (Washington, 1984; Hellawell, 1986). However, the usefulness of even this scoring system was questioned recently by Pinder and Farr (1987) who showed that the average CBS was insensitive to subtle changes in water quality in the River Frome, England, when compared with another biotic index and two diversity indices.

Modifications of the TBI have been used extensively in continental Europe (e.g., France: Tuffery and Verneaux, 1968; Verneaux and Tuffery, 1976; Verneaux et al., 1982; Belgium: De Pauw and Vanhooren, 1983; Denmark: Andersen et al., 1984; Norway: Borgstrøm and Saltveit, 1978). The Indice Biotique of Tuffery and Verneaux (1968) differs from the Trent Biotic Index in the following ways: (1) a larger number of indicator taxa are included; (2) indicator weights are different; (3) single individuals are removed; and (4) sampling is done with a Surber (lotic) or grab (lentic) sampler rather than with a hand net. Verneaux et al. (1982) further modified the method by improving the sampling procedure and including a greater number of indicator taxa. Similarly, the Belgian Biotic Index Method (De Pauw and Vanhooren, 1983) is a slightly modified combination of the Trent Biotic Index and the Indice Biotique.

Lotic ecologists in the United Kingdom have begun using the Biological Monitoring Working Party (BMWP) Score (ISO, 1979; Armitage et al., 1983) to standardize the assessment of water quality. The system uses binary data and relies on taxonomic resolution to the family level only. Pollution-intolerant families gain high scores (e.g., Siphlonuridae = 10), whereas pollution-tolerant families are given low scores (e.g., Chironomidae = 2). The sum of scores of individual families present in a sample yields the site score. Armitage et al. (1983) recognized that using site scores for comparative purposes could be influenced disproportionately by the presence/absence of particular taxa as an artifact of sampling effort. To suppress this effect, each site score was divided by the number of scoring taxa in the sample to give an average score per taxon (ASPT). Armitage et al. (1983) evaluated the sensitivity of BMWP and ASPT and showed that ASPT was less sensitive to sampling effort and seasonal change than BMWP. Both indices presently are being evaluated and used in predictive modeling by British research groups (e.g., Wright et al., 1988). More recently, Wright et al. (1989) and Wright et al. (in press) presented the River InVertebrate Prediction and Classification System (RIVPACS) program. RIVPACS, which uses the probability of capture of taxa as described by Moss et al. (1987), predicts the occurrence of macroinvertebrate fauna at a given site from a small number of environmental variables and is a promising method for environmental monitoring and assessment studies (see below).

3. Chironomid exuviae

Although chironomid midges are reliable indicators of organic pollution (Wilson and McGill, 1977; Ferrington and Crisp, 1989), they are often not included in pollution studies because of the tedious work required in sample processing and identification. The collection of surface-floating chironomid pupal exuviae (Wilson, 1989, and references cited therein) reduces much of the time required to sort and count samples. Furthermore, taxonomic keys for the identification of pupae are available (e.g., Wiederholm, 1986; Langton, 1991).

Several taxa either increased or decreased in response to organic effluents in rivers of the Bristol Avon catchment, England (Wilson, 1989). At intermediate levels of organic enrichment, numerical changes occurred mostly within the established, ambient species complex within each system. However, as the severity of organic pollution increased, between-system similarities increased, ultimately resulting in a predominance of *Chironomus riparius* pupae at high pollution levels. Ferrington and Crisp (1989) suggested that the percent community composition of *Ch. riparius* may be the most reliable approach for detecting sites of severely degraded water quality, but this measure is of limited use for evaluating zones of recovery because *Ch. riparius* populations rapidly decline with modest improvements.

4.3.2.1.2. Organic Enrichment/Eutrophication: Lentic Systems

1. Chironomid indices

Profundal benthic macroinvertebrate communities, particularly larval chironomids, often are used in the classification of freshwater temperate lakes. Thienemann's (1922) now classic lake trophic classification scheme has proved useful over the years, and has undergone several modifications (Brundin, 1949, 1958; Sæther, 1979; Wiederholm, 1980). Early researchers (Thienemann, 1921; Brundin, 1949) considered oxygen concentration to be an important community structuring factor; more recent studies (Sæther, 1979; Wiederholm, 1981; Johnson, 1989) have stressed food availability (as indicated by total phosphorus, chlorophyll-*a*, and algal biovolume). Building on the pioneering studies of Thienemann (1921) and Brundin (1949, 1956), Wiederholm (1976) constructed the Benthic Quality Index (BQI) and Sæther (1979) developed a lake trophic classification scheme using chironomid indicators. The BQI of Wiederholm (1976) consists of five indicator species, with values ranging from a low score for eutrophic lakes where *Ch. plumosus* larvae predominate to the highest score for oligotrophic lakes where *Heterotrissocladius subpilosus* larvae predominate. Sæther (1979) identified 15 lake types (six each in the oligotrophic and eutrophic ranges and three in the mesotrophic range) using profundal chironomid assemblages from nearctic and palearctic lakes. Recent studies (e.g., Kansanen et al., 1984; Meriläinen, 1987; Gerstmeier, 1989) have demonstrated the value of these methods in classifying lake type.

2. Oligochaete indices

Oligochaetes are not used as frequently as chironomid midges in lake classification studies, presumably because of difficulties in species identification. The tubificid *Limnodrilus hoffmeisteri* is known to increase in abundance with organic pollution, so Brinkhurst (1966) suggested that the proportion of this species to other tubificid species could be used as an index of organic pollution. Likewise, tolerant oligochaete species such as *L. hoffmeisteri, Potamothrix hammoniensis,* and *Tubifex tubifex* tend to increase in abundance relative to sediment-dwelling chironomids with increasing organic enrichment, so Wiederholm (1980) suggested that the ratio among oligochaetes and sedentary chironomids (O:C) could be used as an index of pollution. After correcting for depth, a strong relationship was obvious among the O:C index, oligochaete abundance, and lake trophic state measured as chlorophyll-*a* concentration (Wiederholm, 1980). However, the use of overall oligochaete abundance in classification studies of organic pollution remains equivocal. The literature reviewed by Brinkhurst (1974) suggested that no general relationship existed between oligochaete abundance and lake trophic state.

Using tolerant and intolerant oligochaete species assemblages, Ahl and Wiederholm (1977) created an indicator system similar to the chironomid BQI of Wiederholm (1976). Values based on the presence/absence of five species ranged from a low score for oligotrophic lakes where *L. hoffmeisteri* predominated to a high score for eutrophic lakes where *Stylodrilus heringianus* predominated. Lang (1985) suggested that the decrease in relative abundance of the oligotrophic species *Peloscolex velutinus* and *S. heringianus* be used as an indicator of organic enrichment. More recently, Lang (1989) showed that the median relative abundance of oligochaetes characteristic of oligotrophic lakes was related positively to lake trophic state, as measured by total phosphorus and/or primary production. Lang (1989) also showed that sampling location (e.g., depth, slope) was important in lake classification, particularly if the lake was undergoing rapid changes of trophic state.

4.3.2.2. Acidification

Anthropogenic acidification of freshwater ecosystems is a serious environmental problem, particularly in northwestern Europe and eastern North America. The biological effects of acid deposition on the aquatic ecosystem are well-documented (e.g., Wiederholm, 1984b; Økland and Økland, 1986; Gorham, 1989; Stenson and Eriksson, 1989). Hall et al. (1980) and Økland and Økland (1986) suggested that acidification affected aquatic organisms in the following ways: (1) directly through changes in physiology; (2) indirectly by the increase of trace metal concentrations that may be toxic to many organisms; and (3) indirectly through food availability, that is, by reduced primary production and/or reduced bacterial decomposition.

Acidification has been studied extensively during the last decade, but causal linkage of invertebrate distribution patterns to low pH has been difficult. For example, the following different effects of acid stress have been described for species of Ephemeroptera, which are perhaps the best-studied of invertebrate taxa: (1) increased drift following experimentally reduced pH (Hall et al., 1985); (2) reduced growth at low pH (Fiance, 1978); (3) higher respiration rates at high aluminum and high H^+ concentrations (Herrmann and Andersson, 1986); and (4) avoidance of acidic streams during oviposition (Sutcliffe and Carrick, 1973). Furthermore, indirect effects of acidification on food availability (e.g., Hildrew et al., 1984a) and biotic interactions such as interspecific competition and predation (e.g., Hildrew et al., 1984b) have been postulated as potentially important causal mechanisms for explaining mayfly distribution patterns. All of these effects ultimately are manifested at the population level.

Generally, freshwater benthic macroinvertebrate communities become impoverished (i.e., species richness and productivity may decline) as H^+ concentration increases (Wiederholm, 1984b). However, a decrease in acid-sensitive taxa often is accompanied by an increase in the abundance and biomass of species that are quantitatively less significant in circumneutral aquatic systems (Stokes et al., 1989). Certain functional groups (*sensu* Cummins, 1973), such as the scrapers and collectors (e.g., Ephemeroptera), often are more susceptible to the effects of acidification than others, such as shredders and deposit feeders (Smith et al., 1990). Although species richness and possibly productivity may be affected adversely by H^+ stress, overall benthic macroinvertebrate abundance and biomass often are not reliable indicators of acid stress, presumably because of the increased contribution of acid-tolerant taxa.

Although it may be difficult to identify causal links between acidification and macroinvertebrate distribution, several benthic macroinvertebrate groups or species are known to be sensitive to acidification, and these may serve as reliable early warning indicators of acid stress (Appendix 4.1; caveats regarding the use of Appendix 4.1 have been mentioned above).

The Crustacea (particularly several species of crayfish and *Gammarus lacustris*), as well as the Mollusca (except for the Sphaeriidae), are sensitive to low pH. Conversely, certain taxa of Coleoptera, Megaloptera (*Sialis*), Odonata (Zygoptera), Hemiptera (Notonectidae and Corixidae), and Chaoboridae (*Chaoborus*) are relatively tolerant of acidification and often increase in acidified aquatic systems; this is particularly true of taxa released from the pressure of vertebrate predation (Økland and Økland, 1986).

In Scandinavia, the effects of acidification on freshwater systems have been documented thoroughly, and biotic indices have been developed to summarize the effects of high H^+ concentration on aquatic macroinvertebrates (Engblom and Lingdell, 1983; Raddum and Fjellheim, 1984; Raddum et al., 1988; Fjellheim and Raddum, 1990; Lingdell and Engblom, 1990). The index of Engblom and Lingdell (1983) attempts to incorporate laboratory-derived tolerance data with field observations, a commendable approach.

Raddum and Fjellheim (1984) and Fjellheim and Raddum (1990) listed several taxa that may be used as indicators of acid stress in northern Europe; the order Ephemeroptera was the most sensitive group of Insecta. Of 38 mayfly species listed under the acidification part of Appendix 4.1, only 12 are found at pH <4.7; the majority of species disappear from natural waters below pH 5.0. Generally, baetid mayflies are considered to be intolerant (e.g., *Baetis lapponicus, B. macani*), whereas *Leptophlebia marginata* and *L. vespertina* are tolerant to low pH (Engblom and Lingdell, 1983, 1984; Økland and Økland, 1986). Engblom and Lingdell (1984) showed that the lower pH limits observed for Ephemeroptera species in the field corresponded well with pH sensitivities determined in laboratory experiments. Fjellheim and Raddum (1990) suggested that baetid mayflies would be good early warning indicators of acidification, particularly *B. rhodani* because of its cosmopolitan distribution. However, some controversy exists over the use of *B. rhodani* for this purpose because it has been found in streams at pH <4.7 (Engblom and Lingdell, 1983).

Based on a large number of empirical studies, Raddum et al. (1988) presented an index to acid-stressed systems, which is dependent on the presence/absence of indicator organisms. Four categories are recognized: (1) category a, in which species tolerating pH >5.5 are given the number 1; (2) category b, in which species tolerating pH >5.0 are given the number

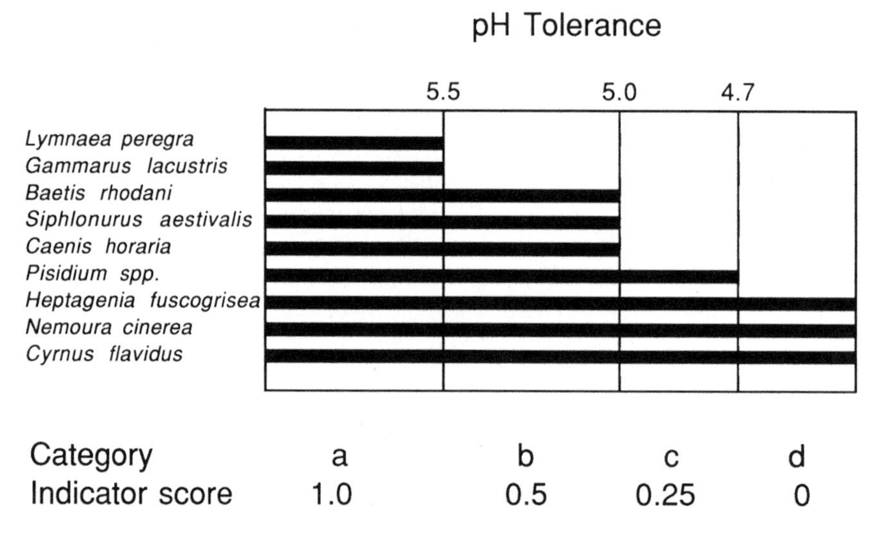

Figure 4.1. Macroinvertebrate indicator organisms and scoring approach used in a biotic index for the classification of acidified, temperate, aquatic systems. Adapted from Fjellheim and Raddum (1990), reproduced by permission of Elsevier Science Publishers Ltd.

0.5; (3) category c, in which species tolerating pH >4.7 are given the number 0.25; and (4) category d, in which species tolerating pH <4.7 are given the number 0 (Fig. 4.1). Accordingly, if a sample contains species of category a, then the site/system is scored as 1 (less acidified). If, on the other hand, species from only category d are found, then the site or system is given a score of 0 (highly acidified). This classification scheme has been approved by many European countries and presently is included in the "International Co-operative Programme on the Assessment and Monitoring of Acidification of Lakes and Rivers" (Anonymous, 1987).

A total of 3,631 Swedish lotic and lentic systems recently were classified using macroinvertebrates (Lingdell and Engblom, 1990). The presence of intolerant taxa indicated that a site was not influenced by acidification or organic pollution. Taxa classified as sensitive to acidification were found at ~60% (n = 2,167) of the sites sampled, whereas at 25% of the sites (n = 922), only acid-tolerant taxa were collected. Seventy-four percent of the sites (n = 2,692) had taxa sensitive or very sensitive to organic pollution. These findings indicated that ≃14% of Sweden's total surface area, or about 20% of Sweden's ≃85,000 lakes, are affected by acidification.

4.3.2.3. *Trace Metals*

The effects of metal pollution on aquatic systems have been studied extensively as a result of widespread contamination by inorganic and organometallic species; these originate from industrial processes, particularly mining activities, and the construction, finishing, and plating of metal objects. Although several studies have documented the lethal effects of heavy metals on aquatic biota, much less is known of sublethal effects and the fate and balance (e.g., metabolization and accumulation) of metal pollutants in aquatic systems (e.g., Wiederholm, 1984b, and see above). In dealing with metal pollutants, field studies, and to some extent laboratory bioassays, often are hampered by the presence of more than one metallic element and the variable toxicity of most metals because of changes in molecular form (chemical speciation) under different environmental conditions (e.g., ambient temperature, oxygen concentration, pH, alkalinity, and organic compounds).

Reduced total abundance and species richness and changes in macroinvertebrate dominance often occur in aquatic systems polluted by heavy metals. Mance (1987) listed the tolerance levels of several macroinvertebrate groups to metal exposure. Generally, insects appear to be less sensitive than gastropods and crustaceans to metal exposure (Fig. 4.2). Invertebrate species-specific tolerance levels are difficult to obtain, however, because trace metal sensitivity often depends on life-history attributes such as size, age (i.e., development stage), and feeding behavior (Wiederholm, 1984b). For example, studies with the isopod *Asellus aquaticus* showed that embryonic development was sensitive to cadmium, whereas juvenile development was more sensitive to copper; moreover, juveniles were more sensitive than adults, and males were more sensitive than females (de Nicola Giudici et al., 1988, and see above). Recent studies indicate that rapid, cost-effective assessment of sublethal contaminant exposure may be acquired by studying macroinvertebrate behavioral responses (e.g., Heinis et al., 1990, and see above).

4.3.2.4. Inorganic Sedimentation

Forestry, mining, and farming activities often are responsible for the release of substantial quantities of inert, particulate substances to aquatic systems (Wiederholm, 1984b). The magnitude of impact and the rate of recovery of macroinvertebrate communities affected by prolonged inorganic pollution often are site-specific and depend on the hydrodynamics of the system. High loading of inorganic matter may affect aquatic macroinvertebrate life histories directly by obstructing respiration, interfering with feeding, and through the loss of habitat or habitat stability; indirect effects may occur because of dilution or altered production of an important food resource (see Wiederholm, 1984b, Hellawell, 1986, Newcombe and MacDonald, 1991, and Ryan, 1991, for recent reviews).

Responses of benthic macroinvertebrates to suspended sediment will depend on both concentration and duration of exposure, which is a response similar to that caused by other environmental contaminants (Newcombe and

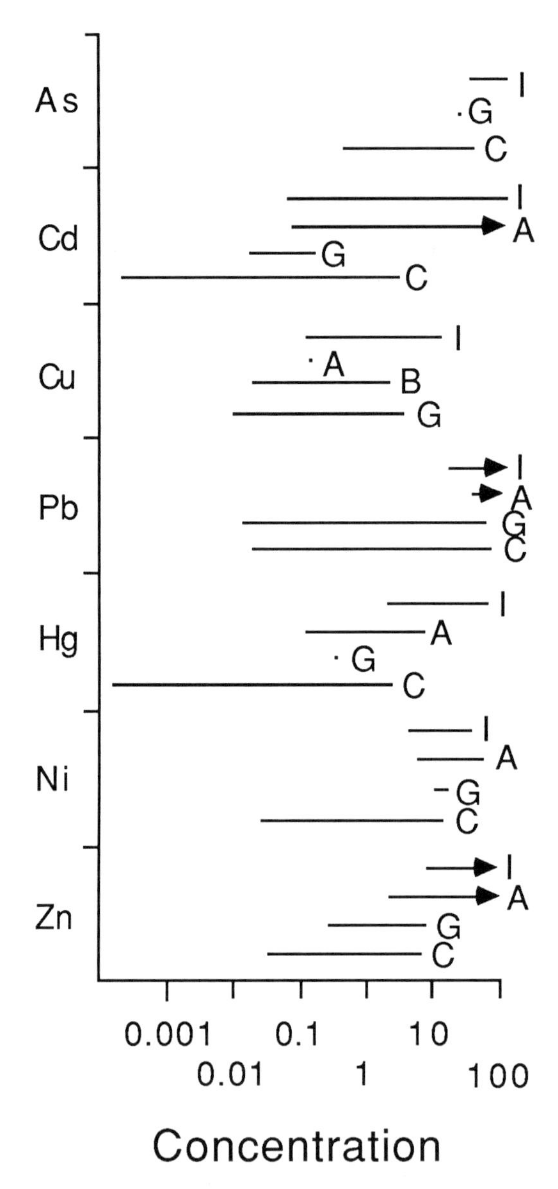

Figure 4.2. Ranges of observed adverse effects of selected metal concentrations (mg/l) for freshwater macroinvertebrates. A = annelids, B = bivalve molluscs, C = crustaceans, G = gastropods, and I = insects. (→ = values >100 mg/l.) Redrawn from Mance (1987), reproduced by permission of Elsevier Science Publishers Ltd.

MacDonald, 1991). Species richness, density, and biomass decrease as inert sediment loading increases (Hellawell, 1986). In general, burrowing, tubicolous oligochaetes and chironomid larvae predominate in habitats where depositing sediments accumulate (Hellawell, 1986). However, invertebrate size, morphology, and behavior are important attributes that determine taxon tolerance (McClelland and Brusven, 1980).

Temporal and spatial changes in macroinvertebrate community composition also must be considered when evaluating any kind of pollution stress. For example, in Uzunov and Kovachev's (1987) study of macroinvertebrate distribution patterns along the River Stuma, Bulgaria, which was polluted by inert suspended solids, the benthic community was depauperate during periods of high loading of suspended materials; several tolerant taxa (mainly beta-mesosaprobionts) predominated. At the upstream reference site, amphipods and mayflies were common; at the first impacted site, tubificids and Diptera dominated. Overall, at sites below the point of sediment discharge, *Tubifex tubifex, Platycnemis pennipes, Heptagenia flava,* and *Asellus aquaticus* were more important during periods of sediment release than after the construction of a sedimentation dam. Following this mitigation, several species were collected that had not been found in studies conducted during the impact period. Immediately after elimination of the impact, opportunistic, passive immigrants such as *Paraleptophlebia submarginata, Baetis scambus,* and *B. tricolor* predominated. However, five years later, these species largely had been replaced by the active immigration of *B. lutheri* and *Ecdyonurus* gr. *venosus,* two species that are indicative of clean-water conditions.

4.3.2.5. *Summary of Univariate Studies*

The examples given above reveal that a great deal of empirical data exist on the responses of macroinvertebrates to pollution gradients. Biotic index and score systems have been constructed for some types of pollution (e.g., organic, H^+ concentration), and these presently are being used throughout western Europe in national monitoring programs (Anonymous, 1987; Newman, 1988). Less is known regarding species-specific responses of macroinvertebrates to several other types of anthropogenic pollution not reviewed here (e.g., chlorinated hydrocarbon, organophosphate, and carbamate pesticides; PCBs). Obviously, before a biotic index can be developed for a specific type of pollution, considerable knowledge concerning how a pollutant affects the biota is necessary. Simply put, predictability of changes in the presence and abundance of indicator populations or species assemblages is a prerequisite to index creation.

4.3.3. *Multivariate Studies*

When species abundance appears to vary monotonically (uniformly) with variation in the environment, a situation that may occur over short sections of an environmental gradient, it is appropriate to use a linear response model to study populations, as was discussed above. However, species often experience the effects of more than one environmental variable simultaneously. Thus, multivariate analysis often is required to account for the observed variation in the abundance or presence/absence of a species (see Norris and Georges, Chapter 7).

The use of multivariate techniques to ordinate indicator taxa (or species assemblages) along environmental gradients is increasing. In their review of gradient analysis, ter Braak and Prentice (1988) listed four data-analysis techniques commonly used by ecologists: (1) *direct* gradient analysis (regression), in which taxa abundance or probability of occurrence is described as a function of measured environmental variables; (2) the *inference* of an environmental variable using the species composition of the community (calibration), that is, the converse of direct gradient analysis; (3) *indirect* gradient analysis, in which community samples are displayed along axes of variation in composition that subsequently can be interpreted in terms of environmental gradients; and (4) *constrained* ordination (multivariate direct gradient analysis), which constructs axes of variation in overall community composition, but does so to optimize their fit to the environmental data provided.

The first two approaches show great promise in advancing the biological indicator concept (see below). In direct gradient analysis, it is possible to predict the presence or abundance of a taxon using environmental variables. Conversely, a species (or species assemblage) may be used to infer water quality, that is, to provide a system classification. The latter is, of course, one of the primary goals of biomonitoring studies.

Freshwater ecologists in Great Britain have used indirect gradient analysis to study the distribution patterns of populations of lotic macroinvertebrates. Indirect ordination procedures such as Principal Components Analysis (PCA) and Detrended Correspondence Analysis (DECORANA) search for major gradients in the species data irrespective of any environmental variables. Furse et al. (1984) and Wright et al. (1984) used DECORANA (Hill, 1979a) to show that factors such as substratum type, depth, slope, discharge, altitude, and total oxidized nitrogen were important between- and within-system correlates in modeling the macrozoobenthos of lotic systems. Similarly, benthic macroinvertebrate distributions in poorly buffered, lotic systems were associated primarily with pH, aluminum, slope, and flow (Ormerod and Edwards, 1987; Weatherley and Ormerod, 1987; Wade et al. 1989).

More recent studies have emphasized the use of direct gradient analysis (e.g., Johnson and Wiederholm, 1989; Dixit et al., 1989; see also Walker, Chapter 9) to study the distribution of taxa along environmental gradients. Direct gradient analysis has a strong foundation in the ecological literature (e.g., ter Braak and Prentice, 1988); taxa occurrences are related directly to environmental variables (Gauch, 1982). For example, Canonical Correspondence Analysis (CCA; ter Braak, 1986), and Two-Way Indicator Species Analysis (TWINSPAN; Hill, 1979b) were used to examine the profundal macrozoobenthos of poorly buffered, nutrient-poor, temperate lakes in Sweden (Johnson, 1989; Johnson and Wiederholm, 1989). These studies showed that lake depth, phytoplankton biovolume, $KMnO_4$ consumption, and pH were important predictors of taxa distributions. Macroinvertebrates usually considered to be indicators of nutrient-poor, well-buffered systems (e.g., *Micropsectra* sp., *Heterotanytarsus apicalis, Monodiamesa* sp., *Gammarus* spp., and *Asellus aquaticus*) aligned positively with lake pH and profundal depth. In contrast, acidophilic taxa such as *Chironomus* spp., *Aeshna grandis*, and *Chaoborus* spp. showed a preference for shallow, brownwater systems with poor buffering capacity and low pH (Johnson and Wiederholm, 1989). In CCA ordination (ter Braak, 1986), arrows point toward the maximum change of an environmental variable, and arrow length is proportional to the variable's importance in the ordination. Thus, by superimposing species or sample plots on environmental plots (i.e., by constructing "biplots"), relationships between specific taxa and selected environmental gradients can be observed readily (Fig. 4.3).

Several caveats should be mentioned in conjunction with these modeling approaches. Although aquatic macroinvertebrates may react to severe but relatively short-term pollution episodes, modeling is often done using average mean water chemistry values rather than ecologically more meaningful factors, such as minimum pH or oxygen conditions or maximum food availability. Furthermore, modeling requires that a number of assumptions be met: normal distribution of explanatory variables, equal covariance among the groups being discriminated, and accurate estimates of the prior probabilities of group membership (Williams, 1983; see also Williams, 1983, for implications of violating these assumptions). The assumptions do not preclude the use of discriminant analysis in an exploratory manner (Williams, 1983).

4.3.3.1. Predictive Modeling

Ordination procedures presently are being used to establish predictive algorithms (*sensu* Armitage et al., 1987; Moss et al., 1987; Ormerod et al., 1988; Wright et al., 1988, 1989; Ormerod and Edwards, 1991) for modeling species occurrence. Predictive modeling appears to be a promising endeavor; for example, the ability to predict the probability of occurrence of a site-specific community allows for comparisons between the expected and observed community and the identification of potential environmental stress (Wright et al., 1988). Moss et al. (1987) modeled macroinvertebrates from unpolluted sites; for most sites, the number and type of taxa recorded were

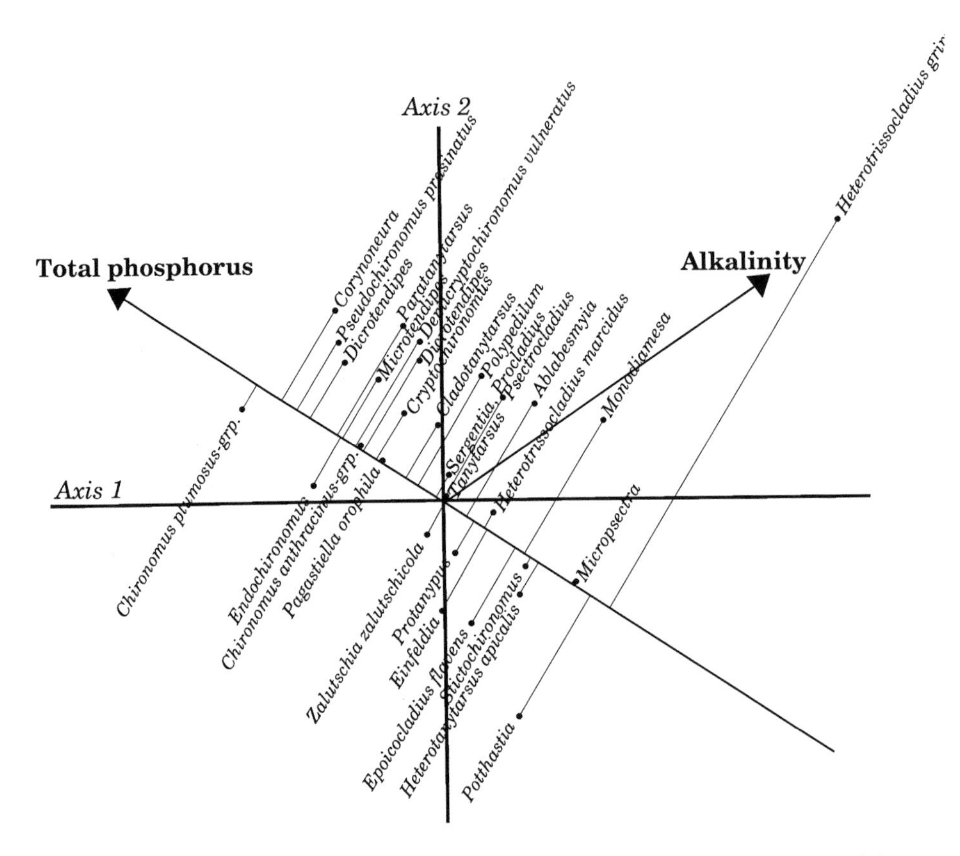

Figure 4.3. Direct ordination (Canonical Correspondence Analysis) of profundal chironomid taxa from 45 Swedish lakes along two environmental gradients. Perpendicular lines between chironomid taxa and the total phosphorus (TP) gradient indicate the approximate ranking of the weighted averages of chironomids with respect to TP. Taxon distance from the center of the ordination is related approximately inversely to taxon tolerance, i.e., peripheral taxa have narrow tolerances. Redrawn from Johnson (1989).

close to those predicted. Similarly, Armitage et al. (1987) used environmental data to assess the effects of stream regulation on families of macroinvertebrates; surprisingly, 22 out of 37 commonly occurring families showed statistically significant trends. Welsh researchers have taken a somewhat different approach in using the Model of Acidification of Groundwaters In Catchments (MAGIC) developed by Cosby et al. (1985) to predict macroinvertebrate assemblages (Ormerod et al., 1988; Weatherley and Ormerod, 1989; Ormerod and Edwards, 1991). MAGIC uses soil, groundwater, and rainfall chemistry data to simulate changes in stream water chemistry. Ormerod and Edwards (1991) used MAGIC to predict stream chemistry under various future scenarios of acid deposition and tested the model's predictive power to assess the relative biological effects of deposition reduction and catchment liming. Ormerod and Edwards (1991) cautioned that although modeling provides insights for determining the future changes in deposition that are necessary for desired biological change, a poor understanding of the relationships between hydrochemical and biological responses to acidification makes model interpretation difficult. Nevertheless, the aforementioned predictive approaches mark novel developments in application of the biological indicator concept.

As previously mentioned, use of a direct gradient approach also is valuable because it allows the probability of occurrence of populations along environmental gradients to be inferred. The majority of methods currently available for direct gradient analysis essentially consider one species at a time (ter Braak, 1986). Plotting species abundance against a single environmental variable is the simplest approach, although it is time-consuming; regression methods are useful in simultaneously studying the effects of multiple environmental variables. However, both of these approaches are impractical when a large number of taxa are involved; hence, the usefulness of CCA (ter Braak, 1986). In this multivariate direct gradient analysis, a taxon is related directly to a set of environmental factors; results are based simultaneously on species abundances and environmental variables. Weighted averaging indicates the center (mode of the unimodal curve) of a species distribution along an environmental gradient (ter Braak and Looman, 1986). Weighted averaging is used as a surrogate for the more computer-intensive, maximum-likelihood Gaussian-model procedure. A taxon's optimum is estimated by taking a weighted average of the values of environmental variables from systems in which the taxon is present. The following formula is used:

$$\mu_k = \frac{\sum_{i=1}^{n} y_{ik} \cdot x_i}{\sum_{i=1}^{n} y_{ik}}$$

where μ_k is the weighted averaging estimate of the optimum, y_{ik} is the abundance of the taxon, and x_i is the value of the environmental variable (ter Braak, 1986). Differences in the weighted averages among species indicate differences in their distributions along an environmental gradient, which may be interpreted as distances between indicator populations. Simply put, a taxon will be most abundant close to its optima, excluding confounding factors such as predation. Once the optima have been estimated or measured directly, the above equation can be rearranged so that the optima can be used to *predict* an environmental variable or system status:

$$x_i = \frac{\sum_{i=1}^{m} y_{ik} \cdot \mu_k}{\sum_{i=1}^{m} y_{ik}}$$

For example, a simple estimate of lake pH is reached through the weighted average of the pH optima of the taxa present.

Ter Braak and van Dam (1989) recently tested various multivariate methods for inferring lake pH from diatoms. Although they found that the maximum-likelihood method was better than all other methods tested, they recommended weighted averaging because it is a good approximation and robust alternative to the maximum-likelihood method. Similarly, Oksanen et al. (1988), who modeled the ecological optima and tolerances of diatoms along a pH gradient in Finnish lakes, found that weighted averaging yielded good predictive estimates. These methods deserve greater consideration in future work with benthic macroinvertebrates (see also Walker, Chapter 9).

4.4. Use of The Biological Indicator Approach

In considering the application of biological indicators to water quality assessment, various situations may call for different or complementary strategies. The use of indicator organisms to monitor water quality changes when causal mechanisms of the disturbance are known or are fairly well-understood (e.g., organic pollution or acidification) has proved fruitful (Hynes, 1960; Raddum and Fjellheim, 1984; Engblom and Lingdell, 1984; Raddum et al., 1988). However, detection of stress when species-specific sensitivities or tolerances are poorly understood or are not available may be best approached by using a less-focused strategy (Rosenberg and Wiens, 1976; Jørgensen, 1978; Pontasch and Brusven, 1988). In situations where overt damage between pollutant exposure and biological effects is less obvious, far more sophisticated, expensive, and in many cases, time-consuming studies may be required. It may be necessary to examine life-history attributes (survivorship, fecundity, recruitment, and population growth rate) in conjunction with knowledge of the ecological consequences of deviations from "normal" physiological and behavioral responses (e.g., Depledge, 1989, and see above). Alternatively, pathological, physiological, genetic, or biochemical (e.g., RNA/DNA ratios, metallothionein and glutathione concentrations, amino acid profiles) indicators may be required to quantify pollutant toxicity to the biological community.

Regardless of the approach taken, confounding factors may mask eventual effects. Although the earliest warnings of environmental deterioration potentially are provided at the organism and suborganism levels, it is doubtful that any single measure can encompass all the biological activities that might influence a population. Simply put, the effects of a pollutant at the organism level may be difficult to isolate from other environmental stressors; well-established cause-and-effect relationships are few. Bayne et al. (1979) suggested, for example, that suborganism, physiological responses, such as growth, may summarize the effects of stressors on biochemical and cytological targets, thereby masking the pathways of causality. Moreover, unless the population is situated at the optima of all ecologically influential environmental gradients at all times of its life history—a highly improbable occurrence—then the population is living partly or wholly under stressed conditions.

If the organism in question already is stressed naturally, then the extra imposition of a pollutant may result in an exaggerated effect. For example, *Gammarus pulex* individuals infected with the acanthocephalan parasite *Pomphorhynchus laevis* were more sensitive (i.e., exhibited greater mortality and lower feeding) to acid stress than uninfected animals (McCahon et al., 1989). However, because the most sensitive organisms already may be stressed naturally, and these organisms are also the most likely to succumb quickly to new stresses, then such organisms may make the most reliable, early warning indicators of a pollutant (e.g., Peterson and Black, 1988). Ultimately, using pollution measures at different levels of biological organization may provide the most cost-effective strategy. Community measures, in contrast to organism and population levels of detection, tend to summarize the magnitude, ecological consequence, and significance of the stressor on the system (see Resh and McElravy, Chapter 5, Resh and Jackson, Chapter 6, and Norris and Georges, Chapter 7).

4.4.1. Life-History Data

The characterization of macroinvertebrate life histories is a major obstacle confronting aquatic biologists because it complicates the evaluation of field-oriented biomonitoring studies. Ignorance or poor understanding of an animal's life history makes for perplexing circumstances when evaluating field data. The existence and importance of strong seasonal variations in life-history patterns, such as emergence and recruitment, must be known to avoid erroneous inferences regarding macroinvertebrate abundance. In other words, knowledge of voltinism (the frequency with which a life cycle is completed) and phenology (the seasonal timing of life-cycle processes) (Butler, 1984) often is necessary for the proper evaluation of field-collected population or demographic data.

Considerable variation occurs in temporal and spatial scales of life-history features such as emergence, feeding and growth, and movements and migrations of aquatic insects (Resh and Rosenberg, 1989). The importance of sampling season on classification and ordination of zoobenthos recently has been examined by Furse et al. (1984), Ormerod (1987), and Johnson et al. (1990). For example, autumn (October) profundal macroinvertebrate communities gave the best measure of lake type in Johnson et al. (1990): 91% of the lakes were classified correctly in TWINSPAN analysis compared with late summer (August) communities in which only 68% of the lakes were classified correctly according to type. Several dipterans (*Chaoborus flavicans, Zalutschia zalutschicola,* and *Sergentia* sp.), which were strong lake-group discriminators in the autumn data set, were collected from fewer lakes in late summer presumably because of emergence; this resulted in a higher percentage of lake misclassifications in the summer data set.

4.4.2. Genetic Variation

The genetic variation of natural populations should be considered when evaluating macroinvertebrate tolerances along pollution gradients. Invertebrates may acclimate physiologically or behaviorally to a pollutant; this resistance would not be inherited. Alternatively, they may develop increased resistance by natural selection (genetic adaptation); this resistance would be inherited. Bürki et al. (1978) showed the existence of two genetically distinct populations of *Chironomus plumosus* along a depth gradient in Lake Murten, Switzerland; one was shorebound at 1–2 m and the other was a profundal population at 10–30 m. Similarly, Pedersen (1986) found differences in genotype frequencies and enzyme polymorphisms for the *Ch. plumosus* populations inhabiting two basins of Lake Tystrup-Bavelse, Denmark. On this basis, Pedersen (1986) argued that each population was genetically specific, without recognizable gene flow. Tolerance to low oxygen conditions may have been the structuring factor in both of these studies. For example, Pedersen (1984) showed that oxygen stress caused differential mortality among three chromosomal genotypes; the heterozygote tolerated anoxia better than either homozygote. Pedersen (1986) then showed that, compared to the Hardy-Weinberg estimate, sites with poor oxygen conditions had the highest frequencies and excesses of the heterozygote genotype.

The evolution of resistance to metal pollutants has been known for some time. Klerks and Weis (1987) found that most populations in polluted areas have some form of increased resistance; however, in most of the studies they reviewed, it was not possible to determine whether metal tolerance was a result of physiological acclimation or genetic adaptation. Furthermore, selection may occur over relatively short time scales. For example, the oligochaete *Limnodrilus hoffmeisteri* developed a genetic resistance to Cd and Ni pollutants in just one to four generations of laboratory experiments (Klerks and Levinton, 1989).

Such findings, if common, will have profound consequences on our understanding of the responses of species to toxic substances. The exposure history of a macroinvertebrate population will have to be taken into account when bioassay studies are performed because the use of a genetically resistant population may lead to erroneous conclusions if results are extrapolated to nonresistant field populations.

4.4.3. Choosing Indicator Species or Species Assemblages

Choice of indicator taxa often is straightforward for well-known disturbances, such as organic pollution and eutrophication. However, for situations lacking empirical data, taxonomic background, or knowledge of pollutant effects on the environment and its biota, greater effort may be required in choosing reliable biological indicators.

If the historical groundwork is missing, indicator species or assemblages still may be identified by studying the distribution of species among community types. This may be accomplished by establishing a baseline data set through regional or national surveys, followed by classification of sites or systems using macroinvertebrate assemblages (Furse et al., 1984; Johnson and Wiederholm, 1989). Indicator taxa can be chosen by using objective, multivariate analyses, if it is assumed that aquatic macroinvertebrate communities are discrete rather than continuous units. The classification of aquatic systems using macroinvertebrates, followed by assessment of the environmental variables that discriminate most strongly among groups and environmental factors, and then integrated ordination of group indicator taxa with environmental variables is one means of objectively selecting indicator species or assemblages. For example, the work of Johnson (1989) on 45 Swedish lakes tended to support earlier, more subjective classification of lake types by profundal chironomids.

Another approach to choosing indicator taxa was offered by Pearson et al. (1983) who argued that indicator species can be identified objectively, without detailed knowledge of the organisms' ecological requirements or pollution tolerances. Selection of these indicator organisms is based on the fact that the distribution of individuals within species follows a recognizable pattern: a few species are represented by many individuals, many species

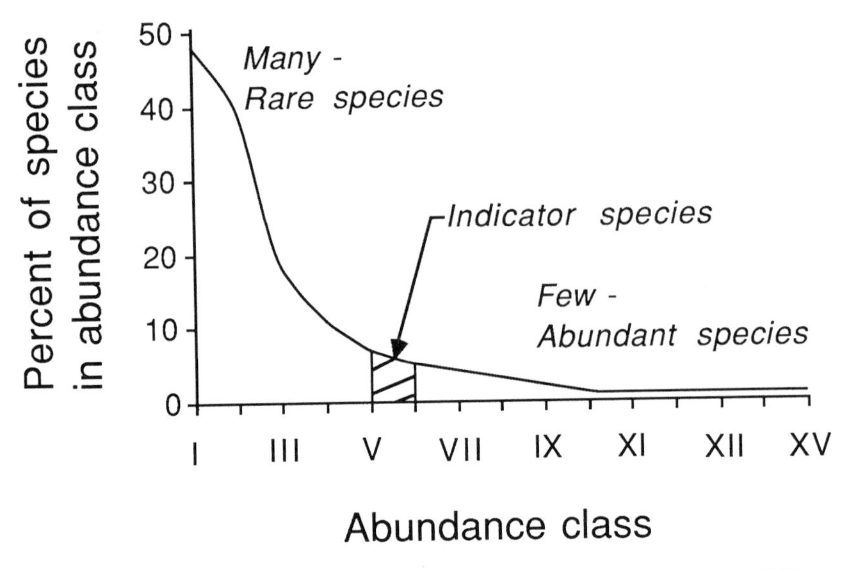

Figure 4.4. Example of a log-normal distribution of individuals among different species in a community. (Actual data plots will = 100%.) According to Pearson et al. (1983), species in abundance classes V and VI may be used as indicator taxa.

are represented by few individuals, and some species are intermediate (e.g., Williams, 1953). A log-normal distribution of individuals among species results from plotting the percentage of species belonging to a particular abundance class against the abundance class (Fig. 4.4). Pearson et al. (1983) proposed using the intermediate abundance classes as pollution indicators (Fig. 4.4). Rare species cannot be used because they may be rare for a variety of reasons other than the effects of pollution. Very abundant taxa, on the other hand, are rejected as indicators because they may have opportunistic characteristics, such as high reproductive capacity and good dispersal mechanisms, rather than being tolerant to the pollutant.

4.5. Conclusions and Future Perspectives

1. Aside from the potential indicator value of changes in ion concentrations in crayfish because of acid stress, biochemical and physiological indicators have not yet been sufficiently well-developed to be used in routine benthic macroinvertebrate monitoring programs in fresh waters, and probably will not soon replace direct observations of stress at population, community, and ecosystem levels (Giesy, 1988). However, these suborganismal indicators can

help us to understand toxicant-induced effects (i.e., modes of action) and their ecological results. Ultimately, this may improve the predictability of stress effects between toxicants and species, and under different environmental conditions.

2. The use of morphological deformities in freshwater benthic macroinvertebrates to detect the presence of contaminants in an ecosystem is currently a taxonomically restricted, qualitative measure. Therefore, a greater variety of macroinvertebrates needs to be examined, and morphological deformities need to be used in a more quantitative fashion. The experimental production of morphological deformities, and the establishment of dose-response relationships between stressors and deformities are high priorities for future research on this subject.
3. Because behavioral endpoints in pollutant assessment are likely to be of physiological origin, they may provide a sensitive, early warning measure of sublethal toxicity. Laboratory studies have shown that ecologically relevant results can be obtained in short-term behavioral experiments. The short time necessary to complete an assay, combined with the ability to obtain sublethal, ecologically relevant results, makes this approach attractive and cost-effective.
4. In contrast to macroinvertebrate behavioral measures of pollutant effects, field measures of invertebrate life-history variables are somewhat easier to obtain. However, the application of life-history approaches to field-oriented biomonitoring requires more basic knowledge of natural variability.
5. The routine use of freshwater benthic macroinvertebrates as sentinel organisms to monitor ongoing environmental contamination requires considerable understanding of the factors involved in determining final contaminant concentrations in these organisms. These factors include: physical and chemical characteristics of the environment from which the organism originates; chemical form of the contaminant(s) involved; an organism's physiological behavior with regard to the target contaminants; and life-history variables of the organism. Research on sentinel organisms should focus on obtaining this understanding.
6. Biotic indices have been used extensively to evaluate pollution stress in Europe and are used increasingly in North America. They were devised originally to measure the effects of organic pollution, so their application to other forms of pollution or even to complex sewage effluents may pose problems. Emphasis on improving the efficiency, accuracy, precision, and predictive ability of biotic indices and scoring systems is needed. A need also exists to combine

laboratory-derived tolerance data and field observations in the development of biotic indices.

7. Biotic index and score systems work reasonably well along univariate environmental gradients (e.g., changes in O_2 or H^+ concentrations). However, the confounding factors often associated with pollution indicate that a multivariate approach may enhance our understanding of the effects of pollution stress. Predictive modeling is a new and promising direction in pollution analysis, and greater effort should be devoted to the development of this technique. The use of newly developed multivariate procedures, and the predictive algorithms they generate, allows for the identification of potential environmental stress by: (1) calculation of probability of occurrence of expected taxa, which then can be compared to those present; and (2) calculation of measures of distance (= tolerance limits) among taxa along ecologically important environmental gradients. These new methods may de-emphasize the expertise of the individual investigator, thereby making inter-investigator analysis less troublesome. They introduce objectivity into handling of the data matrix, and they can be used at population, species assemblage, or community levels.
8. The next generation of pollution or environmental monitoring studies should consist not only of qualitative and quantitative analyses of benthic macroinvertebrate communities, but also of hypothesis generation through classification, ordination, and model construction; the accuracy of model prediction also must be tested. International standardization of macroinvertebrate identification codes would enhance data transfer and allow for the creation of an international data bank for lentic and lotic macroinvertebrates. This would be a major step toward the establishment of tolerance levels of macroinvertebrates and their ordination along environmental gradients.
9. Last, although the need for applied freshwater ecology is obvious, it is impossible to apply knowledge that one does not have. A sound understanding of macroinvertebrate ecology is a prerequisite to the implementation of a biological approach to ecosystem management.

References

Abaychi, J.K. and Y.Z. Mustafa. 1988. The Asiatic clam, *Corbicula fluminea*: an indicator of trace metal pollution in the Shatt al-Arab River, Iraq. *Environmental Pollution* 54:109–22.

Abbasi, S.A., P.C. Nipaney, and R. Soni. 1985. Environmental consequences of the inhibition in the hatching of pupae of *Aedes aegypti* by mercury, zinc and chromium—the abnormal toxicity of zinc. *International Journal of Environmental Studies* 24:107–14.

Abel, P.D. 1989. *Water Pollution Biology*. Ellis Horwood, Chichester, England.

Adams, J.A. and H.M. Haileselassie. 1984. The effects of polychlorinated biphenyls (Aroclors 1016 and 1254) on mortality, reproduction and regeneration in *Hydra oligactis*. *Archives of Environmental Contamination and Toxicology* 13:493–9.

Adams, T.G., G.J. Atchison, and R.J. Vetter. 1981. The use of the three-ridge clam (*Amblema perplicata*) to monitor trace metal contamination. *Hydrobiologia* 83:67–72.

Ahl, T. and T. Wiederholm. 1977. *Svenska vattenkvalitetskritier. Eutrofierand ämnen*. SNV Report No. 918. Statens Naturvårdsverk, Stockholm, Sweden.

Alikhan, M.A., G. Bagatto, and S. Zia. 1990. The crayfish as a "biological indicator" of aquatic contamination by heavy metals. *Water Research* 24:1069–76.

Allard, M. and P.M. Stokes. 1989. Mercury in crayfish species from thirteen Ontario lakes in relation to water chemistry and smallmouth bass (*Micropterus dolomieui*) mercury. *Canadian Journal of Fisheries and Aquatic Sciences* 46:1040–6.

Andersen, M.M., F.F. Rigét, and H. Sparholt. 1984. A modification of the Trent Index for use in Denmark. *Water Research* 18:145–51.

Anderson, N.H. and K.W. Cummins. 1979. Influences of diet on the life histories of aquatic insects. *Journal of the Fisheries Research Board of Canada* 36:335–42.

Anderson, R.L., C.T. Walbridge, and J.T. Fiandt. 1980. Survival and growth of *Tanytarsus dissimilis* (Chironomidae) exposed to copper, cadmium, zinc and lead. *Archives of Environmental Contamination and Toxicology* 9:329–35.

Anderson, R.V. and J.E. Brower. 1978. Patterns of trace metal accumulation in crayfish populations. *Bulletin of Environmental Contamination and Toxicology* 20:120–7.

Anonymous. 1987. *International Co-operative Programme on Assessment and Monitoring of Acidification of Rivers and Lakes: Programme Manual*. Norwegian Institute for Water Research (NIVA), Oslo, Norway.

Aoki, Y., S. Hatakeyama, N. Kobayashi, Y. Sumi, T. Suzuki, and K.T. Suzuki. 1989. Comparison of cadmium-binding protein induction among mayfly larvae of heavy metal resistant (*Baetis thermicus*) and susceptible species (*B. yoshinensis* and *B. sahoensis*). *Comparative Biochemistry and Physiology* 93C:345–7.

Appelberg, M. 1984. Early development of the crayfish *Astacus astacus* L. in acid water. *Institute of Freshwater Research Drottningholm Report* 61:48–59.

Appelberg, M. 1985. Changes in haemolymph ion concentrations of *Astacus astacus* L. and *Pacifastacus leniusculus* (Dana) after exposure to low pH and aluminium. *Hydrobiologia* 121:19–25.

Armitage, P.D., R.J.M. Gunn, M.T. Furse, J.F. Wright, and D. Moss. 1987. The use of prediction to assess macroinvertebrate response to river regulation. *Hydrobiologia* 144:25–32.

Armitage, P.D., D. Moss, J.F. Wright, and M.T. Furse. 1983. The performance of a new biological water quality score system based on macroinvertebrates over a wide range of unpolluted running-water sites. *Water Research* 17:333–47.

Balloch, D., C.E. Davies, and F.H. Jones. 1976. Biological assessment of water quality in three British Rivers: the North Esk (Scotland), the Ivel (England) and the Taf (Wales). *Water Pollution Control* 75:92–114.

Bascombe, A.D., J.B. Ellis, D.M. Revitt, and R.B.E. Shutes. 1990. The development of ecotoxicological criteria in urban catchments. *Water Science and Technology* 22(10/11):173–9.

Bates, N.M. and J.B. Stahl. 1985. Effect of pH and total ion concentration on growth rate of *Chironomus maturus* larvae (Diptera: Chironomidae) from an acid strip-mine lake. *Transactions of the Illinois State Academy of Science* 78:127–32.

Bayne, B.L., M.N. Moore, J. Widdows, D.R. Livingstone, and P. Salkeld. 1979. Measurement of the responses of individuals to environmental stress and pollution: studies with bivalve molluscs. *Philosophical Transactions of the Royal Society of London* (Series B) 286:563–81.

Bhagyalakshmi, A., P. Sreenivasula Reddy, and R. Ramamurthi. 1983. Muscle nitrogen metabolism of freshwater crab, *Oziotelphusa senex senex,* Fabricius, during acute and chronic Sumithion™ intoxication. *Toxicology Letters* (Amsterdam) 17:89–93.

Bonacina, C., G. Bonomi, and C. Monti. 1987. Population dynamics of *Tubifex tubifex,* first settler in the profundal of a copper and ammonia polluted, recovering lake (L. Orta, North Italy). *Hydrobiologia* 155:305.

Borgmann, U., O. Kramar, and C. Loveridge. 1978. Rates of mortality, growth, and biomass production of *Lymnaea palustris* during chronic exposure to lead. *Journal of the Fisheries Research Board of Canada* 35:1109–15.

Borgmann, U., W.P. Norwood, and I.M. Babirad. 1991. Relationship between chronic toxicity and bioaccumulation of cadmium in *Hyalella azteca. Canadian Journal of Fisheries and Aquatic Sciences* 48:1055–60.

Borgstrøm, R. and S.J. Saltveit. 1978. *Faunaen i elver og bekker innen Oslo kommune. Del II. Bunndyr og fisk i Akerselva, Sognsvannsbekken—Frognerelva, Holmenbekken—Hoffselva og Mærradalsbekken.* Rapport 38. Laboratoriet for Ferskvattens Økologiskt Innlandsfiske, Oslo, Norway.

Brinkhurst, R.O. 1966. The Tubificidae (Oligochaeta) of polluted waters. *Internationale Vereinigung für Theoretische und Angewandte Limnologie Verhandlungen* 16:854–9.

Brinkhurst, R.O. 1974. *The Benthos of Lakes.* Macmillan Press, London.

Brinkhurst, R.O., P.M. Chapman, and M.A. Farrell. 1983. A comparative study of respiration rates of some aquatic oligochaetes in relation to sublethal stress. *Internationale Revue der gesamten Hydrobiologie* 68:683–99.

Brinkhurst, R.O., A.L. Hamilton, and H.B. Herrington. 1968. *Components of the Bottom Fauna of the St. Lawrence, Great Lakes.* Great Lakes Institute, Univ. of Toronto No. PR 33, Toronto.

Brundin, L. 1949. Chironomiden und andere Bodentiere der südschwedischen Urgebirgsseen. *Institute of Freshwater Research Drottningholm Report* 30:1–914.

Brundin, L. 1956. Die bodenfaunistischen Seetypen und ihre Anwendbarkeit auf die

Südhalbkugel. Zugleich eine Theorie der produktionsbiologischen Bedeutung der glazialen Erosion. *Institute of Freshwater Research Drottningholm Report* 37:186–235.

Brundin, L. 1958. The bottom faunistical lake type system and its application to the southern hemisphere. Moreover a theory of glacial erosion as a factor of productivity in lakes and oceans. *Internationale Vereinigung für Theoretische und Angewandte Limnologie Verhandlungen* 13:288–97.

Bürki, E., R. Rothen, and A. Scholl. 1978. Koexistenz von zwei cytologisch verschiedenen Populationen der zuckmücke *Chironomus plumosus* im Murtensee. *Revue Suisse de Zoologie* 85:625–34.

Burrows, I.G. and B.A. Whitton. 1983. Heavy metals in water, sediments and invertebrates from a metal-contaminated river free of organic pollution. *Hydrobiologia* 106:263–73.

Butler, M.G. 1984. Life histories of aquatic insects. In *The Ecology of Aquatic Insects,* eds. V.H. Resh and D.M. Rosenberg, pp. 24–55. Praeger Pubs., New York.

Calabrese, E.J., C.C. Chamberlain, R. Coler, and M. Young. 1987. The effects of trichloroacetic acid, a widespread product of chlorine disinfection, on the dragonfly nymph respiration. *Journal of Environmental Science and Health* A22:343–55.

Camargo, J.A. 1991. Toxic effects of residual chlorine on larvae of *Hydropsyche pellucidula* (Trichoptera, Hydropsychidae): a proposal of biological indicator. *Bulletin of Environmental Contamination and Toxicology* 47:261–5.

Campbell, P.G.C. and P.M. Stokes. 1985. Acidification and toxicity of metals to aquatic biota. *Canadian Journal of Fisheries and Aquatic Sciences* 42:2034–49.

Chandler, J.R. 1970. A biological approach to water quality management. *Water Pollution Control* 69:415–21.

Chapman, P.M. 1986. Sediment quality criteria from the sediment quality triad: an example. *Environmental Toxicology and Chemistry* 5:957–64.

Chapman, P.M., R.N. Dexter, and E.R. Long. 1987. Synoptic measures of sediment contamination, toxicity and infaunal community composition (the Sediment Quality Triad) in San Francisco Bay. *Marine Ecology Progress Series* 37:75–96.

Chutter, F.M. 1972. An empirical biotic index of the quality of water in South African streams and rivers. *Water Research* 6:19–30.

Colborn, T. 1982. Measurement of low levels of molybdenum in the environment by using aquatic insects. *Bulletin of Environmental Contamination and Toxicology* 29:422–8.

Coler, R.A., M.S. Coler, and P.T. Kostecki. 1988. Tubificid behavior as a stress indicator. *Water Research* 22:263–7.

Corkum, L.D. and J.J.H. Ciborowski. 1988. Use of alternative classification in studying broad-scale distributional patterns of lotic invertebrates. *Journal of the North American Benthological Society* 7:167–79.

Correa, M., E.J. Calabrese, and R.A. Coler. 1985a. Effects of trichloroacetic acid, a new contaminant found from chlorinating water with organic material, on dragonfly nymphs. *Bulletin of Environmental Contamination and Toxicology* 34:271–4.

Correa, M. and R. Coler. 1983. Enhanced oxygen uptake rates in dragonfly nymphs (*Somatochlora cingulata*) as an indication of stress from naphthalene. *Bulletin of Environmental Contamination and Toxicology* 30:269–76.

Correa, M., R.A. Coler, and C.-M. Yin. 1985b. Changes in oxygen consumption and nitrogen metabolism in the dragonfly *Somatochlora cingulata* exposed to aluminum in acid waters. *Hydrobiologia* 121:151–6.

Correa, M., R. Coler, C.-M. Yin, and E. Kaufman. 1986. Oxygen consumption and ammonia excretion in the detritivore caddisfly *Limnephilus* sp. exposed to low pH and aluminum. *Hydrobiologia* 140:237–41.

Cosby, B.J., R.F. Wright, G.M. Hornberger, and J.N. Galloway. 1985. Modeling the effects of acid deposition: estimation of long-term water quality responses in a small forested catchment. *Water Resources Research* 21:1591–601.

Costa, H.H. 1970. Effects of some common insecticides and other environmental factors on the heart beat of *Caridina pristis*. *Hydrobiologia* 35:469–80.

Cummins, K.W. 1973. Trophic relations of aquatic insects. *Annual Review of Entomology* 18:183–206.

Cushman, R.M. 1984. Chironomid deformities as indicators of pollution from a synthetic, coal-derived oil. *Freshwater Biology* 14:179–82.

Czarnezki, J.M. 1987. Use of the pocketbook mussel, *Lampsilis ventricosa,* for monitoring heavy metal pollution in an Ozark stream. *Bulletin of Environmental Contamination and Toxicology* 38:641–6.

Darlington, S.T., A.M. Gower, and L. Ebdon. 1987. Studies on *Plectrocnemia conspersa* (Curtis) in copper contaminated streams in south west England. In *Proceedings of the Fifth International Symposium on Trichoptera, Lyon, France, July 21–26, 1986,* eds. M. Bournaud and H. Tachet, pp. 353–7. Series Entomologica, Vol. 39. Junk Pubs., Dordrecht, The Netherlands.

Darville, R.G., H.J. Harmon, M.R. Sanborn, and J.L. Wilhm. 1983. Effect of naphthalene on the hemolymph ion concentrations of *Chironomus attenuatus* and the possible mode of action. *Environmental Toxicology and Chemistry* 2:423–9.

Darville, R.G. and J.L. Wilhm. 1984. The effect of naphthalene on oxygen consumption and hemoglobin concentration in *Chironomus attenuatus* and on oxygen consumption and life cycle of *Tanytarsus dissimilis*. *Environmental Toxicology and Chemistry* 3:135–41.

Davies, I.J. 1989. Population collapse of the crayfish *Orconectes virilis* in response to experimental whole-lake acidification. *Canadian Journal of Fisheries and Aquatic Sciences* 46:910–22.

Day, K.E., J.L. Metcalfe, and S.P. Batchelor. 1990. Changes in intracellular free amino acids in tissues of the caged mussel, *Elliptio complanata,* exposed to contaminated environments. *Archives of Environmental Contamination and Toxicology* 19:816–27.

Day, K.E. and I.M. Scott. 1990. Use of acetylcholinesterase activity to detect sublethal toxicity in stream invertebrates exposed to low concentrations of organophosphate insecticides. *Aquatic Toxicology* 18:101–13.

de Nicola Giudici, M., L. Migliore, C. Gambardella, and A. Marotta. 1988. Effect of chronic exposure to cadmium and copper on *Asellus aquaticus* (L.) (Crustacea, Isopoda). *Hydrobiologia* 157:265–9.

De Pauw, N. and G. Vanhooren. 1983. Method for biological quality assessment of watercourses in Belgium. *Hydrobiologia* 100:153–68.

Depledge, M. 1989. The rational basis for detection of the early effects of marine pollutants using physiological indicators. *Ambio* 18:301–2.

Dermott, R.M. 1985. Benthic fauna in a series of lakes displaying a gradient of pH. *Hydrobiologia* 128:31–8.

Detra, R.L. and W.J. Collins. 1991. The relationship of parathion concentration, exposure time, cholinesterase inhibition and symptoms of toxicity in midge larvae (Chironomidae: Diptera). *Environmental Toxicology and Chemistry* 10:1089–95.

Dickman, M., Q. Lan, and B. Matthews. 1990. Teratogens in the Niagara River watershed as reflected by chironomid (Diptera: Chironomidae) labial plate deformities. *Water Pollution Research Journal of Canada* 24:47–79.

Di Giulio, R.T., P.C. Washburn, R.J. Wenning, G.W. Winston, and C.S. Jewell. 1989. Biochemical responses in aquatic animals: a review of determinants of oxidative stress. *Environmental Toxicology and Chemistry* 8:1103–23.

Dixit, S.S., A.S. Dixit, and J.P. Smol. 1989. Relationship between chrysophyte assemblages and environmental variables in seventy-two Sudbury lakes as examined by canonical correspondence analysis (CCA). *Canadian Journal of Fisheries and Aquatic Sciences* 46:1667–76.

Doherty, F.G., D.S. Cherry, and J. Cairns, Jr. 1987a. Valve closure responses of the Asiatic clam *Corbicula fluminea* exposed to cadmium and zinc. *Hydrobiologia* 153:159–67.

Doherty, F.G., M.L. Failla, and D.S. Cherry. 1987b. Identification of a metallothionein-like, heavy metal binding protein in the freshwater bivalve, *Corbicula fluminea*. *Comparative Biochemistry and Physiology* 87C:113–20.

Dominguez, T.M., E.J. Calabrese, P.T. Kostecki, and R.A. Coler. 1988. The effects of tri- and dichloroacetic acids on the oxygen consumption of the dragonfly nymph *Aeschna umbrosa*. *Journal of Environmental Science and Health* A23:251–71.

Donald, D.B. 1980. Deformities in Capniidae (Plecoptera) from the Bow River, Alberta. *Canadian Journal of Zoology* 58:682–6.

Elder, J.F. and H.C. Mattraw, Jr. 1984. Accumulation of trace elements, pesticides, and polychlorinated biphenyls in sediments and the clam *Corbicula manilensis* of the Apalachicola River, Florida. *Archives of Environmental Contamination and Toxicology* 13:453–69.

Engblom, E. and P.-E. Lingdell. 1983. *Bottenfaunans användbarhet som pH-indikator*. SNV Report No. 1741. Statens Naturvårdsverk, Stockholm, Sweden.

Engblom, E. and P.-E. Lingdell. 1984. The mapping of short-term acidification with the help of biological pH indicators. *Institute of Freshwater Research Drottningholm Report* 61:60–8.

Evans, M.L. 1980. Copper accumulation in the crayfish (*Orconectes rusticus*). *Bulletin of Environmental Contamination and Toxicology* 24:916–20.

Farris, J.L., S.E. Belanger, D.S. Cherry, and J. Cairns, Jr. 1989. Cellulolytic activity as a novel approach to assess long-term zinc stress to *Corbicula*. *Water Research* 23:1275–83.

Farris, J.L., J.H. Van Hassel, S.E. Belanger, D.S. Cherry, and J. Cairns, Jr. 1988.

Application of cellulolytic activity of Asiatic clams (*Corbicula* sp.) to in-stream monitoring of power plant effluents. *Environmental Toxicology and Chemistry* 7:701–13.

Ferrington, L.C., Jr. and N.H. Crisp. 1989. Water chemistry characteristics of receiving streams and the occurrence of *Chironomus riparius* and other Chironomidae in Kansas. In Advances in chironomidology. Proceedings of the Tenth International Chironomid Symposium, Debrecen, Hungary, July 25–28, 1988. Part 2. Faunistics, population dynamics, ecology, production and community structure, ed. G. Dévai. *Acta Biologica Debrecina Oecologica Hungarica* 3:115–26.

Fiance, S.B. 1978. Effects of pH on the biology and distribution of *Ephemerella funeralis* (Ephemeroptera). *Oikos* 31:332–9.

Fjellheim, A. and G.G. Raddum. 1990. Acid precipitation: biological monitoring of streams and lakes. *Science of the Total Environment* 96:57–66.

Flannagan, J.F. 1974. Influence of trisodium nitrilotriacetate on the mortality, growth, and fecundity of the fresh-water snail (*Helisoma trivolvis*) through four generations. *Journal of the Fisheries Research Board of Canada* 31:155–61.

Flannagan, J.F., W.L. Lockhart, D.G. Cobb, and D. Metner. 1978. Stonefly (Plecoptera) head cholinesterase as an indicator of exposure to fenitrothion. *Manitoba Entomologist* 12:42–8.

Foster, R.B. and J.M. Bates. 1978. Use of freshwater mussels to monitor point source industrial discharges. *Environmental Science and Technology* 12:958–62.

France, R.L. 1987a. Calcium and trace metal composition of crayfish (*Orconectes virilis*) in relation to experimental lake acidification. *Canadian Journal of Fisheries and Aquatic Sciences* 44 (Supplement 1):107–13.

France, R.L. 1987b. Reproductive impairment of the crayfish *Orconectes virilis* in response to acidification of Lake 223. *Canadian Journal of Fisheries and Aquatic Sciences* 44 (Supplement 1):97–106.

France, R.L. and B.D. LaZerte. 1987. Empirical hypothesis to explain the restricted distribution of *Hyalella azteca* (Amphipoda) in anthropogenically acidified lakes. *Canadian Journal of Fisheries and Aquatic Sciences* 44:1112–21.

France, R.L. and P.M. Stokes. 1987. Life stage and population variation in resistance and tolerance of *Hyalella azteca* (Amphipoda) to low pH. *Canadian Journal of Fisheries and Aquatic Sciences* 44:1102–11.

Friedrich, G. 1990. Eine Revision des Saprobiensystems. *Zeitschrift für Wasser und Abwasser Forschung* 23:141–52.

Furse, M.T., D. Moss, J.F. Wright, and P.D. Armitage. 1984. The influence of seasonal and taxonomic factors on the ordination and classification of running-water sites in Great Britain and on the prediction of their macro-invertebrate communities. *Freshwater Biology* 14:257–80.

Gardner, W.S., W.H. Miller, III, and M.J. Imlay. 1981. Free amino acids in mantle tissues of the bivalve *Amblema plicata:* possible relation to environmental stress. *Bulletin of Environmental Contamination and Toxicology* 26:157–62.

Gauch, H.G. 1982., Jr. *Multivariate Analysis in Community Ecology*. Cambridge Univ. Press, Cambridge, England.

Gerstmeier, R. 1989. Lake typology and indicator organisms in application to the profundal chironomid fauna of Starnberger See (Diptera, Chironomidae). *Archiv für Hydrobiologie* 116:227–34.

Giesy, J.P. 1988. Clinical indicators of stress-induced changes in aquatic organisms. *Verhandlungen Internationale Vereinigung für Theoretische und Angewandte Limnologie* 23:1610–8.

Giesy, J.P., C.S. Duke, R.D. Bingham, and G.W. Dickson. 1983. Changes in phosphoadenylate concentrations and adenylate energy charge as an integrated biochemical measure of stress in invertebrates: the effects of cadmium on the freshwater clam *Corbicula fluminea*. *Toxicological and Environmental Chemistry* 6:259–95.

Gorham, E. 1989. Scientific understanding of ecosystem acidification: a historical review. *Ambio* 18:150–4.

Graney, R.L., Jr., D.S. Cherry, and J. Cairns, Jr. 1983. Heavy metal indicator potential of the Asiatic clam (*Corbicula fluminea*) in artificial stream systems. *Hydrobiologia* 102:81–8.

Graney, R.L., Jr., D.S. Cherry, and J. Cairns, Jr. 1984. The influence of substrate, pH, diet and temperature upon cadmium accumulation in the Asiatic clam (*Corbicula fluminea*) in laboratory artificial streams. *Water Research* 18:833–42.

Graney, R.L. and J.P. Giesy, Jr. 1986. Effects of long-term exposure to pentachlorophenol on the free amino acid pool and energy reserves of the freshwater amphipod *Gammarus pseudolimnaeus* Bousfield (Crustacea, Amphipoda). *Ecotoxicology and Environmental Safety* 12:233–51.

Graney, R.L. and J.P. Giesy, Jr. 1987. The effect of short-term exposure to pentachlorophenol and osmotic stress on the free amino acid pool of the freshwater amphipod *Gammarus pseudolimnaeus* Bousfield. *Archives of Environmental Contamination and Toxicology* 16:167–76.

Graney, R.L. and J.P. Giesy, Jr. 1988. Alterations in the oxygen consumption, condition index and concentration of free amino acids in *Corbicula fluminea* (Mollusca: Pelecypoda) exposed to sodium dodecyl sulfate. *Environmental Toxicology and Chemistry* 7:301–15.

Hadderingh, R.H., G. van der Velde, and P.G. Schnabel. 1987. The effects of heated effluent on the occurrence and the reproduction of the freshwater limpets *Ancylus fluviatilis* Müller, 1774, *Ferrissia wautieri* (Mirolli, 1960) and *Acroloxus lacustris* (L., 1758) in two Dutch water bodies. *Hydrobiological Bulletin* 21:193–205.

Hall, R.J., R.C. Bailey, and J. Findeis. 1988. Factors affecting survival and cation concentration in the blackflies *Prosimulium fuscum/mixtum* and the mayfly *Leptophlebia cupida* during spring snowmelt. *Canadian Journal of Fisheries and Aquatic Sciences* 45:2123–32.

Hall, R.J., C.T. Driscoll, G.E. Likens, and J.M. Pratt. 1985. Physical, chemical, and biological consequences of episodic aluminum additions to a stream. *Limnology and Oceanography* 30:212–20.

Hall, R.J., G.E. Likens, S.B. Fiance, and G.R. Hendrey. 1980. Experimental acidification of a stream in the Hubbard Brook Experimental Forest, New Hampshire. *Ecology* 61:976–89.

Hamilton, A.L. and O.A. Sæther. 1971. The occurrence of characteristic defor-

mities in the chironomid larvae of several Canadian lakes. *Canadian Entomologist* 103:363–8.
Hamilton, J.D. and J. Timmons. 1980. Effects of mild tannery pollution on growth and emergence of two aquatic insects *Rhithrogena semicolorata* and *Ephemerella ignita*. *Water Research* 14:723–7.
Hamilton, S.J. and P.M. Mehrle. 1986. Metallothionein in fish: review of its importance in assessing stress from metal contaminants. *Transactions of the American Fisheries Society* 115:596–609.
Hare, L. and J.C.H. Carter. 1976. The distribution of *Chironomus* (s.s.)? *cucini* (*salinarius* group) larvae (Diptera: Chironomidae) in Parry Sound, Georgian Bay, with particular reference to structural deformities. *Canadian Journal of Zoology* 54:2129–34.
Hare, L., A. Tessier, and P.G.C. Campbell. 1991. Trace element distributions in aquatic insects: variations among genera, elements, and lakes. *Canadian Journal of Fisheries and Aquatic Sciences* 48:1481–91.
Hargeby, A. and R.C. Petersen, Jr. 1988. Effects of low pH and humus on the survivorship, growth and feeding of *Gammarus pulex* (L.) (Amphipoda). *Freshwater Biology* 19:235–47.
Hart, D.R., P.M. McKee, A.J. Burt, and M.J. Goffin. 1986. Benthic community and sediment quality assessment of Port Hope Harbour, Lake Ontario. *Journal of Great Lakes Research* 12:206–20.
Hatakeyama, S. 1987. Chronic effects of Cd on reproduction of *Polypedilum nubifer* (Chironomidae) through water and food. *Environmental Pollution* 48:249–61.
Havas, M. and T.C. Hutchinson. 1983. Effect of low pH on the chemical composition of aquatic invertebrates from tundra ponds at the Smoking Hills, N.W.T., Canada. *Canadian Journal of Zoology* 61:241–9.
Heinis, F., K.R. Timmermans, and W.R. Swain. 1990. Short-term sublethal effects of cadmium on the filter feeding chironomid larva *Glyptotendipes pallens* (Meigen) (Diptera). *Aquatic Toxicology* 16:73–86.
Hellawell, J.M. 1986. *Biological Indicators of Freshwater Pollution and Environmental Management*. Elsevier, London.
Hemelraad, J., D.A. Holwerda, H.J. Herwig, and D.I. Zandee. 1990a. Effects of cadmium in freshwater clams. III. Interaction with energy metabolism in *Anodonta cygnea*. *Archives of Environmental Contamination and Toxicology* 19:699–703.
Hemelraad, J., D.A. Holwerda, H.J.A. Wijnne, and D.I. Zandee. 1990b. Effects of cadmium in freshwater clams. I. Interaction with essential elements in *Anodonta cygnea*. *Archives of Environmental Contamination and Toxicology* 19:686–90.
Henry, M.G., D.N. Chester, and W.L. Mauck. 1986. Role of artificial burrows in *Hexagenia* toxicity tests: recommendations for protocol development. *Environmental Toxicology and Chemistry* 5:553–9.
Herrmann, J. and K.G. Andersson. 1986. Aluminium impact on respiration of lotic mayflies at low pH. *Water, Air, and Soil Pollution* 30:703–9.
Hildrew, A.G., C.R. Townsend, J. Francis, and K. Finch. 1984a. Cellulolytic decomposition in streams of contrasting pH and its relationship with invertebrate community structure. *Freshwater Biology* 14:323–8.

Hildrew, A.G., C.R. Townsend, J. Francis, and K. Finch. 1984b. Community structure in some southern English streams: the influence of species interactions. *Freshwater Biology* 14:297–310.

Hill, M.O. 1979a. *DECORANA—A FORTRAN Program for Detrended Correspondence Analysis and Reciprocal Averaging*. Cornell Univ., Ithaca, NY.

Hill, M.O. 1979b. *TWINSPAN—A FORTRAN Program for Arranging Multivariate Data in an Ordered Two-Way Table by Classification of the Individuals and Attributes*. Cornell Univ., Ithaca, NY.

Hollett, L., M. Berrill, and L. Rowe. 1986. Variation in major ion concentration of *Cambarus robustus* and *Orconectes rusticus* following exposure to low pH. *Canadian Journal of Fisheries and Aquatic Sciences* 43:2040–4.

Hunter, R.D. 1988. Effects of acid water on shells, embryos, and juvenile survival of *Planorbella trivolvis* (Gastropoda: Pulmonata): a laboratory study. *Journal of Freshwater Ecology* 4:315–27.

Hynes, H.B.N. 1960. *The Biology of Polluted Waters*. Liverpool Univ. Press, Liverpool, England.

Idoniboye-Obu, B. 1977. Bioelectric action potentials of *Procambarus acutus acutus* (Girrard) in serially diluted solution of selected C_6 hydrocarbons in water. *Environmental Pollution* 14:5–24.

Ishizaki, S. and H. Hamada. 1987. Effects of heavy metals on the freshwater snail, *Semisulcospira bensoni*, in a closed mining area. *Japanese Journal of Limnology* 48:91–8.

ISO (International Organization for Standardization). 1979. *Assessment of the Biological Quality of Rivers by a Macroinvertebrate "Score."* ISO/TC 147/SC 5/WG 6 N 5, Draft Proposal.

Johnson, R.K. 1989. Classification of profundal chironomid communities in oligotrophic/humic lakes of Sweden using environmental data. In Advances in chironomidology. Proceedings of the Tenth International Chironomid Symposium, Debrecen, Hungary, July 25–28, 1988. Part 2. Faunistics, population dynamics, ecology, production and community structure, ed. G. Dévai. *Acta Biologica Debrecina Oecologica Hungarica* 3:167–75.

Johnson, R.K. and T. Wiederholm. 1989. Classification and ordination of profundal macroinvertebrate communities in nutrient poor, oligo-mesohumic lakes in relation to environmental data. *Freshwater Biology* 21:375–86.

Johnson, R.K., T. Wiederholm, and L. Eriksson. 1990. The influence of season on the classification and ordination of profundal communities of nutrient poor, oligo-mesohumic Swedish lakes using environmental data. *Internationale Vereinigung für Theoretische und Angewandte Limnologie Verhandlungen* 24:646–52.

Jónasson, P.M. 1972. Ecology and production of the profundal benthos in relation to phytoplankton in Lake Esrom. *Oikos Supplement* 14:1–148.

Jones, W.G. and K.F. Walker. 1979. Accumulation of iron, manganese, zinc and cadmium by the Australian freshwater mussel *Velesunio ambiguus* (Phillipi) and its potential as a biological monitor. *Australian Journal of Marine and Freshwater Research* 30:741–51.

Jørgensen, G.F. 1978. Use of biotic indices as a tool for water quality analysis in rivers. *Internationale Vereinigung für Theoretische und Angewandte Limnologie Verhandlungen* 20:1772–8.

Kabré, G. and C. Chaisemartin. 1986–1987. Variations éco-biochimiques de l'hémolymphe chez quatre populations d'Ecrevisses dans une région minière. *Annales des Sciences Naturelles, Zoologie, Paris* 8:19–23.

Kansanen, P.H., J. Aho, and L. Paasivirta. 1984. Testing the benthic lake type concept based on chironomid associations in some Finnish lakes using multivariate statistical methods. *Annales Zoologici Fennici* 21:55–76.

Kapu, M.M. and D.J. Schaeffer. 1991. Planarians in toxicology. Responses of sexual *Dugesia dorotocephala* to selected metals. *Bulletin of Environmental Contamination and Toxicology* 47:302–7.

Khoruzhaya, T.A. 1989. Prospects for the use of biochemical indicators in biomonitoring of natural waters. *Hydrobiological Journal* 25(5):48–52.

Kingett, P.D. 1985. Genetic techniques: a potential tool for water quality assessment. In *Biological Monitoring in Freshwaters: Proceedings of a Seminar, Hamilton, November 21–23, 1984. Part 1,* eds. R.D. Pridmore and A.B. Cooper, pp. 179–88. Water and Soil Miscellaneous Publication No. 82, National Water and Soil Conservation Authority, Wellington, NZ.

Kitching, R.L., H.F. Chapman, and J.M. Hughes. 1987. Levels of activity as indicators of sublethal impacts of copper contamination and salinity reduction in the intertidal gastropod, *Polinices incei* Philippi. *Marine Environmental Research* 23:79–87.

Kititsyna, L.A. and M.L. Pidgayko. 1974. Production of *Pontogammarus robustoides* in the cooling pond of the Kurakhovka Thermal Power Plant. *Hydrobiological Journal* 10(4):20–6.

Klerks, P.L. and J.S. Levinton. 1989. Rapid evolution of metal resistance in a benthic oligochaete inhabiting a metal-polluted site. *Biological Bulletin* 176:135–41.

Klerks, P.L. and J.S. Weis. 1987. Genetic adaptation to heavy metals in aquatic organisms: a review. *Environmental Pollution* 45:173–205.

Koehn, T. and C. Frank. 1980. Effect of thermal pollution on the chironomid fauna in an urban channel. In *Chironomidae. Ecology, Systematics, Cytology and Physiology. Proceedings of the 7th International Symposium on Chironomidae, Dublin, August 1979,* ed. D.A. Murray, pp. 187–94. Pergamon Press, Oxford, England.

Kolkwitz, R. and M. Marsson. 1909. Ökologie der tierischen Saprobien. Beiträge zur Lehre von de biologischen Gewasserbeurteilung. *Internationale Revue der gesamten Hydrobiologie und Hydrographie* 2:126–52.

Kosalwat, P. and A.W. Knight. 1987. Chronic toxicity of copper to a partial life cycle of the midge, *Chironomus decorus. Archives of Environmental Contamination and Toxicology* 16:283–90.

Lang, C. 1985. Eutrophication of Lake Geneva indicated by the oligochaete communities of the profundal. *Hydrobiologia* 126:237–43.

Lang, C. 1989. Effects of small-scale sedimentary patchiness on the distribution of tubificid and lumbriculid worms in Lake Geneva. *Freshwater Biology* 21:477–81.

Langton, P.H. 1991. *A Key to Pupal Exuviae of West Palaearctic Chironomidae.* P.H. Langton, Huntington, England.

Lauenstein, G.G. and T.P. O'Connor. 1988. Measuring the health of U.S. coastal waters. *Sea Technology* 29(5):29–32.

Lechleitner, R.A., D.S. Cherry, J. Cairns, Jr., and D.A. Stetler. 1985. Ionoregulatory and toxicological responses of stonefly nymphs (Plecoptera) to acidic and alkaline pH. *Archives of Environmental Contamination and Toxicology* 14:179–85.

Lingdell, P.-E. and E. Engblom. 1990. *Rena och oförsurade vatten, finns dom?* SNV Report No. 3708. Statens Naturvårdsverk, Stockholm, Sweden.

Little, E.E., B.A. Flerov, and N.N. Ruzhinskaya. 1985. Behavioral approaches in aquatic toxicity investigations: a review. In *Toxic Substances in the Environment: an International Aspect,* eds. P.M. Mehrle, Jr., R.H. Gray, and R.L. Kendall, pp. 72–98. American Fisheries Society, Bethesda, MD.

Long, E.R. and P.M. Chapman. 1985. A sediment quality triad: measures of sediment contamination, toxicity and infaunal composition in Puget Sound. *Marine Pollution Bulletin* 16:405–15.

Lower, W.R. and R.J. Kendall. 1990. Sentinel species and sentinel bioassay. In *Biomarkers of Environmental Contamination,* eds. J.F. McCarthy and L.R. Shugart, pp. 309–31. Lewis Pubs., Chelsea, MI.

Madigosky, S.R., X. Alvarez-Hernandez, and J. Glass. 1991. Lead, cadmium, and aluminum accumulation in the red swamp crayfish *Procambarus clarkii* G. collected from roadside drainage ditches in Louisiana. *Archives of Environmental Contamination and Toxicology* 20:253–8.

Maki, A.W., K.W. Stewart, and J.K.G. Silvey. 1973. The effects of Dibrom on respiratory activity of the stonefly, *Hydroperla crosbyi,* hellgrammite, *Corydalus cornutus* and the golden shiner, *Notemigonus crysoleucas. Transactions of the American Fisheries Society* 102:806–15.

Malley, D.F., J.D. Huebner, and K. Donkersloot. 1988. Effects on ionic composition of blood and tissues of *Anodonta grandis grandis* (Bivalvia) of an addition of aluminum and acid to a lake. *Archives of Environmental Contamination and Toxicology* 17:479–91.

Mance, G. 1987. *Pollution Threat of Heavy Metals in Aquatic Environments*. Elsevier, New York.

Manly, R. and W.O. George. 1977. The occurrence of some heavy metals in populations of the freshwater mussel *Anodonta anatina* (L.) from the River Thames. *Environmental Pollution* 14:139–54.

Marcucella, H. and C.I. Abramson. 1978. Behavioral toxicology and teleost fish. In *The Behavior of Fish and Other Aquatic Animals,* ed. D.I. Mostofsky, pp. 33–77. Academic Press, New York.

McCahon, C.P., A.F. Brown, M.J. Poulton, and D. Pascoe. 1989. Effects of acid, aluminium and lime additions on fish and invertebrates in a chronically acidic Welsh stream. *Water, Air, and Soil Pollution* 45:345–59.

McClelland, W.T. and M.A. Brusven. 1980. Effects of sedimentation on the behavior and distribution of riffle insects in a laboratory stream. *Aquatic Insects* 2:161–9.

McKee, M.J. and C.O. Knowles. 1989. Temporal and spatial variation of RNA content in nymphs of the mayfly *Stenonema femoratum* (Say). *Environmental Pollution* 58:43–55.

Meriläinen, J.J. 1987. The profundal zoobenthos used as an indicator of the biological condition of Lake Päijänne. *Biological Research Report, University of Jyväskylä* 10:87–94.

Meriläinen, J.J. and J. Hynynen. 1990. Benthic invertebrates in relation to acidity in Finnish forest lakes. In *Acidification in Finland,* eds. P. Kauppi, P. Anttila, and K. Kenttämies, pp. 1029–49. Springer-Verlag, Berlin.

Merlini, M., G. Cadario, and B. Oregioni. 1978. The unionid mussel as a biogeochemical indicator of metal pollution. In *Environmental Biogeochemistry and Geomicrobiology. Vol. 3. Methods, Metals and Assessment. Proceedings of the Third International Symposium on Environmental Biogeochemistry, Wolfenbüttel, West Germany,* ed. W.E. Krumbein, pp. 955–65. Ann Arbor Science Pubs., Ann Arbor, MI.

Metcalfe, J.L. 1989. Biological water quality assessment of running waters based on macroinvertebrate communities: history and present status in Europe. *Environmental Pollution* 60:101–39.

Metcalfe, J.L., M.E. Fox, and J.H. Carey. 1988. Freshwater leeches (Hirudinea) as a screening tool for detecting organic contaminants in the environment. *Environmental Monitoring and Assessment* 11:147–69.

Metcalfe, J.L. and A. Mudroch. 1987. Distribution of arsenic and mercury in zoobenthos from the Shubenacadie River headwater lakes in Nova Scotia. In Proceedings of the Thirteenth Annual Aquatic Toxicity Workshop, Moncton, NB, November 12–14, 1986, ed. J.S.S. Lakshminarayana. *Canadian Technical Report of Fisheries and Aquatic Sciences* No. 1575:85–7.

Milbrink, G. 1983. Characteristic deformities in tubificid oligochaetes inhabiting polluted bays of Lake Vänern, southern Sweden. *Hydrobiologia* 106:169–84.

Millington, P.J. and K.F. Walker. 1983. Australian freshwater mussel *Velesunio ambiguus* (Philippi) as a biological monitor for zinc, iron and manganese. *Australian Journal of Marine and Freshwater Research* 34:873–92.

Mitin, A.V. 1985. The isozyme spectra of hemolymph enzymes in mayfly nymphs as an indicator of industrial pollution. *Hydrobiological Journal* 21(5):89–93.

Moore, J.N., S.N. Luoma, and D. Peters. 1991. Downstream effects of mine effluent on an intermontane riparian system. *Canadian Journal of Fisheries and Aquatic Sciences* 48:222–32.

Moss, D., M.T. Furse, J.F. Wright, and P.D. Armitage. 1987. The prediction of the macro-invertebrate fauna of unpolluted running-water sites in Great Britain using environmental data. *Freshwater Biology* 17:41–52.

Muncaster, B.W., P.D.N. Hebert, and R. Lazar. 1990. Biological and physical factors affecting the body burden of organic contaminants in freshwater mussels. *Archives of Environmental Contamination and Toxicology* 19:25–34.

Mysing-Gubala, M. and M.A. Poirrier. 1981. The effects of cadmium and mercury on gemmule formation and gemmoselere morphology in *Ephydatia fluviatilis* (Porifera: Spongillidae). *Hydrobiologia* 76:145–8.

Naylor, C., L. Maltby, and P. Calow. 1989. Scope for growth in *Gammarus pulex,* a freshwater benthic detritivore. *Hydrobiologia* 188/189:517–23.

Nebeker, A.V. and F.A. Puglisi. 1974. Effects of polychlorinated biphenyls (PCB's) on survival and reproduction of *Daphnia, Gammarus* and *Tanytarsus. Transactions of the American Fisheries Society* 103:722–8.

Nehring, R.B. 1976. Aquatic insects as biological monitors of heavy metal pollution. *Bulletin of Environmental Contamination and Toxicology* 15:147–54.
Nehring, R.B., R. Nisson, and G. Minasian. 1979. Reliability of aquatic insects versus water samples as measures of aquatic lead pollution. *Bulletin of Environmental Contamination and Toxicology* 22:103–8.
Newcombe, C.P. and D.D. MacDonald. 1991. Effects of suspended sediments on aquatic ecosystems. *North American Journal of Fisheries Management* 11:72–82.
Newman, M.C. and A.W. McIntosh. 1982. The influence of lead in components of a freshwater ecosystem on molluscan tissue lead concentrations. *Aquatic Toxicology* 2:1–19.
Newman, M.C. and A.W. McIntosh. 1983a. Lead elimination and size effects on accumulation by two freshwater gastropods. *Archives of Environmental Contamination and Toxicology* 12:25–9.
Newman, M.C. and A.W. McIntosh. 1983b. Slow accumulation of lead from contaminated food sources by the freshwater gastropods, *Physa integra* and *Campeloma decisum*. *Archives of Environmental Contamination and Toxicology* 12:685–92.
Newman, P.J. 1988. *Classification of Surface Water Quality*. Heinemann Professional Publishing, London.
Nikanorov, A.M., A.V. Zhulidov, and N.A. Dubova. 1988. Factors determining magnitude of mercury content in hydrobionts of freshwater ecosystems. *Soviet Journal of Ecology* 19:50–6.
Nikinmaa, M., T. Järvenpää, K. Westman, and A. Soivio. 1983. Effects of hypoxia and acidification on the haemolymph pH values and ion concentrations in the freshwater crayfish (*Astacus astacus* L.). *Finnish Fisheries Research* 5:17–22.
Nriagu, J.O. 1990. Global metal pollution. Poisoning the biosphere? *Environment* 32(7):7–11, 28–33.
Økland, J. and K.A. Økland. 1986. The effects of acid deposition on benthic animals in lakes and streams. *Experientia* 42:471–86.
Oksanen, J., E. Läärä, P. Huttunen, and J. Meriläinen. 1988. Estimation of pH optima and tolerances of diatoms in lake sediment by the methods of weighted averaging, least squares and maximum likelihood, and their use for the prediction of lake acidity. *Journal of Paleolimnology* 1:39–49.
Ormerod, S.J. 1987. The influences of habitat and seasonal sampling regimes on the ordination and classification of macroinvertebrate assemblages in the catchment of the River Wye, Wales. *Hydrobiologia* 150:143–51.
Ormerod, S.J. and R.W. Edwards. 1987. The ordination and classification of macroinvertebrate assemblages in the catchment of the River Wye in relation to environmental factors. *Freshwater Biology* 17:533–46.
Ormerod, S.J. and R.W. Edwards. 1991. Modelling the ecological impact of acidification: problems and possibilities. *Internationale Vereinigung für Theoretische und Angewandte Limnologie Verhandlungen* 24:1738–41.
Ormerod, S.J., N.S. Weatherley, P.V. Varallo, and P.G. Whitehead. 1988. Preliminary empirical models of the historical and future impact of acidification on the ecology of Welsh streams. *Freshwater Biology* 20:127–40.
Pearson, T.H., J.S. Gray, and P.J. Johannessen. 1983. Objective selection of sen-

sitive species indicative of pollution-induced change in benthic communities: 2. Data analyses. *Marine Ecology Progress Series* 12:237–55.

Pedersen, B.V. 1984. The effect of anoxia on the survival of chromosomal variants in the larvae of the midge *Chironomus plumosus* L. (Diptera: Chironomidae). *Hereditas* 101:75–7.

Pedersen, B.V. 1986. On microgeographic differentiations of a chromosomal polymorphism in *Chironomus plumosus* L. from Lake Tystrup-Bavelse, Denmark (Diptera: Chironomidae). *Hereditas* 105:209–19.

Petersen, L.B.-M. and R.C. Petersen, Jr. 1984. Effect of kraft pulp mill effluent and 4,5,6 trichloroguaiacol on the net spinning behavior of *Hydropsyche angustipennis* (Trichoptera). *Ecological Bulletins* 36:68–74.

Peterson, C.H. and R. Black. 1988. Density-dependent mortality caused by physical stress interacting with biotic history. *American Naturalist* 131:257–70.

Pettigrove, V. 1989. Larval mouthpart deformities in *Procladius paludicola* Skuse (Diptera: Chironomidae) from the Murray and Darling rivers, Australia. *Hydrobiologia* 179:111–7.

Phillips, D.J.H. 1980. *Quantitative Aquatic Biological Indicators. Their Use to Monitor Trace Metal and Organochlorine Pollution*. Applied Science Pubs., London.

Phillips, D.J.H. 1991. Selected trace elements and the use of biomonitors in subtropical and tropical marine ecosystems. *Reviews of Environmental Contamination and Toxicology* 120:105–29.

Phillips, D.J.H. and D.A. Segar. 1986. Use of bio-indicators in monitoring conservative contaminants: programme design imperatives. *Marine Pollution Bulletin* 17:10–7.

Pinder, L.C.V. and I.S. Farr. 1987. Biological surveillance of water quality—2. Temporal and spatial variation in the macroinvertebrate fauna of the River Frome, a Dorset chalk stream. *Archiv für Hydrobiologie* 109:321–31.

Pontasch, K.W. and M.A. Brusven. 1988. Diversity and community comparison indices: assessing macroinvertebrate recovery following a gasoline spill. *Water Research* 22:619–26.

Pugsley, C.W., P.D.N. Hebert, and P.M. McQuarrie. 1988. Distribution of contaminants in clams and sediments from the Huron-Erie corridor. II—Lead and cadmium. *Journal of Great Lakes Research* 14:356–68.

Pynnönen, K. 1990. Physiological responses to severe acid stress in four species of freshwater clams (Unionidae). *Archives of Environmental Contamination and Toxicology* 19:471–8.

Rada, R.G., J.E. Findley, and J.G. Wiener. 1986. Environmental fate of mercury discharged into the Upper Wisconsin River. *Water, Air, and Soil Pollution* 29:57–76.

Raddum, G.G. and A. Fjellheim. 1984. Acidification and early warning organisms in freshwater in western Norway. *Internationale Vereinigung für Theoretische und Angewandte Limnologie Verhandlungen* 22:1973–80.

Raddum, G.G., A. Fjellheim, and T. Hesthagen. 1988. Monitoring of acidification by the use of aquatic organisms. *Internationale Vereinigung für Theoretische und Angewandte Limnologie Verhandlungen* 23:2291–7.

Raddum, G.G. and O.A. Sæther. 1981. Chironomid communities in Norwegian lakes with different degrees of acidification. *Internationale Vereinigung für Theoretische und Angewandte Limnologie Verhandlungen* 21:399–405.

Radwan, S., W. Kowalik, and R. Kornijów. 1990. Accumulation of heavy metals in a lake ecosystem. *Science of the Total Environment* 96:121–9.

Rand, G.M. 1985. Behavior. In *Fundamentals of Aquatic Toxicology. Methods and Applications,* eds. G.M. Rand and S.R. Petrocelli, pp. 221–63. Hemisphere Publishing Corporation, Washington, DC.

Reddy, S.L.N. and N.B.R.K. Venugopal. 1990a. Effect of fluoride on acetylcholinesterase activity and oxygen consumption in a freshwater field crab, *Barytelphusa guerini. Bulletin of Environmental Contamination and Toxicology* 45:760–6.

Reddy, S.L.N. and N.B.R.K. Venugopal. 1990b. Fluoride induced changes in protein metabolism in the tissues of freshwater crab *Barytelphusa guerini. Environmental Pollution* 67:97–108.

Reddy, S.L.N., N.B.R.K. Venugopal, and J.V. Ramana Rao. 1989. *In vivo* effects of cadmium chloride on certain aspects of carbohydrate metabolism in the tissues of a freshwater field crab *Barytelphusa guerini. Bulletin of Environmental Contamination and Toxicology* 42:847–53.

Renzoni, A. and E. Bacci. 1976. Bodily distribution, accumulation and excretion of mercury in a fresh-water mussel. *Bulletin of Environmental Contamination and Toxicology* 15:366–73.

Resh, V.H. and D.M. Rosenberg. 1989. Spatial-temporal variability and the study of aquatic insects. *Canadian Entomologist* 121:941–63.

Resh, V.H. and J.D. Unzicker. 1975. Water quality monitoring and aquatic organisms: the importance of species identification. *Journal of the Water Pollution Control Federation* 47:9–19.

Reynoldson, T.B. and J.L. Metcalfe-Smith. 1992. An overview of the assessment of aquatic ecosystem health using benthic invertebrates. *Journal of Aquatic Ecosystem Health 1* (In press).

Reynoldson, T.B., S.P. Thompson, and J.L. Bamsey. 1991. A sediment bioassay using the tubificid oligochaete worm *Tubifex tubifex. Environmental Toxicology and Chemistry* 10:1061–72.

Rincon-Leon, F., G. Zurera-Cosano, and R. Pozo-Lora. 1988. Lead and cadmium concentrations in red crayfish (*Procambarus clarkii,* G.) in the Guadalquivir River marshes (Spain). *Archives of Environmental Contamination and Toxicology* 17:251–6.

Rockwood, J.P., D.S. Jones, and R.A. Coler. 1990. The effect of aluminum in soft water at low pH on oxygen consumption by the dragonfly *Libellula julia* Uhler. *Hydrobiologia* 190:55–9.

Rosenberg, D.M. and A.P. Wiens. 1976. Community and species responses of Chironomidae (Diptera) to contamination of fresh waters by crude oil and petroleum products, with special reference to the Trail River, Northwest Territories. *Journal of the Fisheries Research Board of Canada* 33:1955–63.

Russell, R.W. and F.A.P.C. Gobas. 1989. Calibration of the freshwater mussel, *Elliptio complanata,* for quantitative biomonitoring of hexachlorobenzene and octachlorostyrene in aquatic systems. *Bulletin of Environmental Contamination and Toxicology* 43:576–82.

Ryan, P.A. 1991. Environmental effects of sediment on New Zealand streams: a review. *New Zealand Journal of Marine and Freshwater Research* 25:207–21.
Saber Hussain, M. and K. Jamil. 1989. Bioaccumulation of heavy metal ions and their effect on certain biochemical parameters of water hyacinth weevil *Neochetina eichhorniae* (Warner). *Journal of Environmental Science and Health* B24:251–64.
Saber Hussain, M. and K. Jamil. 1990. Bioaccumulation of mercury and its effect on protein metabolism of the water hyacinth weevil *Neochetina eichhornae* (Warner). *Bulletin of Environmental Contamination and Toxicology* 45:294–8.
Sæther, O.A. 1979. Chironomid communities as water quality indicators. *Holarctic Ecology* 2:65–74.
Salánki, J., K. V.-Balogh, and E. Berta. 1982. Heavy metals in animals of Lake Balaton. *Water Research* 16:1147–52.
Santiago-Fandino, V.J.R. 1983. The effects of nickel and cadmium on the growth rate of *Hydra littoralis* and an assessment of the rate of uptake of ^{63}Ni and ^{14}C by the same organism. *Water Research* 17:917–23.
Scherer, E. and R.E. McNicol. 1986. Behavioural responses of stream-dwelling *Acroneuria lycorias* (Ins., Plecopt.) larvae to methoxychlor and fenitrothion. *Aquatic Toxicology* 8:251–63.
Schott, R.J. and M.A. Brusven. 1980. The ecology and electrophoretic analysis of the damselfly *Argia vivida* Hagen, living in a geothermal gradient. *Hydrobiologia* 69:261–5.
Sheffy, T.B. 1978. Mercury burdens in crayfish from the Wisconsin River. *Environmental Pollution* 17:219–25.
Simpson, K.W. 1980. Abnormalities in the tracheal gills of aquatic insects collected from streams receiving chlorinated crude oil wastes. *Freshwater Biology* 10:581–3.
Smith, M.E., B.J. Wyskowski, C.M. Brooks, C.T. Driscoll, and C.C. Cosentini. 1990. Relationships between acidity and benthic invertebrates of low-order woodland streams in the Adirondack Mountains, New York. *Canadian Journal of Fisheries and Aquatic Sciences* 47:1318–29.
Stenson, J.A.E. and M.O.G. Eriksson. 1989. Ecological mechanisms important for the biotic changes in acidified lakes in Scandinavia. *Archives of Environmental Contamination and Toxicology* 18:201–6.
Stephenson, M. and G.L. Mackie. 1988. Multivariate analysis of correlations between environmental parameters and cadmium concentrations in *Hyalella azteca* (Crustacea: Amphipoda) from central Ontario lakes. *Canadian Journal of Fisheries and Aquatic Sciences* 45:1705–10.
Stinson, M.D. and D.L. Eaton. 1983. Concentrations of lead, cadmium, mercury, and copper in the crayfish (*Pacifasticus leniusculus*) obtained from a lake receiving urban runoff. *Archives of Environmental Contamination and Toxicology* 12:693–700.
Stokes, P.M., E.T. Howell, and G. Krantzberg. 1989. Effects of acidic precipitation on the biota of freshwater lakes. In *Acidic Precipitation. Volume 2: Biological and Ecological Effects,* eds. D.C. Adriano and A.H. Johnson, pp. 273–304. Springer-Verlag, New York.

Sutcliffe, D.W. and T.R. Carrick. 1973. Studies on mountain streams in the English Lake District. I. pH, calcium and the distribution of invertebrates in the River Duddon. *Freshwater Biology* 3:437–62.

Symons, P.E.K. and J.L. Metcalfe. 1978. Mortality, recovery, and survival of larval *Brachycentrus numerosus* (Trichoptera) after exposure to the insecticide fenitrothion. *Canadian Journal of Zoology* 56:1284–90.

ter Braak, C.J.F. 1986. Canonical correspondence analysis: a new eigenvector technique for multivariate direct gradient analysis. *Ecology* 67:1167–79.

ter Braak, C.J.F. and C.W.N. Looman. 1986. Weighted averaging, logistic regression and the Gaussian response model. *Vegetatio* 65:3–11.

ter Braak, C.J.F. and I.C. Prentice. 1988. A theory of gradient analysis. *Advances in Ecological Research* 18:271–317.

ter Braak, C.J.F. and H. van Dam. 1989. Inferring pH from diatoms: a comparison of old and new calibration methods. *Hydrobiologia* 178:209–23.

Tessier, A., P.G.C. Campbell, J.C. Auclair, and M. Bisson. 1984. Relationships between the partitioning of trace metals in sediments and their accumulation in the tissues of the freshwater mollusc *Elliptio complanata* in a mining area. *Canadian Journal of Fisheries and Aquatic Sciences* 41:1463–72.

Tevesz, M.J.S., G. Matisoff, S.A. Frank, and P.L. McCall. 1989. Interspecific differences in manganese levels in freshwater bivalves. *Water, Air, and Soil Pollution* 47:65–70.

Thienemann, A. 1921. Biologische Seetypen und die Gründung einer Hydrobiologischen Anstalt am Bodensee. *Archiv für Hydrobiologie* 13:347–70.

Thienemann, A. 1922. Die beiden Chironomusarten der Tiefenfauna der norddeutschen Seen. Ein hydrobiologisches Problem. *Archiv für Hydrobiologie* 13:609–46.

Thornton, K.W. and J.L. Wilhm. 1975. The use of life tables in demonstrating the effects of pH, phenol, and NaCl on *Chironomus attenuatus* populations. *Environmental Entomology* 4:325–8.

Timmermans, K.R. 1991. Trace metal ecotoxicokinetics of chironomids. Ph.D. dissertation, Univ. of Amsterdam, Amsterdam, The Netherlands.

Tohyama, C., Z.A. Shaikh, K.J. Ellis, and S.H. Cohn. 1981. Metallothionein excretion in urine upon cadmium exposure: its relationship with liver and kidney cadmium. *Toxicology* 22:181–91.

Tooby, T.E. and D.J. Macey. 1977. Absence of pigmentation in corixid bugs (Hemiptera) after the use of the aquatic herbicide dichlobenil. *Freshwater Biology* 7:519–25.

Torreblanca, A., J. Del Ramo, and J. Diaz-Mayans. 1989. Gill ATPase activity in *Procambarus clarkii* as an indicator of heavy metal pollution. *Bulletin of Environmental Contamination and Toxicology* 42:829–34.

Trueman, E.R., J.G. Blatchford, H.D. Jones, and G.A. Lowe. 1973. Recordings of the heart rate and activity of molluscs in their natural habitat. *Malacologia* 14:377–83.

Tuffery, G. and J. Verneaux. 1968. *Méthode de Détermination de la Qualité Biologique des Eaux Courantes. Exploitation Codifiée des Inventaires de la Faune du Fond.* Section Pêche et Pisciculture, Centre National d'Etude Techniques et de

Recherches Technologiques pour l'Agriculture, les Forêts et l'Equipment Rural, Ministère de l'Agriculture, Paris, France.

USEPA (U.S. Environmental Protection Agency). 1973. Appendix II-B. Community structure and diversity indices. In *Water Quality Criteria 1972*, pp. 408–9. EPA-R3-73-033. Environmental Protection Agency, Washington, DC.

Uzunov, Y. and S. Kovachev. 1987. The macrozoobenthos of Struma River: an example of a recovered community after the elimination of a heavy industrial impact with suspended materials. *Archiv für Hydrobiologie Supplement* 76:169–96.

van Frankenhuyzen, K. and G.H. Geen. 1987. Effects of low pH and nickel on growth and survival of the shredding caddisfly *Clistoronia magnifica* (Limnephilidae). *Canadian Journal of Zoology* 65:1729–32.

V.-Balogh, K. 1988a. Comparison of mussels and crustacean plankton to monitor heavy metal pollution. *Water, Air, and Soil Pollution* 37:281–92.

V.-Balogh, K. 1988b. Heavy metal pollution from a point source demonstrated by mussel (*Unio pictorum* L.) at Lake Balaton, Hungary. *Bulletin of Environmental Contamination and Toxicology* 41:910–4.

V.-Balogh, K., D.S. Fernandez, and J. Salánki. 1988. Heavy metal concentrations of *Lymnaea stagnalis* L. in the environs of Lake Balaton (Hungary). *Water Research* 22:1205–10.

V.-Balogh, K. and J. Salánki. 1987. Biological monitoring of heavy metal pollution in the region of Lake Balaton (Hungary). *Acta Biologica Hungarica* 38:13–30.

Vermeer, K. 1972. The crayfish, *Orconectes virilis*, as an indicator of mercury contamination. *Canadian Field-Naturalist* 86:123–5.

Verneaux, J., P. Galmiche, F. Janier, and A. Monnot. 1982. Une nouvelle méthode pratique d'evaluation de la qualité des eaux courantes. Un indice biologique de qualité générale (BIG). *Annales scientifiques de l'Université de Besançon. Biologie animale* (4ème série) 3:11–21.

Verneaux, J. and G. Tuffery. 1976. Une méthode zoologique pratique de détermination de la qualité biologique des eaux courantes. *Annales scientifiques de l'Université de Besançon. Zoologie* 3:79–90.

Wade, K.R., S.J. Ormerod, and A.S. Gee. 1989. Classification and ordination of macroinvertebrate assemblages to predict stream acidity in upland Wales. *Hydrobiologia* 171:59–78.

Walker, I.R., C.H. Fernando, and C.G. Paterson. 1984. The chironomid fauna of four shallow, humic lakes and their representation by subfossil assemblages in the surficial sediments. *Hydrobiologia* 112:61–7.

Warwick, W.F. 1980a. Palaeolimnology of the Bay of Quinte, Lake Ontario: 2800 years of cultural influence. *Canadian Bulletin of Fisheries and Aquatic Sciences* 206:1–117.

Warwick, W.F. 1980b. Pasqua Lake, southeastern Saskatchewan: a preliminary assessment of trophic status and contamination based on the Chironomidae (Diptera). In *Chironomidae. Ecology, Systematics, Cytology and Physiology. Proceedings of the 7th International Symposium on Chironomidae, Dublin, August 1979*, ed. D.A. Murray, pp. 255–67. Pergamon Press, Oxford, England.

Warwick, W.F. 1985. Morphological abnormalities in Chironomidae (Diptera) larvae as measures of toxic stress in freshwater ecosystems: indexing antennal deformities in *Chironomus* Meigen. *Canadian Journal of Fisheries and Aquatic Sciences* 42:1881–1914.

Warwick, W.F. 1988. Morphological deformities in Chironomidae (Diptera) larvae as biological indicators of toxic stress. In *Toxic Contaminants and Ecosystem Health: a Great Lakes Focus,* ed. M.S. Evans, pp. 281–320. John Wiley, New York.

Warwick, W.F. 1989. Morphological deformities in larvae of *Procladius* Skuse (Diptera: Chironomidae) and their biomonitoring potential. *Canadian Journal of Fisheries and Aquatic Sciences* 46:1255–71.

Warwick, W.F. 1991. Indexing deformities in ligulae and antennae of *Procladius* larvae (Diptera: Chironomidae): application to contaminant-stressed environments. *Canadian Journal of Fisheries and Aquatic Sciences* 48:1151–66.

Warwick, W.F., J. Fitchko, P.M. McKee, D.R. Hart, and A.J. Burt. 1987. The incidence of deformities in *Chironomus* spp. from Port Hope Harbour, Lake Ontario. *Journal of Great Lakes Research* 13:88–92.

Warwick, W.F. and N.A. Tisdale. 1988. Morphological deformities in *Chironomus, Cryptochironomus,* and *Procladius* larvae (Diptera: Chironomidae) from two differentially stressed sites in Tobin Lake, Saskatchewan. *Canadian Journal of Fisheries and Aquatic Sciences* 45:1123–44.

Washington, H.G. 1984. Diversity, biotic and similarity indices. A review with special relevance to aquatic ecosystems. *Water Research* 18:653–94.

Weatherley, N.S. and S.J. Ormerod. 1987. The impact of acidification on macroinvertebrate assemblages in Welsh streams: towards an empirical model. *Environmental Pollution* 46:223–40.

Weatherley, N.S. and S.J. Ormerod. 1989. Modelling ecological impacts of the acidification of Welsh streams: temporal changes in the occurrence of macroflora and macroinvertebrates. *Hydrobiologia* 185:163–74.

Weatherley, N.S., S.P. Thomas, and S.J. Ormerod. 1989. Chemical and biological effects of acid, aluminium and lime additions to a Welsh hill-stream. *Environmental Pollution* 56:283–97.

Wiederholm, T. 1976. Chironomids as indicators of water quality in Swedish lakes. *Naturvårdsverkets Limnologiska Undersökningar* 10:1–17.

Wiederholm, T. 1980. Use of benthos in lake monitoring. *Journal of the Water Pollution Control Federation* 52:537–47.

Wiederholm, T. 1981. Associations of lake-living Chironomidae. A cluster analysis of Brundin's and recent data from Swedish lakes. *Schweizerische Zeitschrift für Hydrologie* 43:140–50.

Wiederholm, T. 1984a. Incidence of deformed chironomid larvae (Diptera: Chironomidae) in Swedish lakes. *Hydrobiologia* 109:243–9.

Wiederholm, T. 1984b. Responses of aquatic insects to environmental pollution. In *The Ecology of Aquatic Insects,* eds. V.H. Resh and D.M. Rosenberg, pp. 508–57. Praeger Pubs., New York.

Wiederholm, T., ed. 1986. Chironomidae of the Holarctic region. Keys and diagnoses. Part 2—Pupae. *Entomologica scandinavica* (Supplement) 28:1–482.

Wiederholm, T. and G. Dave. 1989. Toxicity of metal polluted sediments to *Daphnia magna* and *Tubifex tubifex. Hydrobiologia* 176/177:411–7.

Wiederholm, T., A.-M. Wiederholm, and G. Milbrink. 1987. Bulk sediment bioassays with five species of fresh-water oligochaetes. *Water, Air, and Soil Pollution* 36:131–54.

Williams, B.K. 1983. Some observations on the use of discriminant analysis in ecology. *Ecology* 64:1283–91.

Williams, C.B. 1953. The relative abundance of different species in a wild animal population. *Journal of Animal Ecology* 22:14–31.

Wilson, R.S. 1989. The modification of chironomid pupal exuvial assemblages by sewage effluent in rivers within the Bristol Avon catchment, England. In Advances in chironomidology. Proceedings of the Tenth International Chironomid Symposium, Debrecen, Hungary, July 25–28, 1988. Part 2. Faunistics, population dynamics, ecology, production and community structure, ed. G. Dévai. *Acta Biologica Debrecina Oecologica Hungarica* 3:367–76.

Wilson, R.S. and J.D. McGill. 1977. A new method of monitoring water quality in a stream receiving sewage effluent, using chironomid pupal exuviae. *Water Research* 11:959–62.

Witters, H., J.H.D. Vangenechten, S. Van Puymbroeck, and O.L.J. Vanderborght. 1984. Interference of aluminium and pH on the Na-influx in an aquatic insect *Corixa punctata* (Illig.). *Bulletin of Environmental Contamination and Toxicology* 32:575–9.

Wood, C.M. and M.S. Rogano. 1986. Physiological responses to acid stress in crayfish (*Orconectes*): haemolymph ions, acid-base status, and exchanges with the environment. *Canadian Journal of Fisheries and Aquatic Sciences* 43:1017–26.

Woodiwiss, F.S. 1964. The biological system of stream classification used by the Trent River Board. *Chemistry and Industry* 83:443–7.

Wright, J.F., P.D. Armitage, M.T. Furse, and D. Moss. 1988. A new approach to the biological surveillance of river quality using macroinvertebrates. *Internationale Vereinigung für Theoretische und Angewandte Limnologie Verhandlungen* 23:1548–52.

Wright, J.F., P.D. Armitage, M.T. Furse, and D. Moss. 1989. Prediction of invertebrate communities using stream measurements. *Regulated Rivers: Research and Management* 4:147–55.

Wright, J.F., M.T. Furse, P.D. Armitage, and D. Moss. In Press. New procedures for evaluating the conservation interest and pollution status of British rivers based on the macroinvertebrate fauna. *Internationale Vereinigung für Theoretische und Angewandte Limnologie Verhandlungen 25.*

Wright, J.F., D. Moss, P.D. Armitage, and M.T. Furse. 1984. A preliminary classification of running-water sites in Great Britain based on macro-invertebrate species and the prediction of community type using environmental data. *Freshwater Biology* 14:221–56.

Yamamura, M., K.T. Suzuki, S. Hatakeyama, and K. Kubota. 1983. Tolerance to cadmium and cadmium-binding proteins induced in the midge larva, *Chironomus yoshimatsui* (Diptera, Chironomidae). *Comparative Biochemistry and Physiology* 75C:21–4.

Yevtushenko, N.Yu., N.V. Bren, and Yu.M. Sytnik. 1990. Heavy metal contents in invertebrates of the Danube River. *Water Science and Technology* 22(5):119–25.

Zauke, G.-P. 1981. Cadmium in Gammaridae (Amphipoda: Crustacea) of the rivers Werra and Weser: geographical variation and correlation to cadmium in sediments. *Environmental Pollution* (Series B) 2:465–74.

Zhulidov, A.V., V.M. Emets, and A.S. Shevtsov. 1980. Biomonitoring of river pollution by heavy metals in reserves on the basis of studies on metal accumulation in the body of aquatic invertebrates. *Doklady Biological Sciences* 252:233–5.

Appendix 4.1. Tolerances of selected palearctic (P) and nearctic (N) macroinvertebrates to organic pollution and acidification. Organic pollution tolerances are taken from Hellawell (1986) (1) and Friedrich (1990) (2). Acidification tolerances are taken from Fjellheim and Raddum (1990) (3), Meriläinen and Hynynen (1990) (4), Engblom and Lingdell (1983) (5-field; 6-lab.), Walker et al. (1984) (7), and Dermott (1985) (8). Nomenclature has been updated.

	Species Tolerance to Organic Pollution				Species Tolerance to H^+ Concentration			
	Poly-saprobic	Alpha-meso-saprobic	Beta-meso-saprobic	Oligo-saprobic	pH >5.5	5.5 < pH > 5.0	5.0 < pH > 4.7	pH <4.7
Turbellaria								
Crenobia alpina				P2		P3		
Dendrocoelum lacteum			P1,2					
Dugesia lugubris			P1,2					
D. gonocephala				P2				
D. tigrina		P1	P2					
Otomesostoma auditivum						P3		
Planaria torva		P1	P2					
Polycelis felina				P1,2				
P. nigra			P1,2					
P. tenuis			P1,2					
Mollusca								
Acroloxus lacustris			P2					
Alasmindonta costata			N1					
A. undulata			N1					
Amblema plicata			N1					
Amnicola emerginata				N1				
A. limosa				N1				
Ancylus fluviatilis			P2	P1				
Anodonta cygnea			P1,2	N1				

A. gibbosus		N1		
A. grandis		N1		
A. imbecillis		N1		
Anodonta sp.				P3
Aplexa hypnorum		N1		
Bathyomphalus contortus		P2		
Birgella subglobosus			N1	
Bithynia tentaculata		P1,2,N1		
Bythinella austriaca			P1	
Bythinella spp.			P2	
Campeloma decisum			N1	
C. gibbum		N1		
C. integrum		N1		
C. subsolidum		N1		
Carunculina parva			N1	
Cyclonais tuberculata		N1		
Dreissena polymorpha		P2		
Elimia livescens		N1		
E. virginica	N1			
Ferrissia rivularis		N1	N1	
Fossaria obrussa		N1		
Fusconaia flava		N1		
Gyraulus albus		P2		P4
G. arcticus		N1		
G. laevis				P4
Helisoma anceps		N1		
Hydrobia jenkinsi		P2		
Laevapex fuscus		N1		

(*continued*)

Appendix 4.1. (Continued)

	Species Tolerance to Organic Pollution				Species Tolerance to H^+ Concentration			
	Poly-saprobic	Alpha-meso-saprobic	Beta-meso-saprobic	Oligo-saprobic	pH >5.5	5.5 < pH > 5.0	5.0 < pH > 4.7	pH <4.7
Lampsilis gracilis			N1					
L. radiata			N1					
L. teres			N1					
Lasmigona complanata		N1						
L. costata			N1					
Leptodea fragilis				N1				
Lymnaea humilis			N1					
L. ovata		N1						
L. stagnalis			P1					
Margaritifera margaritifera					P3			
Musculium transversum		N1						
M. securis			N1					
Mytilopsis leucophaetus			N1					
Obliquaria reflexa			N1					
Physella acuta		P2	N1					
P. cubensis		N1						
P. fontinalis			P1,2,N1					
P. gyrina			N1					
P. heterostropha		N1						
P. integra		N1						
P. virgata		N1						

Pisidium casertanum	N1					P4,N8
P. conventus						N8
P. fallax		N1				
P. fossarinum			N1			
P. henslowanum		N1				P4
P. hibernicum						P4
P. idahoense	N1					
P. lilljeborgii					P4	N8
P. milium					P4	
P. nitidum						P4,N8
P. personatum			P1			
P. subtruncatum		N1			P4	N8
P. supinum			P1			
P. variabile						N8
P. ventricosum						N8
Pisidium sp.						P3,N8
Planorbella trivolvis	N1	N1				
Planorbis carinatus			N1			
P. panus	N1					
P. planorbis		P1				
Planorbis spp.				P3		
Planorbula armigera	N1					
Pleurocerca acuta		N1				
Promenetus umbilicatellus						N8
Proptera alata		N1	N1			
Pseudosuccinea columella		N1				
Quadrula lachrymosa		N1				
Q. pustulosa		N1				

(continued)

Appendix 4.1. (Continued)

	Species Tolerance to Organic Pollution				Species Tolerance to H^+ Concentration			
	Poly-saprobic	Alpha-meso-saprobic	Beta-meso-saprobic	Oligo-saprobic	pH >5.5	5.5 < pH > 5.0	5.0 < pH > 4.7	pH <4.7
Radix auricularia			P1,N1		P4			
R. peregra			P2,N1		P3	P4		
Rangia cuneata			N1					
Sphaerium corneum			P1,2,N1		P4			
S. notatum		N1						
S. rhomboideum			N1					
S. rivicola			P2,N1					
S. simile			N1					
S. solidum				P1,N1				
S. stamineum			N1					
S. striatinum			N1					
Sphaerium sp.						P3		
Stagnicola caperata			N1					
S. catascopium		N1						
Strophitus undulatus			N1					
Theodoxus fluviatilis			P1	P2				
Truncilla donaciformis				N1				
T. truncata				N1				
Unio crassus				P2,N1				
U. pictorum			P1,2	N1				
U. tumidus			P1,2					
Valvata bicarinata				N1				
V. piscinalis			P2,N1					
V. tricarinata			N1	N1				

Viviparus contectus			N1			
V. georgianus				N1		
V. subpurpureus				N1		
V. viviparus			P2			
Oligochaeta						
Aeolosoma niveum			P1			
A. quaternarium			P1			
A. tenebrarum			P1			
Amphichaeta leydigii			P1			
Aulodrilus americanus					N8	
Branchiura sowerbyi		N1	P1,2			
Chaetogaster crystallinus			P1			
C. diastrophus			P1			
Criodrilus lacuum			P1			
Dero sp.			N1			
Enchytraeidae						P4
Haplotaxis gordioides				P1		
Limnodrilus cervix		N1				
L. claparedianus	P1	N1				
L. hoffmeisteri	P1	N1			P4,N8	
L. maumeensis		N1				
L. udekemianus		P1,N1				
Limnodrilus spp.		P2				
Lumbriculidae						P4
Lumbriculus variegatus		P1,2				
Lumbriculus sp.					N8	
Nais barbata		P1				

(*continued*)

Appendix 4.1. (Continued)

	Species Tolerance to Organic Pollution				Species Tolerance to H^+ Concentration			
	Poly-saprobic	Alpha-meso-saprobic	Beta-meso-saprobic	Oligo-saprobic	pH >5.5	5.5 < pH > 5.0	5.0 < pH > 4.7	pH <4.7
N. bretscheri			P1					
N. communis		P1						
N. elinguis		P1						
N. pseudobtusa				P1				
Nais sp.			N1					
Ophidonais sp.		N1						
Peloscolex ferox		P1					P4	
Potamothrix hammoniensis			P1					
Psammoryctides barbatus		P1						P4
Rhyacodrilus coccineus	P1						N8	
Stylaria lacustris							P4	
Stylaria sp.			N1					
Stylodrilus heringianus							N8	P4
Tubifex tubifex	P1	N1						
Tubifex spp.	P2							P4
Hirudinea								
Dina parva		N1					N8	
Dina sp.			N1					
Erpobdella octoculata		P1,2					P4	
E. punctata		N1						

Gloiobdella elongata	N1						
Glossiphonia complanata	N1	P1,2		P3		P4	
G. heteroclita	P1	P2					
Haemopis marmorata						N8	
H. sanguisuga	P1			P3			
Helobdella stagnalis	P1,N1	P2		P4	P3	N8	
Hemiclepsis marginata		P1					
Hirudo medicinalis	P1						
Macrobdella sp.	N1						
Mooreobdella microstoma	N1						
Piscicola punctata		N1					
Placobdella montifera	N1					N8	
P. ornata		N1					
Theromyzon tessolatum		P1		P3			
Crustacea							
Asellus aquaticus	P1,2				P3		P4
A. intermedius		N1					
Atyaephyra desmarestii		P2					
Cambarus acuminatus			N1				
C. asperimanus			N1				
C. conasaugaensis			N1				
C. diogenes	N1						
C. extraneus			N1				

(continued)

Appendix 4.1. (Continued)

	Species Tolerance to Organic Pollution				Species Tolerance to H^+ Concentration			
	Poly-saprobic	Alpha-meso-saprobic	Beta-meso-saprobic	Oligo-saprobic	pH >5.5	5.5 < pH > 5.0	5.0 < pH > 4.7	pH <4.7
C. floridanus			N1					
C. latimanus			N1					
C. striatus		N1						
Crangonyx pseudogracilis			N1					
C. richmondensis							N8	
Daphnia longispina						P3		
D. magna						P3		
Fallicambarus fodiens		N1						
Faxonella clypeata			N1					
Gammarus fossarum				P2				
G. lacustris					P3			
G. pulex			P2					
G. roeseli			P2					
G. tigrinus			P2					
Gammarus sp.			N1					
Hyalella azteca			N1				N8	
Lepidurus arcticus					P3			
Lirceus sp.			N1					
Orconectes erichsonianus			N1					
O. juvenilis				N1				
O. propinquus			N1					
O. rusticus			N1					

Palaemonetes paludosus		N1					
Proasellus coxalis	P2						
Procambarus acutus	N1						
P. angustatus		N1					
P. barbatus		N1					
P. chacei		N1					
P. enoplosternum		N1					
P. epicyrtus		N1					
P. fallax	N1						
P. howellae		N1					
P. litosternum		N1					
P. lunzi		N1					
P. paeninsulanus		N1					
P. pubescens		N1					
P. spiculifer			N1				
P. troglodytes	N1						
P. versutus			N1				
Ephemeroptera							
Ameletus inopinatus			P1		P3		P5,6
Arthroplea congener				P5			P6
Baetidae						N8	
Baetis alpinus			P1,2				
B. buceratus				P5			
B. digitatus					P5		
B. fuscatus		P2		P3,6	P5		
B. lapponicus				P3,5			
B. macani				P3,5			
B. muticus			P2	P3,6	P5		
B. niger				P3		P5	P6
B. rhodani		P1,2		P3			P5,6

(continued)

Appendix 4.1. (Continued)

	Species Tolerance to Organic Pollution				Species Tolerance to H^+ Concentration			
	Poly-saprobic	Alpha-meso-saprobic	Beta-meso-saprobic	Oligo-saprobic	pH >5.5	5.5 < pH > 5.0	5.0 < pH > 4.7	pH <4.7
B. scambus					P3			
B. subalpinus					P3	P5,6		
B. vagans				N1				
B. vernus		P1	P2		P3,5			
Brachycercus harrisella			P1					
Caenis horaria		P1			P3,6	P4,5		
C. diminuta		N1						
C. luctuosa			P1		P4,5			
C. rivulorum						P5,6		
C. robusta					P5,6			
Callibaetis floridanus		N1						
Centroptilum luteolum			P2			P5	P6	
Cloeon dipterum			P1,2			P4,6	P5	
C. simile			P1,2		P5			
Cloeon sp.							N8	
Ecdyonurus dispar				P1				
E. forcipula				P2				
E. venosus				P1,2				
Electrogena lateralis				P2				
Epeorus sylvicola				P2				
Ephemera danica			P1,2			P5,6		
E. simulans				N1				
E. vulgata						P4,5		

Ephemerella aurivillii				P3		P5,6	
E. ignita		P1,2		P5			P3
E. major			P2				
E. mucronata			P2			P5,6	P3
Ephemerella sp.						N8	
Habroleptoides sp.			P2				
Heptagenia dalecarlica					P5,6		
H. flava		P2					
H. fuscogrisea							P3,5,6
H. joernensis				P5	P6		
H. longicauda	P1						
H. sulphurea		P2			P3	P6	P5
Hexagenia bilineata		N1					
H. limbata			N1				
Leptophlebia marginata		P1					P3,5,6
L. vespertina							P3,5,6
Leptophlebia spp.							P4
Metretopus borealis				P5	P6		
Paraleptophlebia strandii					P5,6		
P. submarginata			P1,2				
Parameletus chelifer					P5,6		
Pentagenia vittgera			N1				
Potamanthus luteus		P2					
Procloeon bifidum				P5,6			
Rhithrogena semicolorata			P1,2				

(*continued*)

Appendix 4.1. (Continued)

	Species Tolerance to Organic Pollution				Species Tolerance to H^+ Concentration			
	Poly-saprobic	Alpha-meso-saprobic	Beta-meso-saprobic	Oligo-saprobic	pH >5.5	5.5 < pH > 5.0	5.0 < pH > 4.7	pH <4.7
Siphlonurus alternatus					P5	P3		
S. aestivalis						P3,5		P6
S. lacustris		P1				P3,5		P6
Stenacron interpunctatum			N1	N1				
Stenonema ares			N1					
S. exiguum				N1				
S. femoratum			N1					
S. fuscum				N1				
S. integrum			N1					
S. modestum				N1				
S. pulchellum			N1					
S. smithae				N1				
S. terminatum				N1				
S. tripunctatum				N1				
Odonata								
Aeshna cyanea			P2					
Aeshnidae							N8	
Anax junius				N1				
Argia apicalis			N1					
A. translata			N1					
Calopteryx maculatum			N1					
C. splendens			P2					

C. virgo		P2			
Coenagrion pulchellum	P1				
Coenagrionidae				N8	
Cordulegaster boltoni			P2		
Cordulia aenea					P4
Cordulia sp.				N8	
Enallagma antennatum		N1			
Enallagma sp.				N8	
Epitheca bimaculata					P4
Gomphus externus			N1		
G. pallidus		N1			
G. plagiatus			N1		
G. spiniceps		N1			
G. vastus		N1			
Hataerina titia			N1		
Ischnura elegans		P1			
I. verticalis	N1				
Lestes viridis		P2			
Libellula lydia		N1			
Libellulinae				N8	
Neurocordulia moesta		N1			
Onychogomphus forcipatus		P2			
Platycnemis pennipes		P2			
Pyrrhosoma nymphula		P2			

(*continued*)

Appendix 4.1. (Continued)

	Species Tolerance to Organic Pollution				Species Tolerance to H^+ Concentration			
	Poly-saprobic	Alpha-meso-saprobic	Beta-meso-saprobic	Oligo-saprobic	pH >5.5	5.5 < pH > 5.0	5.0 < pH > 4.7	pH <4.7
Megaloptera								
Climacia areolaris				N1				
Corydalis cornutus			N1					
Sialis infumata				N1				
S. fulginosa		P1	P2					
S. lutaria		P1	P2					P4
S. sordida								P4
Sialis sp.							N8	
Plecoptera								
Acroneuria abnormis			N1					
A. arida				N1				
A. evoluta				N1				
Allocapnia vivipara			N1					
Amphinemura borealis								P3
A. standfussi								P3
A. sulcicollis				P1				P3
Amphinemura spp.				P2				
Arcynopteryx compacta						P3		
Brachyptera risi				P1,2				P3
B. seticornis				P2				
Capnia atra						P3		
C. bifrons				P1				
C. pygmaea						P3		

Chloroperla spp.		P2		
Dinocras cephalotes		P1,2	P3	
Diura bicaudata		P2	P3	
D. nanseni			P3	
Isoperla bilineata		N1		
I. grammatica	P1		P3	
I. obscura			P3	
Leuctra braueri		P2		
L. digitata				P3
L. fusca				P3
L. hippopus				P3
L. moselyi		P1		
L. nigra		P1,2		P3
Nemoura avicularis				P3
N. cambrica		P1		
N. cinerea				P3
Nemurella picteti		P1		P3
Perla bipunctata		P1		
P. burmeisteriana		P2		
P. marginata		P2		
Perlodes dispar		P1		
Protonemura meyeri		P1		P3
Siphonoperla burmeisteri				P3
S. torrentium		P1		
Taeniopteryx maura	N1			
T. nebulosa		P1		
T. nivalis		N1		P3
Trichoptera				
Adicella reducta				P3
Agapetus fuscipes		P1		

(continued)

Appendix 4.1. (Continued)

	Species Tolerance to Organic Pollution				Species Tolerance to H^+ Concentration			
	Poly-saprobic	Alpha-meso-saprobic	Beta-meso-saprobic	Oligo-saprobic	pH >5.5	5.5 < pH > 5.0	5.0 < pH > 4.7	pH <4.7
Agraylea multipunctata				P1				
Agrypnia obsoleta								P3
Agrypnia sp.								P4
Anabolia laevis		P1						
A. nervosa			P1,2					
Apatania fimbriata				P1				
A. stigmatella						P3		
A. zonella						P3		
Athripsodes aterrimus								P3
A. cinereus			P1					P3
Brachycentrus montanus				P1,2				
B. subnubilus			P2					
Chaetopteryx villosa								P3
Cheumatopsyche lepida			P2					
Cheumatopsyche sp.			N1					
Chimarra obscura				N1				
Crunoecia irrorata			P1	P2				
Cyrnus flavidus								P3,4
C. trimaculatus								P3
Ecnomus tenellus			P2					
Glossosoma					P3			

intermedium						
Glossosoma spp.			P2			
Goera pilosa		P1,2				
Halesus radiatus						P3
Holocentropus dubius						P3,4
Hydropsyche angustipennis	P1			P3		
H. bronta		N1				
H. contubernalis	P1					
H. frisoni			N1			
H. morosa		N1				
H. orris		N1				
H. pellucidula	P1			P3		
H. siltalai			P1,2	P3		
H. simulans			N1			
Hydroptila tineoides		P1				
H. waubesiana			N1			
Ithytrichia lamellaris			P1	P3		
Lasiocephala basalis			P1,2			
Lepidostoma hirtum			P2	P3		
Leptocerus tineiformis			P1			
Leptoceridae					N8	
Limnephilidae					P4,N8	
Limnephilus centralis						P3
L. extricatus						P3
L. flavicornis						P3
L. lunatus						P3
L. rhombicus						P3
L. sericeus		P1				

(*continued*)

Appendix 4.1. (Continued)

	Species Tolerance to Organic Pollution				Species Tolerance to H^+ Concentration			
	Poly-saprobic	Alpha-meso-saprobic	Beta-meso-saprobic	Oligo-saprobic	pH >5.5	5.5 < pH > 5.0	5.0 < pH > 4.7	pH <4.7
L. sparsus			P1					
L. stigma								P3
L. vittatus								P3
Limnephilus sp.							N8	
Macronema carolina				N1				
Micropterna lateralis								P3
Molanna angustata								P3,4
Molannodes tinctus								P3
Mystacides azurea								P3,4
M. longicornis							P4	
Nectopsyche sp.							N8	
Neureclipsis bimaculata								P3
N. crepuscularis				N1				
Notidobia ciliaris								P3
Odontocerum albicorne				P2				
Oligoplectrum maculatum				P2				
Oxyethira spp.								P3
Philopotamidae							N8	
Philopotamus montanus				P1		P3		
Philopotamus spp.				P2				
Phryganea								P4

bipunctata					
P. grandis					P3
Phryganeidae				N8	
Plectrocnemia conspersa					P3
Plectrocnemia spp.		P2			
Polycentropus flavomaculatus	P1				P3
P. irroratus					P3
Polycentropus spp.	P2				
Potamophylax cingulatus					P3
P. latipennis					P3
Potamya flava	N1				
Psychomyia pusilla	P1,2				
Ptilocolepus granulatus		P1,2			
Rhyacophila nubila					P3
R. philopotamoides		P1			
R. tristis		P1			
Rhyacophila spp.		P2			
Sericostoma personatum			P3		
Sericostomatinae		P2			
Silo nigricornis		P1,2			
S. pallipes		P1,2			
S. piceus		P2			
Stenophylax permistus					P3
Tinodes waeneri			P3		
Trichostegia minor		P1			

(*continued*)

Appendix 4.1. (Continued)

	Species Tolerance to Organic Pollution				Species Tolerance to H^+ Concentration			
	Poly-saprobic	Alpha-meso-saprobic	Beta-meso-saprobic	Oligo-saprobic	pH >5.5	5.5 < pH > 5.0	5.0 < pH > 4.7	pH <4.7
Diptera								
Chaoboridae								
Chaoborus americanus							N8	
C. flavicans								P4
C. trivittatus							N8	
Chironomidae								
Ablabesmyia aspera				N1			N8	
A. illinoense		N1						
A. janta			N1					
A. longistyla								P4
A. mallochi				N1				
A. monilis			P1					P4
A. peleensis			N1					
A. phatta								P4
A. rhamphe			N1					
Ablabesmyia sp.								N7
Anatopynia plumipes		P1						
Apsectrotanypus trifascipennis		P1			P4			
Arctopelopia spp.								P4
Axarus scopula				N1				
Brillia longifurca				P1				
B. modesta				P1				
Chaetolabis				N1				

atroviridis							
C. ochreatus				N1			
Chironomus anthracinus				N1			
Ch. attenuatus nom. dub.		N1					
Ch. crassicaudatus		N1					
Ch. decorus			N1				
Ch. pagana				N1			
Ch. plumosus	P1	P2					
Ch. riparius		N1					
Ch. staegeri			N1				
Ch. stigmaterus		N1					
Ch. tentans	P1			N1			
Ch. thummi gr.	P1	P2					P4
Chironomus sp.							N7
Cladopelma collator			N1				
C. viridula							P4
Cladopelma sp.						N8	N7
Cladotanytarsus mancus				P1			
Cladotanytarsus spp.						N8	P4
Clinotanypus nervosus							P4
C. pinguis				N1			
Clinotanypus sp.						N8	
Coelotanypus concinnus			N1				
C. scapularis			N1				
Conchapelopia spp.					P4		
Corynoneura lacustris							P4

(*continued*)

Appendix 4.1. (Continued)

	Species Tolerance to Organic Pollution				Species Tolerance to H^+ Concentration			
	Poly-saprobic	Alpha-meso-saprobic	Beta-meso-saprobic	Oligo-saprobic	pH >5.5	5.5 < pH > 5.0	5.0 < pH > 4.7	pH <4.7
C. celtica			P1					
C. scutellata			P1	N1				
C. taris				N1				
Corynoneura sp.					N7			
Cricotopus absurdus				N1				
C. bicinctus			P1	N1				
C. politus				N1				
C. triannulatus			P1,N1					
C. trifasciatus			N1					
Cryptochironomus defectus gr.								P4
C. fulvus		N1						
C. psittacinus				N1				
Cryptochironomus sp.							N7	
Cryptotendipes emorsus			N1					
Demicryptochironomus vulneratus								P4
Diamesa nivoriunda				N1				
Dicrotendipes fumidus				N1				
D. modestus			N1					P4

D. tritomus	N1						
Dicrotendipes sp.						N8	N7
Endochironomus albipennis					P4		
E. impar				P4			
Endochironomus sp.						N7	
Glyptotendipes amplus		N1					
G. barbipes	N1						
G. lobiferus	N1						
G. meridionalis		N1					
G. paripes	N1						
G. senilis			N1				
Glyptotendipes spp.						N7,8	P4
Goeldichironomus carus	N1						
G. holoprasinus	N1						
Harnischia curtilamellata				P4		N7	
Heterotanytarsus apicalis					P4		
Heterotanytarsus sp.						N8	N7
Heterotrissocladius marcidus			P1				P4
Heterotrissocladius sp.						N7	
Kiefferulus dux	N1						
Labrundinia pilosella			N1				
L. virescens			N1				
Larsia sp.						N8	
Lauterborniella sp.							N7

(*continued*)

Appendix 4.1. (Continued)

	Species Tolerance to Organic Pollution				Species Tolerance to H^+ Concentration			
	Poly-saprobic	Alpha-meso-saprobic	Beta-meso-saprobic	Oligo-saprobic	pH >5.5	5.5 < pH > 5.0	5.0 < pH > 4.7	pH <4.7
Macropelopia nebulosa			P1					
M. notata gr.								P4
Meropelopia americana				N1				
M. flavifrons		N1						
Metriocnemus fuscipes			P1					
M. obscuripes			P1					
Micropsectra atrofasciata				P1				
M. deflecta				N1				
M. nigripila				N1				
Microtendipes britteni				P1				
M. pedellus				N1				
Microtendipes spp.			P1		N7			P4
Nanocladius bicolor				P1				
Orthocladius obumbratus				N1				
Pagastiella orophila								P4
Pagastiella sp.							N7	
Parachironomus arcuatus gr.							P4	
P. pectinatellae			N1					

P. tenicaudatus			N1		
Parachironomus sp.					N7
Paracladopelma nais		N1			
Paratanytarsus dissimilis			N1		
Paratanytarsus sp.				N7,8	
Paratendipes albimanus			N1		
Paratendipes sp.					N7
Pentaneura paramelanops	N1				
Phaenopsectra flavipes				P4	
Phaenopsectra sp.				N8	N7
Polypedilum bicrenatum gr.				P4	
P. convictum gr.				P4	
P. fallax		N1			
P. flavus		N1			
P. illinoense		N1			
P. laetum		P1			
P. nubeculosum			N1		
P. pullum				P4	
P. scalaenum	N1				
P. tritum		N1			
P. vibex			N1		
Polypedilum sp.				N8	N7
Procladius bellus	N1			N8	
P. culiciformis	N1				
P. denticulatus	N1				
P. denticulatus gr.				N8	

(*continued*)

Appendix 4.1. (Continued)

	Species Tolerance to Organic Pollution				Species Tolerance to H^+ Concentration			
	Poly-saprobic	Alpha-meso-saprobic	Beta-meso-saprobic	Oligo-saprobic	pH >5.5	5.5 < pH > 5.0	5.0 < pH > 4.7	pH <4.7
Procladius gr.							N8	P4,N7
Prodiamesa olivacea				N1				
Psectrocladius calcaratus								P4
P. dilatatus				N1				
P. limbatellus gr.								P4
P. psilopterus gr.								P4
P. simulans							N8	
P. sordidellus gr.								P4
Psectrocladius sp.								N7
Psectrotanypus dyari		N1						
P. varius		P1						
Pseudochironomus prasinatus								P4
Pseudochironomus sp.					N7		N8	
Pseudodiamesa branickii				P1				
Rheocricotopus robacki				N1				
Rheotanytarsus exiguus		N1						
Sergentia coracina								P4
Stempellina sp.					N7			
Stempellinella sp.					N7		P8	

Stenochironomus hilaris		N1	N1			
Stenochironomus sp.					N4	
Stictochironomus devinctus			N1			
S. varius			N1			
Stictochironomus spp.					P4,N7	
Tanypus carinatus		N1				
T. punctipennis		N1				
T. stellatus	N1					
Tanytarsus gracilentus			N1			
T. gregarius	N1					
Tanytarsus spp.					N8	P4,N7
Thienemanniella xena			N1			
Thienemannimyia spp.				P4		
Tribelos fuscicorne			N1			
T. intextum						P4
T. jucundum			N1			
Tvetenia calvescens			P1			
T. gracei			P1			
Xenochironomus xenolabis			N1			
Xylotopus par			N1			
Zalutschia zalutschicola						P4
Zalutschia sp.						N7

(*continued*)

Appendix 4.1. (Continued)

	Species Tolerance to Organic Pollution				Species Tolerance to H^+ Concentration			
	Poly-saprobic	Alpha-meso-saprobic	Beta-meso-saprobic	Oligo-saprobic	pH >5.5	5.5 < pH > 5.0	5.0 < pH > 4.7	pH <4.7
Other Diptera								
Anopheles punctipennis				N1				
Atherix ibis				P2				
Bezzia glabra		N1						
Bezzia sp.							N8	
Brachydeutera argentata		N1						
Ceratopogonidae								P4
Cnephia pecuarum				N1				
Culex pipiens		N1						
Eristalini	P2							
Eristalinus aeneus		N1						
E. bastardi		N1						
E. brousi		N1						
Metasyrphus americanus		N1						
Odagmia ornata			P2					
Odontomyia cincta			N1					
Palpomyia tibialis			N1					
Prosimulium hirtipes				P2				
Pseudolimnophila luteipennis				N1				
Psychoda alternata		N1						
Psychoda spp.		P2						

Simulium johannseni			N1	
S. venustum			N1	
S. vittatum		N1		
Stilobezzia antennalis	N1			
Stratiomys discalis	N1			
S. meigeni	N1			
Tabanus atratus	N1			
T. calens			N1	
T. lineola	N1			
T. proximus	N1			
T. sulcifrons			N1	
Telmatoscopus albipunctatus	N1			
Tipula abdominalis			N1	
T. caloptera			N1	
Brachycera				P4
Coleoptera				
Agabus biguttatus		P2		
Ancyronyx variegata			N1	
Berosus sp.	N1			
Brychius elevatus		P2		
Cymbiodyta dorsalis			N1	
Dineutes americanus	N1			
Dryops ernesti		P2		
D. luridus		P2		
Dubiraphia sp.		N1		
Dytiscidae				N8
Elmis latreillei			P2	
E. maugetii			P2	
Esolus angustatus			P2	
E. parallelepipedus			P2	

(continued)

Appendix 4.1. (Continued)

	Species Tolerance to Organic Pollution				Species Tolerance to H^+ Concentration			
	Poly-saprobic	Alpha-meso-saprobic	Beta-meso-saprobic	Oligo-saprobic	pH >5.5	5.5 < pH > 5.0	5.0 < pH > 4.7	pH <4.7
Gonielmis dietrichi			N1					
Gyrinus floridansis		N1						
G. substriatus			P1,2					
Hadrenya minutissima				P2				
Haliplus fluviatilis			P1					
H. laminatus			P2					
H. lineatocollis		P1						
Helichus lithophilus				N1				
Helophorus aquaticus			P2					
H. arvernicus			P2					
Hydraena nigrita				P2				
H. pygmaea				P2				
Hydrobius fuscipes			P1					
Laccophilus maculosus		N1						
Limnebius truncatellus				P2				
Limnius perrisi				P2				
L. volckmari				P2				
Macronychus glabratus				N1				
Microcylloepus pusillus				N1				

Optioservus sp.		N1			
Orectochilus villosus		P2			
Oreodytes rivalis			P2		
Platambus maculatus		P2			
Potamonectes assimilis		P1,2			
P. depressus		P2			
Riolus cupreus		P2			
R. subviolaceus			P2		
Stenelmis crenata			N1		
S. decorata	N1				
Stictotarsus duodecimpustulatus		P2			
Tropisternus lateralis	N1				
T. natator	N1				
Hemiptera					
Belostoma sp.	N1				
Corixa sp.	N1				
Corixidae				N8	
Hesperocorixa sp.	N1				
Hydrometra martini	N1				
Nepa cinera		P1			
Notonecta glauca		P1			
Notonecta sp.				N8	
Other phyla					
Acari, Hydrachnellae					P4
Cordylophora lacustris		N1			
Cristatella mucedo		N1			
Ephydatia fluviatilis		P2			
E. muelleri		P2			

(*continued*)

Appendix 4.1. (Continued)

	Species Tolerance to Organic Pollution				Species Tolerance to H^+ Concentration			
	Poly-saprobic	Alpha-meso-saprobic	Beta-meso-saprobic	Oligo-saprobic	pH >5.5	$5.5 <$ pH > 5.0	$5.0 <$ pH > 4.7	pH <4.7
Fredericella sultana			P2					
Limnesia sp.							N8	
Lophopodella carteri				N1				
Paludicella articulata			P2					
Pectinatella magnifica				N1				
Plumatella emarginata			P2					
P. fungosa			P2					
P. repens			P2,N1					
Sperchonopsis sp.							N8	
Spongilla fragilis			N1					
S. lacustris			P1			P4		
Urnatella gracilis			N1					

5

Contemporary Quantitative Approaches to Biomonitoring Using Benthic Macroinvertebrates

Vincent H. Resh and Eric P. McElravy

5.1. Introduction

In this chapter, we examine quantitative approaches that currently are being used to study the effects of actual or potential disturbances on populations and communities of benthic macroinvertebrates. In the first half of the chapter, we examine the study design, field procedures, and post-sampling procedures and analyses used in 90 quantitative studies of disturbance effects on lentic and lotic benthos. In the second half, we treat two of these topics—the number of sample units to be collected in benthic studies and the taxonomic levels of identification done on organisms collected—in detail, and in particular we examine how they apply to small-scale studies. We then conclude with a series of recommendations.

5.2. Contemporary Quantitative Studies

Forty-five lotic and 45 lentic papers were chosen to assess characteristics of current studies that examine the effects of disturbance on benthic macroinvertebrates. The lotic examples are the same as used by Voshell et al. (1989), whereas the lentic examples came from our files. To be considered for use, a study had to be published in a scientific journal (not the "grey literature"; see Marshall, Chapter 3). Each study was examined in terms of study design (e.g., frequency of sampling, types of controls used), field procedures (e.g., type of sampler, number of sample unit replicates), and post-sampling procedures and analyses (e.g., sorting procedures, inferential statistics used). A similar "contemporary literature" approach has been used by Downing (1984) for selected aspects of lake benthic studies and by Winterbourn (1985) for stream benthic studies.

The streams included in this analysis were all rocky-bottomed. Fifteen percent of the studies were done in small, headwater streams and springs,

Table 5.1. Summary of study design procedures in lotic and lentic habitats. Lotic results are based on table 1 of Voshell et al. (1989).

	Percent of Studies	
	Lotic	Lentic
TYPE OF STUDY	N = 48	N = 45
Comparative evaluation	96	71
Synoptic survey	4	29
CONTROL FOR COMPARATIVE STUDY	N = 46	N = 32
Spatial–same water body	63	26
Spatial–different water body	15	28
Temporal (before and after)	22	46
DURATION OF STUDY	N = 45	N = 45
One-time event	33	7
< 1 yr	38	38
1–2 yr	18	15
≥ 2 yr	11	40
FREQUENCY OF SAMPLING (IF REPEATED)	N = 30	N = 43
1 d–1 wk	13	19
1 wk–1 mo	20	23
1 mo	20	19
1 mo–1 yr	40	25
1 yr	7	7
> 1 yr	0	7

36% in moderately sized streams (2nd to 4th order, 5 to 10 m wide), 28% in rivers (> 5th order), and 13% examined streams of various sizes; stream size could not be determined for 9% of the studies. Of the lentic studies, 31% were done in ponds, 65% in lakes or reservoirs, and 4% in marshes. Fifty-seven percent of these habitats were natural; 43% were man-made. Almost all of the lotic studies were comparative evaluations, whereas 29% of the lentic studies were synoptic surveys (Table 5.1). Comparative evaluations are defined by Weber (1973) and Voshell et al. (1989) as surveys that attempt to determine the effects of specific perturbations by studying the biota with and without the perturbation and with different levels of the perturbation. Synoptic surveys are defined by Weber (1973) and Voshell et al. (1989) as surveys undertaken to determine the kinds and abundances of all organisms in all recognizable habitats in the environment being studied.

5.2.1. *Study Design*

5.2.1.1. *Controls*

Major differences appeared between lotic and lentic systems in the use of controls for comparisons among sites (Table 5.1). Sixty-three percent of lotic studies used a site in the same water body (e.g., a site upstream of the one in question), but this was the least common approach in lentic studies. Perhaps because disturbance often affects the entire lentic system, temporal controls comprised almost one-half (46%) of the studies in lakes, ponds, and marshes. The establishment of suitable spatial and/or temporal controls is an essential component of sampling design for any biomonitoring study and has been discussed by Green (1979), Stewart-Oaten et al. (1986), Resh et al. (1988), and Resh and Rosenberg (1989), among others.

5.2.1.2. Duration

Lentic studies clearly are conducted over longer periods than lotic ones (Table 5.1). This pattern also was found by Resh and Rosenberg (1989) when they examined the duration of 275 lotic studies and 214 lentic studies published from 1983 to 1988. However, McElravy (1988) examined the duration of 172 stream studies and found that although only 20% of stream studies related to impact assessment published from 1980–1984 lasted more than one year, 56% of the papers published from 1985–1987 were more than one year in duration. This suggests that long-term lotic research may be becoming more common, and studies of similar duration in lotic and lentic research may be observed during the 1990s.

5.2.1.3. Sampling Frequency

Little difference was noted between sampling frequency in lotic and lentic systems (Table 5.1), although repeated sampling (cf. single sampling intervals) was a much more common feature among lentic studies (43 studies cf. 30 studies). In the majority of studies, samples were collected at monthly or more frequent intervals. Specific recommendations on sampling frequency are difficult to propose because this feature clearly is a function of study objectives. However, monthly or other fixed-interval sampling may be less valuable than irregularly spaced samples, depending on the variables of interest (Resh, 1979b; Waters, 1979). For many monitoring studies, annual or semi-annual sampling may be all that a budget will permit. In these cases, choice of sampling time is critical and should follow logically from the questions being asked. Depending on the question of interest, sampling often is done just before emergence of insects, during periods of maximum growth, or prior to physical events such as floods.

5.2.2. Field Procedures

5.2.2.1. Number of Replicates

Lotic and lentic studies were compared with regard to the number of replicate sample units collected (or "sample size") in Table 5.2. In the majority

In shallow streams, stratification into riffles and pools and then by substrate size in riffles is the most obvious approach and has been done for decades to reduce intersample variation (e.g., Allen, 1959). In both types of systems, the objective of stratification is to reduce variability of estimates, thereby reducing required sample sizes and increasing "sensitivity" of the study to detection of environmental change.

Related to stratification is the practice of selecting only one site for each treatment effect (or control) being examined; the investigator then collects replicate sample units within each of these single sites. Problems arising from such a procedure have been discussed at great length under the heading of "pseudoreplication" by Hurlbert (1984) as well as Allan (1984), Hawkins (1986), Stewart-Oaten et al. (1986) and others. A thorough discussion of this problem and suggestions for corrections in study designs are provided in Cooper and Barmuta, Chapter 11.

5.2.2.3. Sampling Device

In the 90 studies examined, the Surber and other enclosed samplers such as the Hess, T, and portable box primarily were used in lotic systems, whereas dredges and grabs, followed by corers, made up the majority of lentic samplers used (Table 5.2). The Surber sampler (38%), hand net (28%), and box or cylinder samplers (22%) were the most common devices in the lotic studies surveyed by Winterbourn (1985), whereas grabs and dredges accounted for 71% of all lentic samplers used in the survey by Downing (1984).

The design of sampling devices has received more attention in lentic research than in lotic research. Traditional lentic-sampling methods depend heavily on sampler function because of the need for remote sampling (except in the use of SCUBA). In addition, lentic researchers have been able to use the sophisticated samplers developed for marine benthic habitats. In shallow lotic habitats, the operator can sample directly, and generally accepted semiquantitative or qualitative substitutes, such as kick nets, have long been available.

Sampling devices that are appropriate for a range of lotic and lentic habitats are listed in Hellawell (1978) and Merritt et al. (1984) and demonstrated in Resh et al. (1990). Sources of bias that may influence sampler operation are presented in Resh (1979b). Downing (1984) provided a detailed analysis of lake sampling devices; a similar analysis should be done for stream samplers.

Choice of sampling device is a critical aspect of study design and investigators must keep in mind that although cost, availability, and other logistical factors are important, sampler accuracy must be the primary consideration. Confidence that the device being used is actually sampling the intended universe in a representative manner is essential. Although an appropriate sampler probably has been described for most habitats (see habitat tables in Merritt et al., 1984), and a plethora of devices are available, investigators need to consider nontraditional devices when established ones are likely to result in significant error, and to take advantage of technological developments. This may require construction of a new site- or study-specific device such as the rock outcrop community sampler (ROCS) used by Voshell et al. (1989).

Surprisingly few of the 90 studies analyzed used emergence traps (Table 5.2), even though insect emergence is a major topic in both lotic and lentic research (Resh and Rosenberg, 1989). Davies (1984) has written an excellent review of sampling devices used to study emergence. Also, the use of drift nets is common in stream studies, but little attention has been given to the appropriate design of drift studies (Voshell et al., 1989). Allan and Russek's (1985) study is an exception; it and a review of invertebrate drift by Brittain and Eikeland (1988) are useful papers to consult in this regard. However, it is more difficult than generally believed to use drift to assess impact because macroinvertebrates will drift under normal conditions, when small or large disturbances occur, and regardless of the stress involved. Finally, artificial substrates were used infrequently in many of the lotic and lentic studies surveyed. Their use has been extensively reviewed by Rosenberg and Resh (1982). Voshell et al. (1989) recently concluded that, with the exception of rock-filled baskets (which are representative artificial substrates, *op. cit.*), little reason exists to substitute artificial substrates for direct sampling of natural substrates in small to medium-sized streams; artificial substrates remain useful in large rivers or lakes where direct sampling is difficult.

5.2.2.4. *Mesh Sizes*

Mesh size is a component of the net used during sampling and the screens used in subsequent sorting. Of the 45 lotic studies examined here, 301–400 μm mesh sizes were most commonly used (27%), and 50% of the studies used mesh sizes of 301–500 μm (Table 5.2). Only 18% of all stream studies used mesh sizes larger than 600 μm. These are slightly larger than the mesh sizes most commonly used (201–300 μm) in the stream studies reviewed by Winterbourn (1985), but they are smaller than those reported by respondents to a questionnaire on benthic sorting practices (Resh et al., 1985) in which the modal mesh size of the samplers used was 590 μm. Of the 45 lentic studies examined, mesh sizes of 101–200 μm were most commonly used (23%), and 41% of the studies done in lentic systems used mesh sizes from 101–300 μm (Table 5.2). This is considerably smaller than the mesh sizes that Downing (1984) reported to be most common in his analysis of lentic studies (450–600 μm).

Because sampling with fine meshes requires greater time in removing animals from sediment, coarser nets often have been chosen; of course, this makes absolute population estimates suspect (Downing, 1984). Coarse mesh sizes may be allowable, however, if a comparative study design maintains a consistent (and appropriate) mesh size. For studies using absolute data, Voshell et al. (1989) suggested the following points in considering the tradeoff between sorting time and accuracy: (1) use as fine a mesh size as possible; (2) clearly state mesh size in published results; and (3) interpret results knowing that some taxa and age classes probably passed through the mesh used, and that live animals (e.g., Chironomidae; Storey and Pinder, 1985) may require a finer mesh size for retention than dead or preserved animals.

5.2.3. Post-Sampling Procedures

5.2.3.1. Sample Modification

In both the lotic and lentic studies surveyed, a small percentage of the researchers indicated that only selected sample units were processed, sample units were combined, or that selected sample units were subsampled (Table 5.3). In contrast, the respondents to Resh et al.'s (1985) questionnaire indicated that from 47% to 62% (depending on the sampler used) regularly selected only certain sample units (or parts of these sample units) for sorting and subsequent analysis. The difference may be due to a failure to report these methods in the 90 studies examined, but it is likely that cost-saving techniques were applied more often by the respondents to Resh et al. (1985), who were largely consulting or industry biologists, compared with the mostly academic or research biologists who did the 90 studies examined in the current review.

5.2.3.2. Sorting Facilitation

Clear differences were found between lotic and lentic studies in terms of procedures used to aid in the sorting of samples (Table 5.3). For example, sieving was done in 87% of lentic studies, but only 18% of lotic studies; elutriation, flotation, and staining were used less often than sieving, and unaided sorting (as "no sorting facilitation") was more common in lotic studies than in lentic studies (Table 5.3). In contrast, Resh et al. (1985) found that sieves were used in the majority of lentic and lotic studies, but elutriation, flotation, and stains also were used less commonly than sieving by the questionnaire respondents.

Resh (1979b, table 2) discussed sorting procedures that may result in sample bias and suggested ways of reducing these problems. The benefit of sorting-facilitation procedures has been mentioned often in the literature, but the cost-saving advantages that these can offer should be incorporated into

Table 5.3. Summary of contemporary laboratory procedures for processing benthic samples from lotic and lentic studies.

	Percent of Studies	
	Lotic	Lentic
SAMPLE MODIFICATION AFTER COLLECTION	N = 48	N = 45
All sample units processed	85	93
Selected sample units processed	2	0
Sample units were combined	10	7
At least a portion of the sample units subsampled	8	4
Processing unknown, or sample units not collected	4	0
SORTING FACILITATION PROCEDURES USED	N = 45	N = 45
Sieving (excluding elutriation)	18	87
Elutriation	4	2
Behavioral sorting	0	2
Flotation	11	24
Staining	4	24
Centrifugation	2	0
No sorting facilitation	60	16
FINAL SORTING PROCEDURES	N = 45	N = 45
A. Sorted live	13	16
Sorted preserved	64	62
Unknown live or preserved	22	22
B. Sorted with magnification	16	20
Sorted without magnification, or unknown	89	80

a far greater number of studies. Clearly, if benthic surveys are to become an integral part of environmental assessments, cost-saving techniques will have to be more commonly adopted. Perhaps consultants have made greater use of these techniques because cost savings in time-consuming sample processing can increase profits. Table 5.4 indicates that, following field processing, the average time to sort any of the sample types shown is ≈3 h. Using information from the questionnaire responses, it appears the use of stains, sieves, flotation, and/or elutriation can provide a significant reduction in processing time. Because the time required to take additional samples is only ≈5% of the time required to process them (Resh and Price, 1984), improvements in processing efficiency can greatly reduce costs of benthic sampling. Perhaps even more important, cost-efficient sample processing can allow increased numbers of replicates to be collected, sorted, and used in analysis.

Application of techniques such as staining may not increase time savings but may increase accuracy. In the Resh et al. (1985) questionnaire re-

Table 5.4. Time required to sort various types of benthic samples. NA = no information provided. Based on Resh et al. (1985).

	Surber, Hess, Portable Box Samplers	Ekman, Ponar, and Peterson Grab Samplers	Drift Net Samplers	Floating Multiplate Samplers	Rock-Filled Basket on Substratum
Mean time required to handpick sample	**3.2 h (0.3–11.4 h)	**2.65 h (0.1–10.9 h)	*3.3 h (0.1–16.7 h)	*3.5 h (0.4–21 h)	**3.6 h (1.1–11.8 h)
Mean % time savings using elutriation or flotation, per sample	**36.4% (25–50%)	*38.3% (11–50%)	NA	25.8% (0–50%)	*38.2% (16.7–50%)
Mean % time saved using sieves, per sample	37.5% (25–50%)	**45.3% (14–100%)	11.9% (0–23.8%)	*15.4% (0–50%)	*18.9% (0–50%)
Mean % time saved using stains	*18.4% (10–50%)	*40.6% (14.3–75%)	11.1% (0–33.3%)	21.8% (0–50%)	31.9% (20–50%)

*Based on 5–9 questionnaire responses.
**Based on 10 or more questionnaire responses.

sponses, rose bengal was the most common stain used. Additional comments on the processing of samples are presented in Resh et al. (1984) and Winterbourn (1985). This is one practical area in which major contributions to benthic research can be made by sheer ingenuity.

5.2.4. Taxonomic Levels of Identification

Levels of taxonomic identification (e.g., family, genus) used in the 90 studies examined varied widely between lotic and lentic systems, as well as among taxonomic groups (Table 5.5). Differences reflect the diversity and occurrence of faunal groups in lentic and lotic systems, as well as the current state of taxonomic knowledge of these groups. In lotic systems, insect groups (except Odonata) were identified most commonly to genus, although the Ephemeroptera, Plecoptera, and Simuliidae (Diptera), as well as two non-insect groups (Platyhelminthes and Crustacea) were identified to species in approximately one-third or more of the lotic studies (Table 5.5). Nematodes, annelids, and water mites (Hydracarina) were identified most often to family level or above.

Most of the insect groups in the lentic studies also were identified to genus; many lentic, noninsect taxa frequently were identified to the species level. This is due in large part to the common occurrence of these taxa in lakes and to a tendency of lentic studies, more than stream studies, to concentrate on a particular taxonomic group rather than on the whole benthic community.

5.2.5. Data Analysis Used

The 90 lotic and lentic studies showed similarities and differences in terms of benthic measures used and types of analyses performed. Taxa richness was reported for more lotic (49%) than lentic (29%) studies (Table 5.6), perhaps reflecting taxonomic difficulties with the typical Chironomidae and Oligochaeta fauna in the lentic benthos, and concentration on a few groups rather than the whole community in lentic studies. In contrast, a high proportion of both lotic (76%) and lentic (78%) studies estimated density, perhaps because enumeration is the easiest analytical technique. Not surprisingly, the majority of studies (70–72%) in both systems used the traditional $\alpha = 0.05$ in inferential statistics, and analysis of variance (ANOVA) was the most common statistical test used. Slightly more kinds of statistical tests were done in lotic systems, and even though similar proportions of lotic (38%) and lentic (40%) papers used indices, a greater variety of these were used in lotic systems. Shannon's Diversity Index was most commonly used (35% and 38% in lotic and lentic studies, respectively). Among the reports examined, only a small proportion (lotic: 4%, lentic: 15%) indicated that a statistician was consulted during the research (Table 5.6), although statistical

Table 5.5. Summary of taxonomic level of identification in contemporary lotic and lentic studies. The level of the majority of identifications in a systematic group (e.g., order, phylum) determines level of that group for a particular study. (N) = number of studies reporting identifications in that group.

	Percent of Studies	
	Lotic	Lentic
INSECTS		
Ephemeroptera		
Order	6	17
Family	14	4
Genus	43	38
Species	37	42
(N)	(35)	(24)
Odonata		
Order	29	24
Family	27	10
Genus	14	38
Species	29	28
(N)	(14)	(21)
Hemiptera		
Order	0	9
Family	0	36
Genus	0	27
Species	0	27
(N)	(0)	(11)
Megaloptera		
Order	13	0
Genus	62	33
Species	25	67
(N)	(8)	(3)
Plecoptera		
Order	10	50
Family	7	0
Genus	47	0
Species	36	50
(N)	(30)	(2)
Trichoptera		
Order	3	21
Family	11	5
Genus	61	42
Species	25	32
(N)	(36)	(19)

(*continued*)

Table 5.5. (Continued)

	Percent of Studies	
	Lotic	Lentic
Coleoptera		
Order	6	14
Family	26	14
Genus	48	57
Species	20	14
(N)	(31)	(14)
Chironomidae		
Family	31	28
Subfamily	17	0
Genus	50	46
Species	2	26
(N)	(36)	(39)
Simuliidae		
Family	14	0
Genus	55	0
Species	31	0
(N)	(29)	(0)
Other Diptera		
Family	29	19
Genus	61	54
Species	10	27
(N)	(31)	(26)
NON-INSECTS		
Platyhelminthes		
Class	0	20
Family	33	0
Genus	33	20
Species	33	60
(N)	(6)	(5)
Nematoda		
Phylum	75	90
Family	8	0
Genus	17	0
Species	0	10
(N)	(12)	(10)
Annelida		
Class	19	29
Family	43	9
Genus	19	3
Species	19	60
(N)	(21)	(35)

(*continued*)

Table 5.5. (Continued)

	Percent of Studies	
	Lotic	Lentic
Crustacea		
Order	32	8
Family	5	0
Genus	32	16
Species	32	76
(N)	(19)	(25)
Hydracarina		
"Superfamily"	44	20
Family	11	20
Genus	44	20
Species	0	40
(N)	(9)	(5)
Mollusca		
Phylum	20	0
Class	13	13
Family	7	6
Genus	47	31
Species	13	50
(N)	(15)	(16)
TAXONOMIST CONSULTED	24	21
(N)	(45)	(45)
SPECIMENS PLACED IN REPOSITORY	4	0
(N)	(45)	(45)

advice may have been obtained but not reported by the authors. A detailed discussion of statistical analyses in biomonitoring is presented in Norris and Georges, Chapter 7, and Cooper and Barmuta, Chapter 11.

5.3. Some Design Considerations in Small-Scale Quantitative Studies

Several excellent, detailed treatments on design and sampling techniques for benthic macroinvertebrate studies are available. Table 5.7 summarizes topics discussed in these treatments, and Norris and Georges (Chapter 7) and Cooper and Barmuta (Chapter 11) also deal with this topic. Our orientation in discussing the two topics selected for detailed treatment—number of replicate sample units to be collected and taxonomic levels of identification—has been to emphasize design considerations for an audience that generally has been ignored: those workers conducting studies of disturbances having limited scope and objectives (e.g., surveys of three to five stations along a single stream), and limited budgets. Much has been written regard-

Table 5.6. Summary of contemporary data analysis and reporting methods in lotic and lentic studies.

	Percent of Studies	
	Lotic	Lentic
TAXA RICHNESS USED	N = 45	N = 45
	49	29
ENUMERATION PROCEDURES USED	N = 45	N = 45
Report absolute density	76	78
Report absolute biomass	9	29
Report relative abundance	27	11
Report presence-absence	2	4
Report relative density	7	16
STATISTICAL ANALYSIS	N = 45	N = 45
Reports do no statistical analysis; data (may include means) reported in tables/graphs	33	29
Reports do descriptive statistics (e.g., Std. Dev.) on all or part of the data	7	44
Reports do inferential or noninferential statistics	60	69
OF STUDIES THAT USE INFERENTIAL STATISTICS ON ALL OR PART OF THE DATA	N = 27	N = 29
$\alpha = 0.05$	70	72
$\alpha > 0.05$	11	3
$\alpha < 0.05$	0	7
α not specified	19	17
OF THE STUDIES USING INFERENTIAL STATISTICS, PROCEDURES USED WERE	N = 35	N = 41
ANOVA	37	37
ANCOVA	0	2
t-test	17	22
Probit analysis	3	0
Parametric regression/correlation analysis	6	17
Chi square	9	7
Ecological Community Analysis	3	0
Kruskal-Wallace	6	0
Mann-Whitney U test	9	10
Wilcoxon test	0	2
Spearman rank correlation	9	2
Friedman's rank analysis	3	0
OF STUDIES USING NONINFERENTIAL TECHNIQUES, PROCEDURES USED WERE	N = 8	N = 5
Cluster analysis (similarity measure)	38	29
Cluster analysis (distance measure)	25	14
Principal Component Analysis	12	14
Polar ordination	25	0
Other ordination	0	43

(*continued*)

Table 5.6. (Continued)

	Percent of Studies	
	Lotic	Lentic
PROPORTION OF PAPERS USING INDICES	N = 45	N = 45
	38	40
OF REPORTS USING INDICES, PROCEDURES USED WERE	N = 37	N = 24
Shannon's Diversity (incl. "Evenness")	35	38
Simpson's Diversity	11	4
Brillouin's Diversity	5	0
Margalef's Diversity	3	0
Pielou's Evenness	3	0
McIntosh's Diversity	3	0
MacNaughton's Dominance	3	0
Berger-Parker Dominance	3	0
Williams Index	3	0
Other "diversity" index	0	8
Biotic Condition Index	3	0
Chutter's Biotic Index	3	0
Jaccard Similarity Index	3	0
Sokal's Distance Index	5	0
% Similarity Index	0	21
Czekanowski's Similarity Index	0	8
Other similarity index	8	4
Autotrophic Index	3	0
Bosch Fidelity Index	3	0
Constancy Index	3	0
Misc. biotic indices	0	13
Other type of index	0	4
STATISTICIAN CONSULTED	N = 45	N = 45
	4	15

ing the design of large-scale projects and, in fact, this discussion applies to them as well. However, these large projects usually have specialists, or at least a consulting statistician or taxonomist, available for advice on experimental design. In contrast, smaller projects often do not have such inputs.

5.3.1. Number of Sample Units

The number of sample units taken in benthic macroinvertebrate surveys is probably the most widely discussed issue in benthic experimental design, but it is also the one about which most confusion still exists. The collection of sample units is undertaken to accomplish either one (or both) of two objectives in a benthic biomonitoring program. The first objective is to estimate the value of a certain benthic measure (e.g., mean density of individuals) with a desired degree of precision and risk of error. This is the

Table 5.7. Topics discussed by selected reports dealing with the design and analysis of benthic research and biomonitoring. An asterisk indicates detailed treatment.

Author	Lentic	Lotic	Survey Design-Stratification/Sampling Considerations	Study Design/Frequency of Studies	Sampling Devices	Sorting Procedures	Data Storage and Retrieval/Quality Control
APHA (1989)	X	X	X		X	X	
Collins and Resh (1989)	X		X	X	X		X
Comiskey and Brandt (1982)	X	X	X*		X	X	X*
Cuff and Coleman (1979)	X		X*				
Cummins (1975)		X	X			X	
Dawson and Hellenthal (1986)	X	X					X*
Downing (1984)	X				X*	X	
Elliott (1977)	X	X	X				
Green (1979)	X	X	X	X			
Hellawell (1978)		X	X		X*	X	
Hellawell (1986)	X	X	X		X		
Kathman (1984)		X	X			X	X
McIntire et al. (1984)	X		X	X		X	
Merritt et al. (1984)	X	X	X		X		
Norris and Georges (1986)		X	X				X
Ortal and Ritman (1985)		X					X*
Peckarsky (1984)		X	X		X		
Prepas (1984)	X	X	X				
Resh (1979b)		X	X		X	X	
Resh et al. (1984)		X			X*	X	X
Resh et al. (1985)		X			X	X	
Resh et al. (1990)	X	X			X		
Southwood (1978)	X	X	X		X		
Tetra-Tech (1987)	X				X	X	X
Voshell et al. (1989)		X	X	X	X		
Weber (1973)	X	X	X		X	X	
Winterbourn (1985)		X	X		X	X	

objective of many surveys and population studies at single sites. The second objective is to determine if a given degree of change in a particular benthic measure has occurred among several sites or times, again with a certain level of risk of error. Error is defined here as concluding that two or more sample means are from different populations (groups) when in fact they are from the same population (a Type I error), or concluding that all means are derived from the same population when in fact they are not (a Type II error).

Replicate sample units are required to provide a measure of variability, and thus define the precision of estimates of benthic measures (e.g., mean density). The number of sample units required for a given level of precision for a particular mean value is determined from sample size statistics (e.g., formulae in Elliott, 1977); Needham and Usinger (1956), Chutter (1972), Resh (1979b), and many others have discussed this topic in detail. The important question to be answered during the design phase of a biomonitoring program is how to determine the number of sample units to be taken so that a desired level of precision will be achieved. (An exception to this is studies that use a sequential decision plan, in which case the number of sample units required will be determined as the data become available during the sorting process; this topic will be discussed later.)

Many authors have discussed how the number of replicate sample units (sample size) is determined based on pilot studies or reconnaissance surveys (Table 5.7), and these reviews and most statistics texts provide appropriate sample size formulae. However, most sampling designs that are used in small-scale studies involve a predetermined number of sample units, based on budget considerations. This number is usually far less than the recommended number of sample units, particularly when high degrees of precision ($< \pm$ 20% of the mean) are desired. Although this approach has been singled out by some critics as the worst way to determine sample size, for many studies it may be the only way of budgeting limited resources. So, if a predetermined number of sample units is required, how many should it be, and what are the limitations of taking small numbers of replicate sample units?

The following factors influence how many sample units should be collected to estimate the mean of a benthic measure: (1) the size of the mean; (2) the degree of aggregation; and (3) the desired precision (Resh, 1979b). First, low values of means require larger numbers of sample units than high values of means, assuming (2) and (3) are the same (see discussion and figs. 2a–2c in Norris and Georges, 1986). Thus, taking a small-sized quadrat is useful in increasing sample unit replicates, but not if mean values are low and zero counts become common. Second, the greater the degree of aggregation, the greater the number of sample units required if mean density is constant. Third, the higher the desired precision, the greater the number of sample units necessary (but as a squared function, so a twofold change in precision requires a fourfold increase in sample units).

If a population under study has high variability because of low density or a high degree of aggregation, relatively large numbers of sample units will be required. It should be remembered that aggregation may result from physical and biotic factors, but also may be related to how well the sampling universe conforms to the specific population universe.

Canton and Chadwick (1988) and Voshell et al. (1989) concluded that although large numbers of sample units are required to estimate most benthic measures, six sample units generally will provide estimates of ±40% of the mean total number of individuals (with 95% confidence intervals) in a community. In conservation studies, where an estimate of the density of a particular species is desired, precise estimates of the mean would be an appropriate goal. For studies of disturbance, however, it is the *difference* between means over time or at different sites that is the more meaningful measure. This involves a different approach than estimating specific population or community measures with a given level of precision.

5.3.2. *Detecting Differences Between Means*

As noted previously, the objectives of many impact-oriented benthic biomonitoring programs may be satisfied by the collection of replicate sample units, which are then used to determine whether a significant change has occurred in the mean values of a particular benthic measure between two or more sites or times. The problem of determining the number of replicates required to detect a given difference between means with a known chance of error is related to precision, but it has received less attention than the question of estimating population measures. Sample sizes for detecting differences between means can vary greatly. For example, Schwenneker and Hellenthal (1984) reported that a range of from three to 1,560 sample units would be required to estimate a 100% change in density for individual populations of benthic macroinvertebrates from a midwestern stream. Allan (1984) calculated that from 11 to 9,591 sample units were necessary to detect a 100% difference between means, depending on the benthic measure used and the size of the mean. The majority of the required number of sample units reported by these authors is far greater than the sample sizes used in the 90 studies examined in Table 5.2.

5.3.2.1. *Formulae Used*

Estimates of precision can be expressed in terms of confidence intervals that have a particular significance value, and these can be compared to provide a statistical test of significance of differences between several sites or times. Procedures with increased power, such as t-tests and ANOVA, also can be used (Norris and Georges, 1986). A procedure for determining the number of replicate sample units (using an ANOVA design) to detect a difference between any two means is given in Sokal and Rohlf (1981, p. 263); their formula is repeated here:

$$n \geq 2(\sigma/\delta)^2\{t_{\alpha[\nu]} + t_{2(1-P)[\nu]}\}^2$$

where n = number of replicate sample units, σ = the true standard deviation, δ = the smallest true difference to be detected, ν = degrees of freedom of the ANOVA error mean square [for a groups and n replications per group, $\nu = a(n - 1)$ for a balanced design], α = significance level, $P = 1 - \beta$, or the statistical power desired, and t = values from a two-tailed t-table with probabilities $\alpha_{[\nu]}$, and $2(1 - P)_{[\nu]}$. This formula, which incorporates Student's t, requires *a priori* knowledge of the coefficient of variation ($CV = SD/\bar{X}$) of the benthic measure of interest and is solved by iteration. The user selects the degree of difference in means to be detected (i.e., the sensitivity of the analysis to change across sites or times) and specifies the significance for both Type I (= α) and Type II (= β) errors. Use of the formula requires that the data meet the assumptions of ANOVA (e.g., normality, homogeneity of variances) and, thus, it is run with transformed data when required (Norris and Georges, 1986).

5.3.2.2. A Case Study

Three to five replicates are commonly used in benthic biomonitoring studies in streams (Table 5.2). In order to examine what differences in mean values of selected benthic measures are statistically detectable (i.e., the "sensitivity" of the study to change) when using ANOVA designs with these small numbers of replicates, Sokal and Rohlf's (1981) formula was applied to four commonly used measures (number of taxa, total number of individuals, Simpson's Diversity Index, and total number of a numerically dominant species, the mayfly *Baetis tricaudatus*) calculated from samples collected from a single site in a northern California stream, Hunting Creek, Napa County (see Resh and Jackson, Chapter 6, for more information on this stream), over a six-year period. In using the Sokal and Rohlf formula, the following assumptions were made: (1) the variability of estimates of means, as measured by the CV, is constant across time intervals; (2) the median value of CV for each mean across time represents, with reasonable certainty, the actual CV of each benthic measure; (3) Simpson's Diversity Index is normally distributed (Patil and Taillie, 1976); (4) because the variance and mean for both number of taxa and Simpson's Index at Hunting Creek showed no significant relationship, transformation was not required for these variables (cf. Fig. 7.3 in Norris and Georges, Chapter 7); and (5) the logarithmic transformation of counts of individuals and *B. tricaudatus* is an adequate correction for violations of ANOVA assumptions for these variables.

5.3.2.3. *Effects of Study Design on Detecting Disturbance: Number of Replicates*

First, the effect of sample size on the ability of a biomonitoring study to detect changes in mean values of commonly used benthic measures was examined. This was done by simulating a study using the Hunting Creek data in which the number of years being compared (which could represent different sites in other streams) was held constant at two, while the number of replicate sample units was varied from three to 10. In each case, the percent difference in mean values statistically detectable between the two years for each sample size (with a 95% chance of being correct at a 5% level of significance) was calculated.

In the case of this simulated two-year study, a 56% difference in mean values of number of taxa could be detected with five replicates (Fig. 5.1A). A similar difference (73%) was observed for Simpson's Diversity (Fig. 5.1C). However, number of individuals and number of *B. tricaudatus* were substantially more variable. With five replicates, it was only possible to detect an increase of 237% or a decrease of 174% in back-transformed (geometric) means for number of individuals (Fig. 5.1B), whereas a 976% increase or a 339% decrease in back-transformed means would be required to detect a difference in *B. tricaudatus* between years (Fig. 5.1D).

For all four of these benthic measures, increasing the number of replicates above five resulted in little reduction in the minimum detectable differences between means (Fig. 5.1A–D). A substantial reduction in minimum detectable differences, however, is observed for all measures when the number of replicates is increased from two to three. The variability of *B. tricaudatus* is so high that only drastic changes in mean density (e.g., >200%) are detectable with even 10 replicates (Fig. 5.1D), a problem also apparent to a lesser degree with number of individuals (Fig. 5.1B).

5.3.2.4. *Effects of Study Design on Detecting Disturbance: Number of Years*

Second, the effect of increasing the number of years (or sites in other studies) on the ability of the analysis to detect changes in means was examined with the Hunting Creek data by keeping the sample size constant at five replicate sample units, but setting the number of years examined at the single Hunting Creek site at two or six. When the median value of the CV is used, little improvement in the resolution of the monitoring program occurred using six years compared with two years (e.g., for number of taxa,

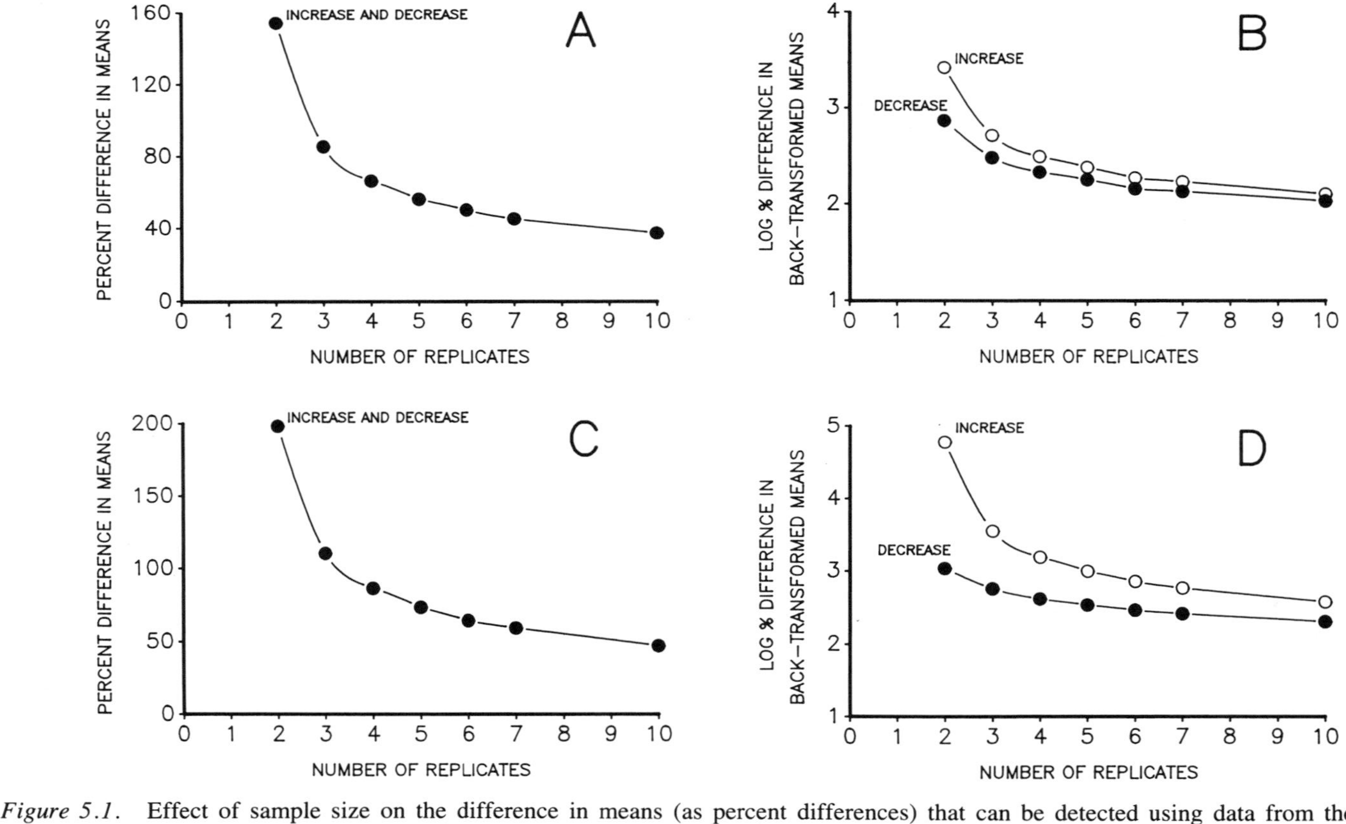

Figure 5.1. Effect of sample size on the difference in means (as percent differences) that can be detected using data from the Hunting Creek biomonitoring study described in the text. Values are calculated from the equation of Sokal and Rohlf (1981, p. 263), using median values of means and coefficients of variation (CV). A: Mean number of taxa. B: Back-transformed (geometric) mean number of individuals. C: Mean Simpson's Diversity. D: Back-transformed (geometric) mean number of *Baetis tricaudatus*.

Table 5.8. Percent difference between two means detectable, with a 95% chance of being correct at a 5% level of significance, for four benthic measures at Hunting Creek, in ANOVA designs with two or six years of data, five replicate sample units, and three estimates of the coefficient of variation (CV).

	No. of Taxa	No. of Indi-viduals	Simpson's Diversity	No. of *Baetis tricaudatus*
A. 2 yr				
MEDIAN CV				
Times/divide factor	1.56	1.21*	1.73	1.74*
Detectable increase, % difference	56	237**	73	976**
Detectable decrease, % difference	56	174**	73	339**
MAXIMUM CV				
Times/divide factor	2.28	1.63*	2.92	3.91*
Detectable increase, % difference	128	3,720**	192	> 10^6**
Detectable decrease, % difference	128	850**	192	1,766**
MINIMUM CV				
Times/divide factor	1.12	1.04*	1.30	1.21*
Detectable increase, % difference	12	28**	30	96**
Detectable decrease, % difference	12	27**	30	78**
B. 6 yr				
MEDIAN CV				
Times/divide factor	1.51	1.19*	1.66	1.67*
Detectable increase, % difference	51	200**	66	762**
Detectable decrease, % difference	51	152**	66	300**
MAXIMUM CV				
Times/divide factor	2.16	1.57*	2.74	3.64*
Detectable increase, % difference	116	2,600**	174	432,958**
Detectable decrease, % difference	116	728**	174	1,500**
MINIMUM CV				
Times/divide factor	1.10	1.04*	1.27	1.19*
Detectable increase, % difference	11	25**	27	88**
Detectable decrease, % difference	11	25**	27	70**

*Based on transformed data.
**Based on back-transformed data.

a 56% difference in means is detectable with two years of data; for six years of data the difference required for significance is 51%) (Table 5.8). The percent differences in back-transformed means required for detection of a significant difference between years for number of individuals and number of *B. tricaudatus* remain high enough (even with data from six years) that only catastrophic changes in the biotic community likely would be detected with these variables (Table 5.8).

5.3.2.5. *Effects of Variability on the Ability to Detect Disturbance*

The importance of the variability of a particular estimate of a benthic measure (e.g., as measured by the CV) relative to the magnitude of change that is required to detect a difference between years in Hunting Creek is clear when the minimum or maximum values of the CV are substituted for median values in the Sokal and Rohlf formula (Table 5.8). For five replicates, substantial reductions in differences between means are detectable under minimum CV conditions when compared with those under maximum or median CVs (Table 5.8), for both two years and six years of data. For example, using number of taxa for two years, a 56% difference in means is detectable statistically if the median value of CV (21.2%) applies. If the minimum CV (4.4%) was applicable, however, a 12% difference in means would be significant. Conversely, a substantial increase in detectable mean differences in number of taxa (to 128%) would result if the maximum CV (48.3%) was used as an estimate of variability.

At this point, it is important to ask: How often do detectable differences in these measures occur in the absence of impact? Such "natural" variability can be a confounding factor in any impact assessment or biomonitoring program (e.g., Norris and Georges, 1986; Stewart-Oaten et al., 1986; Resh et al., 1988; McElravy et al., 1989; Resh and Rosenberg, 1989).

The chances of differences due to natural variability occurring in Hunting Creek, and being detected with five replicates, can be estimated by examining the frequency distribution of differences in means from all possible two-year periods (given for each measure in Fig. 5.2A–D) and comparing these frequencies with the detectable differences reported in Fig. 5.1. Using the detectable percent differences in means for two years and five replicates (values shown on Fig. 5.2A–D), the chance that natural variability alone will produce a statistically significant difference in means was estimated by summing the frequency of all categories greater than each detectable percent difference value (as given on Fig. 5.2A–D). In Hunting Creek, the chance of detecting a difference due to annual variability alone is at least 22% for number of taxa, at least 23–35% for number of individuals, at least 5% for Simpson's Diversity, and at least 20–30% for numbers of *B. tricaudatus*. Thus, we see that a detectable difference in mean values will often (20–35% of the time) occur, simply as a result of annual variability, over any two-year period in this stream for three of the four benthic measures examined.

Statistically significant differences may be found in the absence of anthropogenic impact with even fewer numbers of sample units. McElravy et al. (1989) took only four sample units in a design that exaggerated variability (two collected in riffles, two in pools) each year for seven years. They were able to demonstrate statistically significant reductions in species richness and

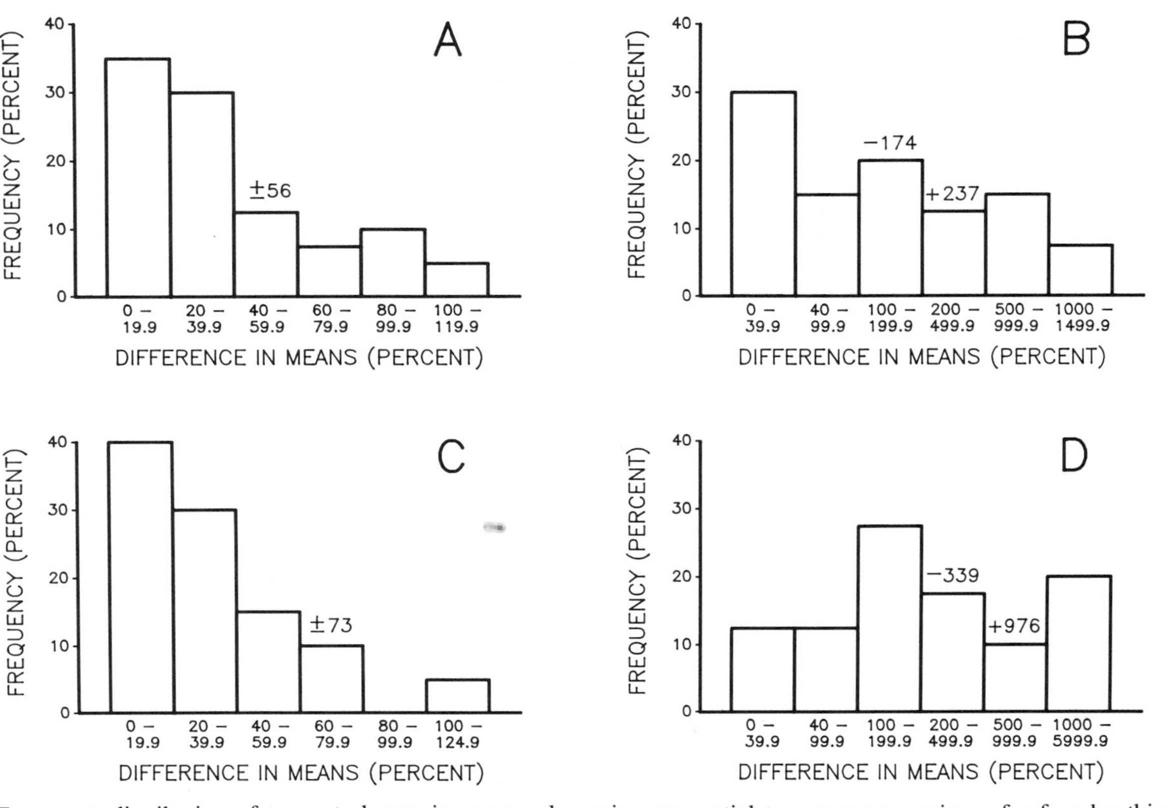

Figure 5.2. Frequency distribution of percent change in mean value using sequential two-year comparisons for four benthic measures in Hunting Creek 1984–1989. Values given in each histogram represent the minimum percent change that can be detected using the equation of Sokal and Rohlf (1981, p. 263), with five replicate sample units and the median value of the coefficient of variation. A: Number of taxa. B: Number of individuals. C: Simpson's Diversity. D: Number of *Baetis tricaudatus*.

in the value of Simpson's Diversity Index during a severe drought and in years with above average rainfall.

If differences between impacted and unimpacted conditions can be determined *a priori* (e.g., by using economic or biological criteria), then sequential decision plans represent an alternative to approaches that use fixed sample sizes. With this method of statistical inference, the number of observations required is not determined in advance, but rather sample units are collected and analyzed separately until a decision can be made that meets specified levels of risk and error (Resh and Price, 1984). Because sample analysis stops when the minimum information needed to make an unimpacted/impacted classification is obtained, fewer sample units often may be examined with sequential decision plans compared with procedures using fixed sample sizes. Jackson and Resh (1988) demonstrated that sequential decision plans using estimates of taxa richness, species diversity, and density saved time (50–79%) over conventional methods and produced 95% accuracy of unimpacted/impacted classifications. Jackson and Resh (1989) provided step-by-step examples of the creation and use of sequential decision plans, and Resh et al. (1988) described their use in a stream biomonitoring program.

If conventional methods are required, two approaches can be used to increase sample size without increasing costs. First, take smaller sample units (e.g., Elliott, 1977; Resh, 1979b; Downing, 1984, 1989; Morin, 1985), but remember that as sample means decrease the required number of sample units increases, that zero counts can create statistical nightmares, and that an edge effect becomes more important with smaller-sized sample units. Second, carefully choose certain study populations (Voshell et al., 1989) having characteristics that permit a large number of sample units to be collected inexpensively (Resh, 1979b). Of course, this last point requires that the population chosen reflects the biomonitoring goals of the study. Alternative strategies might include limiting the sampling universe through stratification (e.g., by habitat type) to reduce variability among replicates while maintaining the goals of the biomonitoring program (Resh, 1979b; Voshell et al., 1989).

5.3.3. Taxonomic Considerations

Species-level identifications have been considered to be a critical aspect of biomonitoring studies in the past (Resh and Unzicker, 1975), but how is this issue currently viewed? To address this question, 34 recent papers were examined that dealt, at least in part, with this issue. From this analysis, it was apparent that 18 of the papers emphasized the importance of species-level identifications, nine recommended use of higher taxa under appropriate circumstances, four suggested the use of both species and higher-level identification depending on the objectives of the study, and three dealt with the importance of accuracy of identifications in biomonitoring work.

5.3.3.1. *Species-Level Identifications Are Necessary*

Authors who felt that species-level identifications were required for biomonitoring followed three general lines of reasoning to support their position. First, the concept of the species as the basic biological unit is widely accepted (e.g., Rosenberg et al., 1979; Herricks, 1984), and this level is assumed to have the greatest information content as a result of studies on individual populations (e.g., specific habitat preference, life-history features, etc.). Use of higher levels can result in a loss of a portion of this information (Waterhouse and Farrell, 1985), because of differences among related species that form the higher levels (e.g., Learner et al., 1978; Resh and Grodhaus, 1983; Krieger, 1984; Hall and Ide, 1987; Pinder, 1989). These differences may have resulted from speciation mechanisms that emphasized ecological specialization (Cranston, 1990). The consequences of this information loss depend on the nature of the environmental gradient being examined (e.g., type and downstream concentration of a pollutant), the response (e.g., as abundances) of the organisms sampled along that gradient (Waterhouse and Farrell, 1985), and the principal objectives of the monitoring program (Furse et al., 1984). Lack of species-level information can decrease sensitivity and reduce the ability of a study to detect more subtle changes (e.g., Howmiller and Scott, 1977; Krieger, 1984), prevent application of the indicator species concept (Sæther, 1980; Herricks and Cairns, 1982; Resh et al., 1983), result in possible misinterpretation of bioassay data (e.g., Millemann et al., 1984), or cause difficulty in validating ecological interpretations for assessment of impact (e.g., Benke et al., 1981; Herricks and Cairns, 1982). Second, identifications at taxonomic levels above species can result in systematic underestimation of certain kinds of biological (e.g., diversity) indices (Hughes, 1978; Resh, 1979a). Third, use of higher taxa can interfere with subsequent analysis (Resh and Unzicker, 1975; Lenat and Penrose, 1980). For example, multivariate procedures usually are done with species data to increase performance (Brinkhurst, 1985).

5.3.3.2. *Higher Taxonomic Levels Are Acceptable*

Authors who recommended use of levels above species recognized that such use represented a compromise between a desire for the increased information content and either its unavailability (i.e., larval keys and descriptions do not exist), and/or the cost (time, required expertise) of obtaining it (e.g., ISO, 1984; Evans, 1988). The use of higher-level categories such as genus, family, or even order was justified as follows: (1) to detect instances of gross pollution that have dramatic effects on benthos (e.g., Howmiller and Scott, 1977; Hellawell, 1978; Osborne et al., 1980; Furse et al., 1984; Armitage et al., 1987); (2) to provide an "early warning" of potential problems or changes in the benthos (Herricks and Cairns, 1982) as an indication of the need for more detailed study; (3) when analysis procedures such as species diversity indices and other multiple taxa measures, which are not as sensitive to information loss, are the main objective of the monitoring study and are used to simplify the presentation of results (e.g., Kaesler and Herricks, 1979; Osborne et al., 1980; Waterhouse and Farrell, 1985; Pinder, 1989); and (4) when the taxa included in a higher level are consistent in their response to environmental change (e.g., most Plecoptera; Baumann, 1979).

The many improved keys now available allow nonspecialists to produce reliable identifications at the family and genus levels. Accuracy of taxonomic identifications is essential to any biomonitoring study, but for macroinvertebrates, the problem is greatest at the species level. For many macroinvertebrate groups, identification to species is difficult because of their small size, a lack of adequate species-level keys and descriptions, and the need for rearing immature forms to the adult stage (e.g., Merritt et al., 1984). However, reliability of species identifications can be increased through maintenance of up-to-date library-reference and voucher collections, communication with other workers (Evans, 1988), and use of computerized databases of habitat information that can provide an initial screening procedure (Hellenthal, 1982; Dawson and Hellenthal, 1986).

Interestingly, $<1/4$ of the papers examined in this survey reported consulting a taxonomist, and only 4% reported sending specimens to a recognized depository (Table 5.5). Access to specimens is an important consideration for future studies because taxonomic concepts of macrobenthic organisms will continue to change. However, even if experienced personnel and adequate literature are available, the cost of species identifications can be quite high if special preparation is required (e.g., clearing and mounting Chironomidae on slides) and if such species occur frequently in samples. Rosenberg et al. (1979) further discuss the need for proper curation and storage of specimens following large-scale surveys, and Cranston (1990) notes the importance of voucher specimens when species-level taxa are recognized but not named during a study.

In examining the importance of species identifications in biomonitoring, the question of the required level of taxonomy still can be asked. The answer, it seems, is "It depends"—it depends on the purpose of the study, the level of sensitivity required, the type of index or analysis being used, and the particular group of organisms of primary interest.

5.4. Recommendations for Future Biomonitoring Studies

Study designs, field procedures, and methods of analysis that are currently being used in quantitative biomonitoring studies with macrobenthos have been described in this chapter. We end this discussion with a listing of suggestions (some old but worth repeating) that may enhance the assessment of disturbance effects, and thus improve the role of biomonitoring in the environmental assessment process:

1. Spatial and temporal variability in the absence of disturbance, which often occur concurrently (Resh and Rosenberg, 1989), must be understood before disturbance effects can be distinguished from natural variability. Separation of these two sources of variability, as opposed to homogenizing their effects, often provides an explanation for underlying factors causing the patterns that are observed.
2. Both accuracy and precision must be considered during the design phase of any biomonitoring study. Conclusions regarding impact too often have been based on significant differences in main-effect means that really resulted from the influence of either covariate factors or sampling bias, or both. With small sample sizes, the ability to detect differences between means of benthic measures over space or time (usually *the* objective in a biomonitoring study) can be enhanced by using more powerful techniques such as ANOVA (as compared to, for example, estimates of means and confidence intervals). Set precision and error risk rates appropriate to the questions asked; $P = 0.05$ is not the only level available.
3. A useful approach to habitat stratification in streams could involve a two-step procedure. First, choose sample areas of comparable substrate size within a riffle, and second, within these areas, choose plots of comparable hydraulic stress (Statzner et al., 1988). The measurement of hydraulic parameters is now easy, fast, and effective (Statzner and Müller, 1989). However, be aware that as stratification increases, extrapolation to other habitats is more difficult. Likewise, low mean densities or highly contagious distributions that result when the sampling universe and population universe do not coincide may create analysis problems.
4. Some studies may be conducted more appropriately using approaches other than fixed-interval sampling or fixed sample size (see Resh, 1979b, and Resh and Jackson, Chapter 6, for examples). Often, however, substantial amounts of reconnaissance data are required in using these approaches.
5. Retain voucher specimens and prepare a reference collection that will be placed in a depository, such as an established university department or government museum. Such institutions also should expect some compensation for this service.

6. Scientists need to communicate better to legislators that attempts to standardize biomonitoring procedures and analyses (such as basing impact on changes in values of the Shannon Diversity Index) actually may inhibit the ability to carry out effective impact assessments, especially when these procedures are no longer recognized by the scientific community as accurate or useful in the biomonitoring process.
7. Scientists working in biomonitoring and environmental assessment need to continue to employ imaginative thinking in both the design of studies as well as specific methods used. Benthic biologists need to take advantage of technological advancements in developing new sampling devices, laboratory procedures, and methods of analysis. This includes using nontraditional sampling devices such as air-lift samplers when these are obviously appropriate (e.g., in rivers >1 m in depth), as well as using inexpensive computers and microelectronics to assist in operating such samplers. Some day, wouldn't it be useful to have a reliable measure, akin to chlorophyll-*a* in algal productivity studies, which is fast, easy to perform, and produces results that can be consistently interpreted?

Acknowledgments

We thank R. Voshell for permission to use data from Voshell et al. (1989) in Tables 5.1 and 5.2 and for providing the 45 lotic references used in that paper. We also thank J. Jackson for useful discussions, L. Bergey for assistance with the lentic literature search and the compilation of data, and R. Voshell and J. Downing for comments on this manuscript.

References

Allan, J.D. 1984. Hypothesis testing in ecological studies of aquatic insects. In *The Ecology of Aquatic Insects,* eds. V.H. Resh and D.M. Rosenberg, pp. 484–507. Praeger Pubs., New York.

Allan, J.D. and E. Russek. 1985. The quantification of stream drift. *Canadian Journal of Fisheries and Aquatic Sciences* 42:210–5.

Allen, K.R. 1959. The distribution of stream bottom fauna. *Proceedings of the New Zealand Ecological Society* 6:5–8.

APHA (American Public Health Association). 1989. Benthic macroinvertebrates. In *Standard Methods for the Examination of Water and Wastewater,* 17th ed., pp. 10–95 to 10–113. American Public Health Association, Washington, DC.

Armitage, P.D., R.J.M. Gunn, M.T. Furse, J.F. Wright, and D. Moss. 1987. The use of prediction to assess macroinvertebrate response to river regulation. *Hydrobiologia* 144:25–32.

Baumann, R.W. 1979. Nearctic stonefly genera as indicators of ecological parameters (Plecoptera: Insecta). *Great Basin Naturalist* 39:241–4.

Benke, A.C., G.E. Willeke, F.K. Parrish, and D.L. Stites. 1981. *Effects of Urbanization on Stream Ecosystems*. Environmental Resources Center, Georgia Institute of Technology Report ERC 07–81, Atlanta, GA.

Brinkhurst, R.O. 1985. Review of *Methods for the Study of Marine Benthos*, 2nd ed., eds. N.A. Holme and A.D. McIntyre. *Canadian Journal of Fisheries and Aquatic Sciences* 42:1445–6.

Brittain, J.E. and T.J. Eikeland. 1988. Invertebrate drift—a review. *Hydrobiologia* 166:77–93.

Canton, S.P. and J.W. Chadwick. 1988. Variability in benthic invertebrate density estimates from stream samples. *Journal of Freshwater Ecology* 4:291–7.

Chutter, F.M. 1972. A reappraisal of Needham and Usinger's data on the variability of a stream fauna when sampled with a Surber sampler. *Limnology and Oceanography* 17:139–41.

Collins, J.N. and V.H. Resh. 1989. *Guidelines for the Ecological Control of Mosquitoes in Non-Tidal Wetlands of the San Francisco Bay Area*. California Mosquito and Vector Control Association Inc. and the University of California Mosquito Research Program, Sacramento, CA.

Comiskey, C. and C. Brandt. 1982. *Marine ecosystem monitoring. Quantitative Impact Assessment Final Report*, Appendix A. Science Advisory Board, U.S. Environmental Protection Agency, Washington, DC.

Cranston, P.S. 1990. Biomonitoring and invertebrate taxonomy. *Environmental Monitoring and Assessment* 14:265–73.

Cuff, W. and N. Coleman. 1979. Optimal survey design: lessons from a stratified random sample of macrobenthos. *Journal of the Fisheries Research Board of Canada* 36:351–61.

Cummins, K.W. 1975. Macroinvertebrates. In *River Ecology*, ed. B.A. Whitton, pp. 170–98. Univ. of California Press, Berkeley, CA.

Davies, I.J. 1984. Sampling aquatic insect emergence. In *A Manual on Methods for the Assessment of Secondary Productivity in Fresh Waters*, 2nd ed., eds. J.A. Downing and F.H. Rigler, pp. 161–227. IBP Handbook 17. Blackwell Scientific Pubs., Oxford, England.

Dawson, C.L. and R.A. Hellenthal. 1986. *A Computerized System for the Evaluation of Aquatic Habitats Based on Environmental Requirements and Pollution Tolerance Associations of Resident Organisms*. EPA/600/S3–86/019. Environmental Research Laboratory, U.S. Environmental Protection Agency, Corvallis, OR.

Downing, J.A. 1979. Aggregation, transformation, and the design of benthos sampling programs. *Journal of the Fisheries Research Board of Canada* 36:1454–63.

Downing, J.A. 1984. Sampling the benthos of standing waters. In *A Manual on Methods for the Assessment of Secondary Productivity in Fresh Waters*, 2nd ed., eds. J.A. Downing and F.H. Rigler, pp. 87–130. IBP Handbook 17. Blackwell Scientific Pubs., Oxford, England.

Downing, J.A. 1989. Precision of the mean and the design of benthos sampling programmes: caution revised. *Marine Biology* 103:231–4.

Elliott, J.M. 1977. Some methods for the statistical analysis of samples of benthic invertebrates, 2nd ed. *Freshwater Biological Association Scientific Publication* No. 25:1–156.

Evans, D.L. 1988. The need for taxonomic accuracy. *Florida Benthological Newsletter* 2(3):5–6.

Furse, M.T., D. Moss, J.F. Wright, and P.D. Armitage. 1984. The influence of seasonal and taxonomic factors on the ordination and classification of running-water sites in Great Britain and on the prediction of their macro-invertebrate communities. *Freshwater Biology* 14:257–80.

Green, R.H. 1979. *Sampling Design and Statistical Methods for Environmental Biologists*. John Wiley, New York.

Hall, R.J. and F.P. Ide. 1987. Evidence of acidification effects on stream insect communities in central Ontario between 1937 and 1985. *Canadian Journal of Fisheries and Aquatic Sciences* 44:1652–7.

Harper, P.P. and L. Cloutier. 1986. Spatial structure of the insect community of a small dimictic lake in the Laurentians (Québec). *Internationale Revue der gesamten Hydrobiologie* 71:655–85.

Hawkins, C.P. 1986. Pseudo-understanding of pseudoreplication: a cautionary note. *Bulletin of the Ecological Society of America* 67:184–5.

Hellawell, J.M. 1978. *Biological Surveillance of Rivers*. National Environment Research Council and Water Research Centre, Stevenage, England.

Hellawell, J.M. 1986. *Biological Indicators of Freshwater Pollution and Environmental Management*. Elsevier, London.

Hellenthal, R.A. 1982. Using aquatic insects for the evaluation of freshwater communities. In *Acquisition and Utilization of Aquatic Habitat Inventory Information. Proceedings of a Symposium, Portland, OR, October 28–30, 1981*, ed. N.B. Armantrout, pp. 347–54. Western Division, American Fisheries Society, Bethesda, MD.

Herricks, E.E. 1984. Aspects of monitoring in river basin management. *Water Science and Technology* 16(5–7):259–74.

Herricks, E.E. and J. Cairns, Jr. 1982. Biological monitoring. Part III—Receiving system methodology based on community structure. *Water Research* 16:141–53.

Howmiller, R.P. and M.A. Scott. 1977. An environmental index based on relative abundance of oligochaete species. *Journal of the Water Pollution Control Federation* 49:809–15.

Hughes, B.D. 1978. The influence of factors other than pollution on the value of Shannon's diversity index for benthic macro-invertebrates in streams. *Water Research* 12:359–64.

Hurlbert, S.H. 1984. Pseudoreplication and the design of ecological field experiments. *Ecological Monographs* 54:187–211.

ISO (International Organization for Standardization). 1984. *Water Quality—Assessment of the Water and Habital Quality of Rivers by a Micro-Invertebrate "Score."* ISO/TC 147/SC 5/WG 6 N 40, Draft Proposal ISO/DP 8689.

Jackson, J.K. and V.H. Resh. 1988. Sequential decision plans in monitoring benthic macroinvertebrates: cost savings, classification accuracy, and development of plans. *Canadian Journal of Fisheries and Aquatic Sciences* 45:280–6.

Jackson, J.K. and V.H. Resh. 1989. Sequential decision plans, benthic macroinvertebrates, and biological monitoring programs. *Environmental Management* 13:455–68.

Kaesler, R.L. and E.E. Herricks. 1979. Hierarchical diversity of communities of aquatic insects and fishes. *Water Resources Bulletin* 15:1117–25.

Kathman, R.D. 1984. *Freshwater Benthic Invertebrates. Collection, Identification, Analysis*. Unpublished report prepared for Weyerhaeuser Canada, Ltd., Kamloops, BC.

Krieger, K.A. 1984. Benthic macroinvertebrates as indicators of environmental degradation in the southern nearshore zone of the central basin of Lake Erie. *Journal of Great Lakes Research* 10:197–209.

Learner, M.A., G. Lochhead, and B.D. Hughes. 1978. A review of the biology of British Naididae (Oligochaeta) with emphasis on the lotic environment. *Freshwater Biology* 8:357–75.

Lenat, D.R. and D.L. Penrose. 1980. Discussion: "Hierarchical diversity of communities of aquatic insects and fishes," by Roger L. Kaesler and Edwin E. Herricks. *Water Resources Bulletin* 16:361–2.

McElravy, E.P. 1988. Temporal variability in abundance of aquatic insects: a comparison of temperate and tropical environments. Ph.D. dissertation, Univ. of California, Berkeley, CA.

McElravy, E.P., G.A. Lamberti, and V.H. Resh. 1989. Year-to-year variation in the aquatic macroinvertebrate fauna of a northern California stream. *Journal of the North American Benthological Society* 8:51–63.

McIntyre, A.D., J.M. Elliott, and D.V. Ellis. 1984. Introduction: design of sampling programmes. In *Methods for the Study of Marine Benthos,* 2nd ed., eds. N.A. Holme and A.D. McIntyre, pp. 1–26. Blackwell Scientific Pubs., Oxford, England.

Merritt, R.W., K.W. Cummins, and V.H. Resh. 1984. Collecting, sampling, and rearing methods for aquatic insects. In *An Introduction to the Aquatic Insects of North America,* 2nd ed., eds. R.W. Merritt and K.W. Cummins, pp. 11–26. Kendall/Hunt Publishing, Dubuque, IA.

Millemann, R.E., W.J. Birge, J.A. Black, R.M. Cushman, K.L. Daniels, P.J. Franco, J.M. Giddings, J.F. McCarthy, and A.J. Stewart. 1984. Comparative acute toxicity to aquatic organisms of components of coal-derived synthetic fuels. *Transactions of the American Fisheries Society* 113:74–85.

Morin, A. 1985. Variability of density estimates and the optimization of sampling programs for stream benthos. *Canadian Journal of Fisheries and Aquatic Sciences* 42:1530–4.

Needham, P.R. and R.L. Usinger. 1956. Variability in the macrofauna of a single riffle in Prosser Creek, California, as indicated by the Surber sampler. *Hilgardia* 24:383–409.

Norris, R.H. and A. Georges. 1986. Design and analysis for assessment of water quality. In *Limnology in Australia,* eds. P. De Deckker and W.D. Williams, pp. 555–72. Junk Pubs., Dordrecht, The Netherlands.

Ortal, R. and S. Ritman. 1985. Israel inland water ecological data base. In *The Role of Data in Scientific Progress,* ed. P.S. Glaeser, pp. 73–6. Elsevier, New York.

Osborne, L.L., R.W. Davies, and K.J. Linton. 1980. Use of hierarchical diversity indices in lotic community analysis. *Journal of Applied Ecology* 17:567–80.

Patil, G.P. and C. Taillie. 1976. Ecological diversity: concepts, indices and applications. *Proceedings of the International Biometry Conference* 9:383–411.

Peckarsky, B.L. 1984. Sampling the stream benthos. In *A Manual on Methods for the Assessment of Secondary Productivity in Fresh Waters,* 2nd ed., eds. J.A. Downing and F.H. Rigler, pp. 131–60. IBP Handbook 17. Blackwell Scientific Pubs., Oxford, England.

Pinder, L.C.V. 1989. Biological surveillance of chalk-streams. *Freshwater Biological Association 57th Annual Report,* pp. 81–92. Freshwater Biological Association, Ambleside, England.

Prepas, E.E. 1984. Some statistical methods for the design of experiments and analysis of samples. In *A Manual on Methods for the Assessment of Secondary Productivity in Fresh Waters,* 2nd ed., eds. J.A. Downing and F.H. Rigler, pp. 266–335. IBP Handbook 17. Blackwell Scientific Pubs., Oxford, England.

Resh, V.H. 1979a. Biomonitoring, species diversity indices, and taxonomy. In *Ecological Diversity in Theory and Practice,* eds. J.F. Grassle, G.P. Patil, W. Smith, and C. Taillie, pp. 241–53. International Co-operative Publishing House, Fairland, MD.

Resh, V.H. 1979b. Sampling variability and life history features: basic considerations in the design of aquatic insect studies. *Journal of the Fisheries Research Board of Canada* 36:290–311.

Resh, V.H., J.W. Feminella, and E.P. McElravy. 1990. *Sampling Aquatic Insects.* Videotape. Office of Media Services, Univ. of California, Berkeley, CA.

Resh, V.H. and G. Grodhaus. 1983. Aquatic insects in urban environments. In *Urban Entomology: Interdisciplinary Perspectives,* eds. G.W. Frankie and C.S. Koehler, pp. 247–76. Praeger Pubs., New York.

Resh, V.H., J.K. Jackson, and E.P. McElravy. 1988. The use of long-term ecological data and sequential decision plans in monitoring the impact of geothermal energy development on benthic macroinvertebrates. *Internationale Vereinigung für Theoretische und Angewandte Limnologie Verhandlungen* 23:1142–6.

Resh, V.H., G.A. Lamberti, E.P. McElravy, J.R. Wood, and J.W. Feminella. 1984. *Quantitative Methods for Evaluating the Effects of Geothermal Energy Development on Stream Benthic Communities at The Geysers, California.* California Water Resources Center Contribution No. 190. California Water Resources Center, Univ. of California, Davis, CA.

Resh, V.H. and D.G. Price. 1984. Sequential sampling: a cost-effective approach for monitoring benthic macroinvertebrates in environmental impact assessments. *Environmental Management* 8:75–80.

Resh, V.H. and D.M. Rosenberg. 1989. Spatial-temporal variability and the study of aquatic insects. *Canadian Entomologist* 121:941–63.

Resh, V.H., D.M. Rosenberg, and J.W. Feminella. 1985. The processing of benthic samples: responses to the 1983 NABS questionnaire. *Bulletin of the North American Benthological Society* 2:5–11.

Resh, V.H., D.M. Rosenberg, and A.P. Wiens. 1983. Emergence of caddisflies (Trichoptera) from eroding and non-eroding shorelines of Southern Indian Lake, Manitoba, Canada. *Canadian Entomologist* 115:1563–72.

Resh, V.H. and J.D. Unzicker. 1975. Water quality monitoring and aquatic organisms: the importance of species identification. *Journal of the Water Pollution Control Federation* 47:9–19.

Rosenberg, D.M., H.V. Danks, J.A. Downes, A.P. Nimmo, and G.E. Ball. 1979. Procedures for a faunal inventory. In Canada and Its Insect Fauna, ed. H.V. Danks. *Memoirs of the Entomological Society of Canada* 108:509–32.

Rosenberg, D.M. and V.H. Resh. 1982. The use of artificial substrates in the study of freshwater benthic macroinvertebrates. In *Artificial Substrates,* ed. J. Cairns, Jr., pp. 175–235. Ann Arbor Science Pubs., Ann Arbor, MI.

Sæther, O.A. 1980. The influence of eutrophication on deep lake benthic invertebrate communities. *Progress in Water Technology* 12:161–80.

Schwenneker, B.W. and R.A. Hellenthal. 1984. Sampling considerations in using stream insects for monitoring water quality. *Environmental Entomology* 13:741–50.

Sokal, R.R. and F.J. Rohlf. 1981. *Biometry. The Principles and Practice of Statistics in Biological Research,* 2nd ed. W.H. Freeman, New York.

Southwood, T.R.E. 1978. *Ecological Methods with Particular Reference to the Study of Insect Populations,* 2nd ed. Chapman and Hall, London.

Statzner, B., J.A. Gore, and V.H. Resh. 1988. Hydraulic stream ecology: observed patterns and potential applications. *Journal of the North American Benthological Society* 7:307–60.

Statzner, B. and R. Müller. 1989. Standard hemispheres as indicators of flow characteristics in lotic benthos research. *Freshwater Biology* 21:445–59.

Stewart-Oaten, A., W.W. Murdoch, and K.R. Parker. 1986. Environmental impact assessment: "pseudoreplication" in time? *Ecology* 67:929–40.

Storey, A.W. and L.C.V. Pinder. 1985. Mesh-size and efficiency of sampling of larval Chironomidae. *Hydrobiologia* 124:193–7.

Tetra Tech, Inc. 1987. *Recommended Protocols for Sampling and Analyzing Subtidal Benthic Macroinvertebrate Assemblages in Puget Sound.* Unpublished report prepared for the U.S. Environmental Protection Agency, Region 10–Office of Puget Sound, Seattle, WA. Contract No. 68–03–1977.

Voshell, J.R., Jr., R.J. Layton, and S.W. Hiner. 1989. Field techniques for determining the effects of toxic substances on benthic macroinvertebrates in rocky-bottomed streams. In *Aquatic Toxicology and Hazard Assessment: 12th Volume,* eds. U.M. Cowgill and L.R. Williams, pp. 134–55. American Society for Testing and Materials Special Technical Publication 1027. American Society for Testing and Materials, Philadelphia, PA.

Waterhouse, J.C. and M.P. Farrell. 1985. Identifying pollution related changes in chironomid communities as a function of taxonomic rank. *Canadian Journal of Fisheries and Aquatic Sciences* 42:406–13.

Waters, T.F. 1979. Influence of benthos life history upon the estimation of secondary production. *Journal of the Fisheries Research Board of Canada* 36:1425–30.

Weber, C.I., ed. 1973. *Biological Field and Laboratory Methods for Measuring the Quality of Surface Waters and Effluents.* EPA–670/4–73–001. U.S. Environmental Protection Agency, Cincinnati, OH.

Winterbourn, M.J. 1985. Sampling stream invertebrates. In *Biological Monitoring in Freshwaters: Proceedings of a Seminar, Hamilton, November 21–23, 1984, Part 2,* eds. R.D. Pridmore and A.B. Cooper, pp. 241–58. Water and Soil Miscellaneous Publication No. 83, National Water and Soil Conservation Authority, Wellington, NZ.

6

Rapid Assessment Approaches to Biomonitoring Using Benthic Macroinvertebrates

Vincent H. Resh and John K. Jackson

6.1. Introduction

Over the past three decades, the use of benthic macroinvertebrates in water-quality monitoring programs in North America has undergone two transitions. Through the 1960s, use of qualitative approaches, such as correlating the presence, absence, or approximate relative abundance of certain macroinvertebrates with preestablished classifications of environmental quality, was emphasized. This approach was influenced by the almost century-old European "Saprobien system" for assessing the pollution status of lotic habitats and by the lake typology concept of characterizing the trophic status of lentic waters by the organisms present (see Johnson et al., Chapter 4). The first transition occurred in the 1970s when the emphasis shifted toward quantitative approaches that typically included calculation of diversity indices, formal hypothesis testing that required replicate sample units, and detailed statistical analyses.

In recent years, however, a second transition has occurred: renewed interest in the use of qualitative techniques, primarily because of the high cost of quantitative approaches. This shift resulted in the development of what are generally called "rapid assessment approaches." The purpose of applying rapid assessment approaches is to identify water quality problems associated with both point and nonpoint source pollution and to document long-term regional changes in water quality. A similar approach also is being taken to assess status of fish communities (Plafkin et al., 1989).

In Europe, the use of qualitative techniques has been predominant for decades, with recent efforts focusing on standardization of the biotic indices and scoring systems used (Metcalfe, 1989). The current application of qualitative approaches on both continents has reduced the intensity of study necessary at individual sites, relative to that required with quantitative approaches such as those described by Resh and McElravy in Chapter 5. This permits a greater number of sites to be examined.

Rapid assessment approaches are somewhat analogous to using thermometers in assessing human health; easily obtained values are compared to a threshold that is considered to be "normal" (cf. a human body temperature of 98.6° F or 37° C). Of course, the key questions in biomonitoring are: What population and community measures are biologically relevant (the thermometers)? What are the thresholds against which they are being compared (the normal body temperature)? How much of a deviation from the threshold is a sign of "ill health"?

In this chapter, a variety of rapid assessment approaches that are currently used in North America and Europe are described. The accuracy of some of the water quality measures used in these programs is then examined by comparing their performance with data collected from impacted and unimpacted streams. Next, the results of that analysis are combined with information from different protocols currently in use, and other criteria, to illustrate how an assessment procedure can be developed for a region. Finally, topics requiring further research are identified.

6.2. Elements of Rapid Assessment Approaches

Rapid assessment involves sampling and analysis approaches that are designed to fulfill two objectives. First, effort (and cost) is reduced in assessing environmental conditions at a site, relative to that needed in quantitative approaches. This can be achieved in several ways: (1) the number of habitats sampled and replicate sample units taken per habitat are reduced; (2) less silt and particulate debris are collected, which makes sorting easier and faster; (3) only a fraction of the animals collected are considered, which means fewer have to be identified; or (4) specimens are identified to family or even higher levels.

A second objective of rapid assessment approaches is to summarize the results of site surveys in a way that they can be understood by nonspecialists such as managers, other decision-makers, and the concerned public. This is done by using analysis measures that express results as single scores, as well as by placing the scores obtained in categories of environmental quality based on regional background data.

The success of any rapid assessment approach ultimately depends on the ability to detect impacted and unimpacted conditions. Therefore, efforts to reduce costs must not be carried to the point that information used in the analysis does not adequately represent the site examined. Likewise, the analysis and summarization should not be so simplified that impact-related conditions are not detected.

6.3. Rapid Assessment Measures

Selection of the most appropriate population and community measures for use in rapid assessment approaches is a critically important, as well as a controversial, decision. A wide variety of data measures has been used in rapid assessment protocols as a basis for assessing whether impact has occurred; many of these are described in Appendix 6.1. "Protocol" is used in this chapter to include: the appropriate habitat, sampling design and collection methods, enumeration methods, taxa used and level of taxonomic analysis done, and analysis measure(s) used to determine the degree of impact on a benthic community. These measures have been divided into five categories: richness, enumerations, community diversity and similarity indices, biotic indices, and functional feeding group measures (Table 6.1).

Measures of *richness* describe the number of distinct, specified taxonomic units (e.g., family, species) in a collection or at a site; richness is a component and estimate of community structure. Macroinvertebrate taxa richness may not really be a measure of macroinvertebrate species richness because it is based on specimens identified to the lowest taxonomic levels with available keys (usually genus), rather than on nominal species, which often requires rearing of adult specimens and taxonomic expertise for accurate identifications. Sometimes species are separated by perceived differences and are given designations (sp. A, sp. B, etc.); these groups may, or may not, correspond to distinct species. Separation of different life stages of the same species (early compared to late instars, or late instars compared to pupae) into different taxa would result in overestimation of taxa richness; more often, however, similar-appearing species are not separated, which results in underestimations of taxa richness (see Resh and McElravy, Chapter 5, for further discussion of this topic).

Enumerations range from counts of all organisms collected to estimates of relative abundances of different taxonomic groups (i.e., number of individuals in certain orders, families, or species, or numerically dominant taxa in these groupings). Essentially no taxonomic effort is required for total number of individuals; relative abundance requires distinctions based on the group under consideration (e.g., number of individuals for a given order, family, or species).

Measures of *community diversity* combine richness and enumerations in a summary statistic. These measures of community structure usually require taxonomic distinctions at the species level (or at some higher taxonomic level of macroinvertebrate richness as discussed above). Total number of taxa provides a richness component in calculating the value of diversity indices; the number of individuals per taxon provides an evenness component (Washington, 1984). Few of the dozens of diversity indices that have been proposed are regularly used; the most commonly used is Shannon's Index.

Table 6.1. Selected examples of measures used in rapid assessment protocols.

Measure and Description of Calculation	Rationale Behind Use of Measure	Comments
A. Richness measures		
Number of taxa (= taxa richness)—All macroinvertebrates are separated into presumed species groups; the number of distinct taxa is counted; species-level discriminations are made.	Taxa richness generally decreases with decreasing water quality (Weber, 1973; Resh and Grodhaus, 1983).	In practice, many taxa are identified to genus, and species are distinguished as sp. 1, sp. 2, etc. Most likely, this results in an underestimate of taxa richness because, with the exception of low-diversity habitats, probably few lentic and no lotic habitats can be characterized so completely. Identification to species also allows use of indicator organism concept (see Johnson et al., Chapter 4).
Number of EPT taxa (= EPT richness)—All Ephemeroptera, Plecoptera, and Trichoptera are separated from the other macroinvertebrates by order; the number of distinct taxa is then counted. An ability to distinguish taxa within these three orders is required. As above, species-level discriminations are made, but identifications are sometimes not done.	This is based on the observation that, in general, the majority of taxa in these three orders are pollution sensitive (Lenat, 1988).	Generic-level identification keys are available for each order for North America (Edmunds et al., 1976; Wiggins, 1977; Stewart and Stark, 1988) and Europe. Species-level keys are only available for, at most, about one-half of North American EPT taxa and about two-thirds of European taxa. Presence/absence of EPT taxa also has been proposed for use by Plafkin et al. (1989).
Number of families (= family richness)—All macroinvertebrate families are separated and number of families are counted. Family-level identifications are required.	As with number of taxa, number of families decreases with decreasing water quality.	Different taxonomic treatments have different numbers (and composition) of families in an order (e.g., Day, 1956, 3 mayfly families; Edmunds et al., 1976, 16 mayfly families); this requires some standardization.
Niche Occupant Forms (Mason, 1979)—This approximates number of taxa, but is based on number of discernible taxa that can be distinguished by a novice biologist. Counts are made of numbers of different forms—not individuals—thereby expediting preliminary surveys.	As with number of taxa, number of Niche Occupant Forms decreases with decreasing water quality.	Different life stages of the same taxon may be counted as different taxa. Data obtained cannot be used for other measures (e.g., biotic indices).

B. Enumerations		
Number of individuals (= total abundance)—All macroinvertebrate specimens are counted. No identifications are made.	Under certain types of stresses, standing crops (numbers or biomass) may increase or be reduced (Weber, 1973; Resh and Grodhaus, 1983).	Low benthic macroinvertebrate abundance is part of Plafkin et al. (1989), Protocol 1 (see Appendix 6.1).
Ratio of EPT abundance to Chironomidae abundance—All specimens of Ephemeroptera, Plecoptera, and Trichoptera are determined to order and counted; number of Chironomidae are determined. An ability to distinguish these four groups is required.	Perhaps it is that Chironomidae are perceived to be pollution-tolerant relative to pollution-sensitive Ephemeroptera, Plecoptera, and Trichoptera. Compared with a nonstressed habitat, a stressed habitat reflects an imbalance between these groups.	Habitat-specific conditions (e.g., substrate, temperature) also can influence relative abundance of these groups. Also, at the species level, the variety of pollution tolerance of the Chironomidae is very high. This measure can also be greatly affected by mesh size (larger mesh, fewer Chironomidae).
Ratio of individuals in numerically dominant taxa to total number of individuals (= % dominant taxa)—Number of individuals in the numerically dominant taxa is distinguished and counted, as is total number of individuals. An ability to distinguish and identify which individuals comprise the numerically dominant taxa is required.	A community dominated by relatively few species would indicate environmental stress (Plafkin et al., 1989), and a high percent contribution by a single taxon indicates community imbalance (Bode, 1988).	Some unstressed habitats also are dominated by a few taxa. Bode (1988) has used tolerance levels with this measure as part of a classification procedure. Plafkin et al. (1989, Protocol 2; see Appendix 6.1) apply this measure at the family level.
Ratio of non-dipterans to total number of individuals (= % non-dipterans)—Number of non-Diptera are counted and divided by total number of individuals. An ability to distinguish Diptera from other organisms is required.	This reflects presumed lower pollution tolerance of non-dipteran groups compared with that of the Chironomidae, Ceratopogonidae, and other dipteran families.	Habitat-specific conditions (e.g., substrate, temperature) also can influence relative abundance of these groups.
C. Community diversity and similarity indices		
Shannon's Index—Calculated as $\sum_{i=1}^{s} p_i \log_2 p_i$, where p_i is the proportion of individuals in the ith species. No identifications are required but all specimens must be separated to species level; numbers in each species are counted.	Species diversity decreases with decreasing water quality.	See Norris and Georges, Chapter 7, for a discussion of diversity indices.
Coefficient of Community Loss (Courtemanch and Davies, 1987)—Calculated as: total number of taxa at the reference site minus total number of taxa at the impacted site/number of taxa common to both sites.	Communities will become more dissimilar as stress increases.	This index depends on presence/absence data. Plafkin et al. (1989) indicate that this index provides better discrimination between sites than the Jaccard Coefficient or the Pinkham-Pearson Index.

(*continued*)

Table 6.1. (Continued)

Measure and Description of Calculation	Rationale Behind Use of Measure	Comments
Jaccard Coefficient (based on Jaccard 1912; Plafkin et al., 1989)—Calculated as: number of species common to both sites divided by the sum of the number of species common to both sites, the number of species present at the reference site but not the study site, and the number of species present at the study site but not the reference site.	Communities will become more dissimilar as stress increases.	This index also depends on presence/absence data. Spatial variation can complicate the interpretation of values obtained using any similarity measure.
Pinkham-Pearson Community Similarity Index (Pinkham and Pearson, 1976)—Calculated as: $B = 1/K \sum_{i=1}^{k} \frac{\text{minimum } (x_{ia}, x_{ib})}{\text{maximum } (x_{ia}, x_{ib})}$ where K is the number of comparisons, x_{ia}, x_{ib} is the number of individuals in the ith species at the reference site (a) or the study site (b).	Communities will become more dissimilar as stress increases.	In addition to presence/absence data, this index also incorporates abundance and compositional information.
D. Biotic indices		
Belgian Biotic Index Method (De Pauw and Vanhooren, 1983)—Taxa are identified to predetermined levels (usually family or genus) and comprise the "systematic units" for special faunistic groups (e.g., Plecoptera, cased Trichoptera). A biotic score is based on total number of systematic units and number of units in different faunistic groups.	This index reflects richness of the benthic macroinvertebrate community and gives weighted scores reflecting richness in various indicator groups.	This index combines the scoring procedure of the Indice Biotique from France (Tuffery and Verneaux, 1968) and the sampling procedure of the Trent Biotic Index from Great Britain (Woodiwiss, 1964).
Biotic Condition Index (Winget and Mangum, 1979; F.A. Mangum, U.S. Forest Service Intermountain Region Aquatic Ecosystem Analysis Lab, Provo, UT, personal communication)—Calculated as [Community Tolerance Quotient (predicted)/CTQ (actual)] × 100, where CTQ (predicted) is based	Four physical and chemical variables (i.e., total alkalinity, sulfate, substrate size, stream gradient) that are correlated significantly with benthic community structure are used to predict unimpacted benthic community structure.	Applies only to fauna of the United States west of the Mississippi River and to some eastern U.S. ecosystems. This is the only North American application of abiotic conditions to biotic measures that we have found.

on physical and chemical data collected at a site and CTQ (actual) is the sum of the tolerance quotients for each species/total number of species. CTQ (predicted) involves a data base of 164 collections in 20 streams. Specific-, family-, and generic-level identifications are used.		
Biotic Index (Chutter, 1972; Hilsenhoff, 1987)—Calculated as $\Sigma\ n_i t_i/N$ where n_i is the number of individuals of a genus or species, t_i is the tolerance value of that taxon, and N is the number of organisms in a sample. Specimens are identified to genus or species and numbers of each taxon are counted; tolerance values are obtained from published tables. Depending on tolerance values, different levels of identification are required. *Family Biotic Index* (Hilsenhoff, 1988a) is calculated as above, but n is the number of individuals of a family and t is the tolerance value of that family. For this index, family-level identifications and enumerations are required.	This index weights the relative abundance of each taxon in terms of its pollution tolerance in determining a community score.	For Wisconsin, Hilsenhoff (1988b) has developed seasonal correction factors for the Biotic Index. Application of tolerance values to other areas may require modifications for faunal or tolerance differences.
BMWP Score (e.g., Wright et al., 1988)—Calculated as a total ($\Sigma\ t_i$) or as average score per taxon ($\Sigma\ t_i/n$) where t_i is the tolerance score for that taxon (i.e., family) and n is the total number of families. Impact is assessed as a proportional change: score observed/score expected. Expected score is derived from a data base of 370 sites.	This index summarizes presence/absence and tolerance of families. Several physical and chemical parameters that are correlated significantly with benthic community structure are used to predict unimpacted benthic community structure.	Developed for Great Britain; modifications in tolerance values and predictive relationships may be necessary for use in other areas.
Florida Index (Ross and Jones, 1979)—Calculated as (2 × Class I taxa) + Class II taxa, in which species, genera, or families are classified into one of 5 classes, based on values assigned to categories determined from Florida surveys.	This index also weights pollution tolerance of different taxa and is based on an earlier index (Beck, 1954, 1955).	Unlike some biotic indices, this index does not use relative abundance, only richness.
Indicator organism presence/absence (Courtemanch, 1989)—Fourteen lake types in Maine have been categorized using chlorophyll-a, phosphorus, and	A consistent relationship between trophic status of lakes and chironomid assemblages permits assignment of a given lake to predetermined categories.	Trophic status of lakes may reflect levels of organic enrichment.

(continued)

Table 6.1. (Continued)

Measure and Description of Calculation	Rationale Behind Use of Measure	Comments
mean depth, with a specific chironomid assemblage associated with each lake type. Chironomids collected in profundal lake samples are identified to genus or species and a dichotomous key to lake types is used for classification.		
ISO Score (ISO, 1984)—Calculated as the sum of the tolerance scores for families present (Total ISO Score), or as the Total ISO Score divided by number of families (Mean ISO Score). Change is then measured as [(new score − previous score)/previous score] × 100. Family-level identifications are required.	Family-level identifications can be done reliably, and families differ in their pollution tolerance, particularly to organic pollution.	Family-level identification is considered a compromise between the greater information content of species identifications and the difficulties of taxonomic accuracy.
Saprobic Index (Zelinka and Marvan, 1961; DIN, 1989)—S is calculated as $\sum_{i=1}^{n} s_i a_i g_i \Big/ \sum_{i=1}^{n} a_i g_i$, where n is the number of taxa in the saprobic list, s is the saprobic valency, a is abundance (estimated as 1 of 7 abundance classes), and g is the indicator value of each taxon. DIN applicability is based on standard error (SM < 0.2) calculated as SM $\pm \sqrt{\sum_{i=1}^{n} (s_i - s)^2 a_i g_i / (n-1) \sum_{i=1}^{n} a_i g_i}$ and the sum of the abundance estimates ($\Sigma a_i \geq 15$), calculated as $\sum_{i=1}^{n} a_i$.	The saprobic valency accounts for the fact that a species is found in multiple saprobic zones; the indicator value is a weighting factor that "stresses the good indicators and suppresses the bad ones" (Sládeček, 1979, p. 3–5).	Hynes (1960), Washington (1984), Metcalfe (1989) and many others have discussed drawbacks to this index. These criticisms could apply to all biotic indices thus far proposed.

E. Functional feeding-group measures		
Ratio of shredders to total number of individuals—Samples are collected from leaf packs or other accumulations of coarse particulate organic matter (CPOM). Specimens are identified and numbers of each taxon counted; those in the shredder functional group usually are determined using the tables in Merritt and Cummins (1984). Generic-level identifications are required. Simplified keys to functional groups are given in Cummins and Wilzbach (1985). See text for further explanation.	Shredder organisms and their microbial food base are sensitive to toxicants and to modifications of the riparian zone (Plafkin et al., 1989).	Some researchers have questioned the applicability of functional designations when applied at the generic level, among ages of a specific taxon, in different regions, etc. Indices clearly depend on accurate designation to functional feeding groups, which is usually dependent on accurate identifications.
Ratio of scrapers to collector-filterers—Specimens are identified and numbers of each taxon counted; those in the scraper and collector-filterer functional groups are determined as above.	These functional groups reflect available food resources. Dominance of collector-filterers may reflect organic enrichment.	Some researchers have questioned the applicability of functional designations when applied at the generic level, among ages of a specific taxon, in different regions, etc. Indices clearly depend on accurate designation of functional feeding groups, which is usually dependent on accurate identifications. Some scraper guilds may be composed of pollution-tolerant taxa (e.g., physid snails; M. Barbour, EA Engineering, Science, and Technology, Inc., Sparks, MD, personal communication).
Ratio of trophic specialist to generalist—Within a functional feeding group, individual taxa are classified as either specialists (restricted to a specific food source) or generalists (able to exploit a broader range of food sources) (Maine Department of Environmental Protection, 1987).	This is based on the assumption that trophic generalists are more pollution-tolerant and thus become numerically dominant in response to environmental stress.	It is difficult to ascertain specialist-generalist categories.
F. Combination indices		
Invertebrate Community Index (Ohio EPA, 1987)—Calculated as the sum of 10 individual measures that are scored individually, based on background data from 232 Ohio reference sites covering five ecoregions. The 10 measures include: (1) total number of taxa	Inclusion of 10 measurements minimizes drawbacks of depending on only one measure. Measurements 1–9 are based on artificial substrate samples, whereas 10 is based on a qualitative sample. All individual measurements reflect rationales provided above regarding pollution-sensitivity and tolerance.	As with other indices based on geographically limited databases, application of scoring criteria outside of these five ecoregions or with different collecting techniques would be limited without recalibration of the indices.

(continued)

Table 6.1. (Continued)

Measure and Description of Calculation	Rationale Behind Use of Measure	Comments
(2) total number of Ephemeroptera taxa (3) total number of Trichoptera taxa (4) total number of Diptera taxa (5) percent of Ephemeroptera (6) percent of Trichoptera (7) percent of the tribe Tanytarsini of the Chironomidae (8) percent of other dipterans and noninsects (9) percent of tolerant organisms (from a specified list) (10) total number of qualitative Ephemeroptera, Plecoptera, and Trichoptera taxa. Measures 1–4, 6, and 8–10 are scored according to size of drainage area.		
Mean Biometric Score (Shackleford, 1988)—Calculated as the sum of seven individual measures that are scored individually based on background data from reference sites. The seven measures include: (1) dominance in common (= the number of the five most abundant taxa common to both reference and study sites) (2) common taxa index (= the number of taxa common to both reference and study sites/maximum number of taxa found at reference or study site)	This score is a combination of community diversity, indicator organism, and functional group approaches. Individual measurements reflect rationales provided above.	Calculation of this score appears to require presence of comparable reference sites.

(3) quantitative similarity index (= Σ minimum p_{ia}, p_{ib}, where a is the reference site and b is the study site and p_{ia}, p_{ib} is the relative abundance at the reference and study sites, respectively, of the different species) (4) taxa richness as percent change between reference and study sites (5) indicator assembly index = $0.05(\%EPT_b/\%EPT_a + \%CA_a/\%CA_b)$, where %EPT is relative abundance of Ephemeroptera, Plecoptera, and Trichoptera; %CA is relative abundance of Chironomidae and Annelida; subscript a is the reference site and subscript b is the study site (6) missing taxa (= the number of EPT genera present at the reference site but absent at the study site) (7) functional group percent similarity [= as in (3) above but uses relative abundance of functional groups]		
Biological Condition Score (Plafkin et al., 1989)—Calculated from summing the scores given to each of eight "metrics." (1) taxa richness (2) biotic index (3) ratio of scrapers to collector-filterers (4) ratio of EPT to chironomid abundance (5) percent contribution of dominant taxon (6) number of EPT taxa (7) Coefficient of Community Loss (8) ratio of shredders to total number of individuals	Each "metric," which reflects group tolerances, community structure, or community function, is individually evaluated in terms of percent similarity with reference sites.	Level of taxonomy varies with the different protocols (see Appendix 6.1).

Community similarity indices are used to compare community structure in space (e.g., among different sites) or over time (e.g., from year to year). Similar levels of taxonomic discernment among the communities being comared is implicit in their use. Some community similarity indices stress richness (e.g., Jaccard Index) or both richness and abundance (e.g., Pinkham-Pearson Index). Diversity and community similarity indices are discussed at length by Norris and Georges in Chapter 7.

Biotic indices use preestablished water-quality tolerance values for taxa (families, genera, or species) that have been collected and identified. The relative abundances of taxa, weighted by tolerance values, sometimes may be included in the calculation of a biotic index. Washington (1984) and Metcalfe (1989) described the history and development of different biotic indices. Some indices have been modified to form other indices. For example, in the United Kingdom over a 15-year period, the Trent Biotic Index led to Chandler's Score, which led to the Biological Monitoring Working Party (BMWP) Score, which led to the Modified BMWP Score (Metcalfe, 1989, fig. 1). In other cases, indices have required little modification. For example, the biotic index developed in South Africa (Chutter, 1972) was modified for Wisconsin, U.S.A., simply by changing tolerance values for the local fauna and excluding selected invertebrate taxa (Hilsenhoff, 1987, 1988a).

Functional feeding group measures are community measures that are based on the morphological structures and behaviors responsible for food acquisition by given species at a site. Apparently, some discrepancy exists as to how functional group designations currently are made and how they were intended to be made. Functional groups, as currently used in the "ecological data tables" of Merritt and Cummins (1984), reflect trophic levels (i.e., herbivores, detritivores, carnivores) and are based on digestive tract analysis (Cummins, 1988). However, Cummins (1988) recently stated that the intent was to base functional-group distinctions on mouthpart morphology and means of food acquisition, such as described by Cummins and Wilzbach (1985). Clearly, problems can arise when these designations are used interchangeably. If the functional feeding group concept is to be applied to biomonitoring, then this distinction must be considered (Cummins, 1988) and perhaps further clarified.

6.4. Rapid Assessment Protocols in Current Use

Information on rapid assessment protocols was obtained primarily by contacting researchers involved in the use of these procedures. Although a literature search was conducted (see Marshall, Chapter 3), the majority of protocols appeared as agency and commission reports that were not generally located by conventional searching techniques. Scores of approaches and modifications have been proposed by provincial and state government agencies in North America and water authorities and regulatory associations in Europe (Appendix 6.1). A range of approaches used is presented here, rather than a comprehensive listing. Certainly, new approaches will appear and be used, and some of those presented in Appendix 6.1 likely will be modified or even abandoned. In any event, this is the most rapidly changing topic in contemporary biomonitoring.

Thirty rapid assessment protocols are presented in Appendix 6.1. Eighty percent of these apply to streams, 10% to lakes, and 10% to both streams and lakes. The bias toward lotic protocols has several probable explanations: most of these approaches have been developed in the last decade, a period when benthic research in streams increased and surpassed benthic research in lakes (Resh and Rosenberg, 1989, fig. 1); streams are logistically easier to sample than lakes; most regions in which these protocols are being developed have greater areas of potentially impacted stream habitat than lake habitat; and perhaps as a result of enforcement of discharge requirements, lentic pollution may be perceived as less of a problem than lotic pollution. Undoubtedly, other reasons also could be offered.

The protocols listed in Appendix 6.1 are described in terms of field procedures used (sampling device and habitat sampled), sample processing (proportion examined, taxonomic level of identifications), and analyses (measures examined). A discussion of each of these follows.

6.4.1. Sampling Devices

A variety of sampling devices is recommended for use but kick nets (or similar devices such as D- and A-frame nets) are by far the most common ones. An obvious advantage of kick nets is that they are inexpensive, portable, and can be used in a diversity of shallow habitats. Mesh sizes for these nets tend to be coarse; however, because rapid assessment approaches generally do not use absolute densities, and small specimens are difficult to identify anyway, this is not considered a problem. Some protocols suggest the use of the Surber sampler; in these protocols, choice of a Hess or other box-type samplers may be a preferable alternative (Voshell et al., 1989).

The sampling devices listed above are of little use in medium- to large-sized streams or rivers where water depths often exceed 1 m; thus, deep-water habitats often are excluded from monitoring programs. Air-lift samplers (e.g., Drake and Elliott, 1983), suction samplers (e.g., Gale and Thompson, 1975), dredges, grab samplers, or representative artificial substrates (Rosenberg and Resh, 1982) may be useful in applying rapid assessment approaches to deep-water habitats.

In rapid assessment protocols, if a sampling device is used to collect more than one sample unit it is to maximize the diversity of subhabitats sampled rather than to take replicate sample units for statistical analyses of hypotheses (see Resh and McElravy, Chapter 5, Norris and Georges, Chapter 7, and Cooper and Barmuta, Chapter 11, for further discussion of this latter approach). However, additional sample units sometimes are collected to obtain a required threshold number of organisms for later analysis.

6.4.2. *Habitat*

The protocols usually specify that all habitats should be sampled, or they specify that only riffle or run habitats should be sampled. Riffle/run areas probably are suggested because of high taxa richness and abundance that occur there, and because this approach provides some habitat stratification and therefore reduced sample variance for intersite comparisons. Little attempt to stratify beyond riffles is apparent (but see Kelso, 1987, and Lenat, 1988, in Appendix 6.1).

In Chapter 5, Resh and McElravy suggested that hydraulic stress, measured using the hemispheres of Statzner and Müller (1989), may be used for habitat stratification and to reduce intersite variability in lotic systems. This approach is equally valid in rapid assessment approaches.

6.4.3. *Proportion Examined*

Most protocols require that the entire sample be examined; however, at least some tendency toward subsampling is apparent. Some protocols have established predetermined "counts" (e.g., 50, 100, 250, 300 individuals), and still others specify that proportions (e.g., 3/8) of the whole sample should be examined. The adequacy of different "counts" has been examined by Hilsenhoff (1977, 1982), who found that 100 individuals is a satisfactory subsample. Of course, bias toward large, mobile, or colorful specimens must be avoided. One lake protocol that is based on lower counts (e.g., 50 individuals; Courtemanch, 1989) may be related to the lower taxonomic richness of Chironomidae in lakes and the time required to prepare specimens for identification. Examination of low numbers of specimens may be more appropriate for lakes than streams because of the reduced richness and patchiness of lentic systems.

6.4.4. *Taxonomic Level*

Most protocols have generic/species level identifications as their goal for analysis, but others specify the family level as sufficient. In some protocols, only taxa of Ephemeroptera, Plecoptera, and Trichoptera (EPT) are identified; in others, only taxa with an "indicator value" are identified, or taxa are separated into "species groups" but no identifications are actually done.

The choice of taxonomic level represents a compromise between the desire for increased information content and the resulting usefulness of species-level identification, and the cost (in time and required expertise) of obtaining it (see Resh and McElravy, Chapter 5, for a detailed discussion of this topic).

6.4.5. Measures Used

The protocols presented in Appendix 6.1 involve calculation of from one to 12 measures, but about one-half of the protocols involve only one or two measures. Over 85% of the protocols include some measure based on established environmental tolerances of organisms, such as a biotic index or numbers of "tolerant" species. A measure of taxa richness (sometimes as EPT richness or number of families) and/or some form of enumeration were included in 67% of the protocols. Calculation of diversity and similarity indices were part of 26% of the protocols. Functional feeding group measures were the least commonly used (<25%).

When only one or two measures are calculated, impact is assessed directly from them, but when multiple measures are used impact is either evaluated in terms of each individual measure, or the individual measures are combined into a compound measure that is used in the evaluation. To assess whether a site is impacted, about two-thirds of the protocols use preestablished, regionwide criteria from previous studies, and about one-third use background information from that site or a nearby reference site for comparison.

One aspect of rapid assessment approaches that is not clear from the descriptions is the frequency of assessing a particular site (yearly, semiannually) or the timing of sampling (see discussion of seasonality below). Plafkin et al. (1989) recommended specific times as best for different climatological regions, and Cummins (1988) has discussed seasonality in terms of the use of functional group analysis.

6.5. Comparison with Quantitative Assessments

Several trends are obvious if the 30 rapid assessment protocols in Appendix 6.1 are compared with quantitative approaches discussed by Resh and McElravy (Chapter 5) in terms of sampling devices, habitats sampled, proportion of sample examined, level of taxonomy, and measures used. First, kick nets are less commonly used in quantitative assessments than they are in rapid assessment protocols. Second, all habitats are sampled in many of the rapid assessment protocols, but in few quantitative assessments. Third, in both quantitative and rapid assessment approaches, most of the samples are examined in their entirety; however, the tendency for specified numbers (i.e., a subsample) to be examined in some rapid assessment protocols is not at all apparent in the quantitative studies. Fourth, in almost all the quantitative studies, the majority of identifications in each order were to genus and species, whereas some reliance on family-level identifications occurs in the rapid assessment protocols. Fifth, rapid assessment approaches rely heavily on use of some biotic index (over 80% of the protocols), whereas quantitative studies do not (only 3% of total). Sixth, and perhaps most fundamental, statistical tests are used in the majority of quantitative studies to assess whether impact has occurred, whereas such tests are generally absent in the rapid assessment protocols.

6.6. Determining the Accuracy of Rapid Bioassessment Measures

Many measures have been proposed for use in water quality monitoring programs over the past several decades, and many new measures have been developed specifically for rapid assessment approaches. But how accurate are these measures? Accuracy, the ability of a measure to reflect actual field conditions, can take two forms: the measure indicates that impact has occurred when, in fact, it has occurred (avoidance of Type I error); or, the measure does not indicate impact when, in fact, it has not occurred (avoidance of Type II error). (For further discussion of Type I and II errors in benthic biomonitoring studies, see Norris and Georges, Chapter 7, and Cooper and Barmuta, Chapter 11.)

In this section, several of these measures are analyzed in terms of their accuracy in detecting impacted and unimpacted conditions in several northern California coastal streams. A similar approach also could be used to ascertain the applicability of rapid assessment measures in other regions.

6.6.1. Sites Examined

Data from two streams in the California Coast Range were used to determine which rapid assessment measures could be used to detect impact when a known impact had occurred (i.e., avoiding Type I error). The first data set consisted of benthic macroinvertebrate samples collected annually in April, 1984–1986, from Knoxville Creek, a second-order stream in Napa County, California. In 1986, an accidental overflow from a treatment pond occurred in this stream, which lowered pH (normally 7.5–7.7) to 2.5 for several days in February and to 3.5 for several days in March. Several potentially toxic metals exceeded permit thresholds (As: 1.4x, Cr: 2.2x, Cu: 1.9x, Hg: 34x). The absence of measurable precipitation had caused the stream to become intermittent during the previous summer (a common condition in the Mediterranean climate of this area, with wet winters and dry summers), so the stream was additionally stressed. Benthic data were available for two pre-impact years and one post-impact year, and were collected in riffle habitats using a Surber sampler.

The second data set was from Little Geysers Creek, a second-order stream in The Geysers Known Geothermal Resource Area in Sonoma County, California. This stream receives chronic thermal input from naturally occurring geothermal springs (Lamberti and Resh, 1985). Annual temperature difference between thermally influenced and normal sites is 9°C. Benthic macroinvertebrate data were obtained from rock-filled baskets that were placed in the thermally impacted and unimpacted portions of this stream (Resh et al., 1984).

In the two examples above, the acid spill represented a short-term impact that was examined temporally; samples were collected about two months after the overflow occurred. In contrast, the thermal comparison was a chronic impact that was examined spatially; impact was occurring at the time of sampling. Furthermore, acidic conditions at Knoxville Creek were not visible, whereas thermal effects at Little Geysers Creek were apparent year round.

Accuracy in terms of a measure not indicating impact when impact did not occur (i.e., avoiding Type II error) was determined by examining two second-order sites at another stream in the California Coast Range (Hunting Creek, Napa County, California) over a four-year period (1984–1987), and by examining two years of pre-impact data at Knoxville Creek (1984–1985). As under impacted conditions, these temporal comparisons were based on Surber sampler collections, restricted to riffles, that were made in mid-April of each year.

Two other analyses were performed at the unimpacted sites. Seasonal comparisons were made at the end of the wet season (April) and at the end of the dry season (August) at a second-order site in Hunting Creek over a three-year period (1985–1987). Spatial comparisons were made between the two second-order Hunting Creek sites over a four-year period (1984–1987).

6.6.2. Determining Accuracy

Two steps were used to examine the accuracy of rapid assessment measures. First, to ascertain whether a measure correctly reflected actual conditions, the percent similarity for that measure was calculated using mean values for the impact and control years or sites (where low similarity would be expected) or for unimpacted years or sites (where high similarity would be expected). Percent similarity was used because it probably will become the evaluation approach adopted for many rapid assessment measures in the United States (Plafkin et al., 1989). Comparisons of percent similarities for the various measures in the impacted year (or site) with values of the unimpacted years (or sites), or between unimpacted years, were calculated either as impact/control × 100 or control/impact × 100; the choice depended on the expected direction of change in the value of the measure when impact occurred (e.g., taxa richness would decrease with impact). Thus, a percent similarity value of <100% in, for example, taxa richness would be expected with impact; a value >100% would be expected if the estimate of taxa richness for the impact year exceeded that for the control year.

As the second step, analysis of variance or t-tests were used to determine whether the change in actual values was statistically significant. Enumerations were log transformed; percentages were arc-sine transformed. Each sample unit, within a site or a year, served as a replicate for statistical analysis; if $P \leq 0.003$ (a conservative correction for procedure-wise error), a Student-Newman-Keuls test was done to elucidate which differences were significant.

6.6.3. *Accuracy of Measures*

Two trends were apparent when the accuracy of each of the measures was examined in terms of the percent similarity of the compared measures and the statistical significance of the differences between the compared measures (Table 6.2). First, more of the rapid assessment measures indicated impact in the chronically thermally impacted stream than in the acutely acid-impacted stream (Table 6.2, Columns I and II). Second, statistically significant differences (i.e., indications of impact based on ANOVA or t-tests) in the absence of impact (inaccurate conclusions, Type II error) were more common for the seasonal temporal comparison within a site (Column VI) and the spatial comparison between two second-order sites (Column VII) than in the annual temporal comparisons within a site (Table 6.2, Columns III–V). This trend illustrates the importance of seasonal differences and of carefully selecting reference sites for comparison.

For *richness measures,* impact was consistently indicated by low percent similarities when sites or years were compared at the impacted sites (Table 6.2A, Columns I and II); likewise, statistically significant differences in impacted sites or years were detected with all richness measures. Statistically significant differences in the absence of impact were found only for number of EPT taxa in the spatial comparison (Table 6.2A, Column VII). Percent similarities of 70% to 80% in richness measures seem to include the variation found for these measures in the unimpacted sites; richness similarities were generally <60% in the two impacted conditions examined. Clearly, richness measures were quite accurate in both detecting impact when it occurred and not indicating impact when it did not occur.

Of the *community indices,* the reduction in similarity for all indices was statistically significant in response to chronic thermal stress (Table 6.2B, Column II), but only Margalef's Index provided a statistically significant reduction in similarity in response to the acid impact (Table 6.2B, Column I). Except for seasonal comparisons, unimpacted years or sites had high similarities (Table 6.2B, Columns III–VII), but these overlapped with values observed under impacted conditions for Simpson's and Shannon's Index (Table 6.2B, Column I).

The *Family Biotic Index* showed lower similarities under impacted conditions (<80%; Table 6.2C, Columns I and II) than under unimpacted conditions (>90%; Table 6.2B, Columns III–VII), and reductions under impacted conditions were statistically significant.

For *enumerations,* number of individuals can increase or decrease under impacted conditions, so both possibilities were considered in the analysis (Table 6.2D, superscripts a,b). Percent similarities for this measure were highly variable and overlapped under impacted and unimpacted conditions. Percent similarities of number of EPT individuals and the ratio of EPT to total number of individuals (% EPT) were low, but the decrease was only statistically significant in the thermally impacted stream (Column II). However, the values for the acid-impacted stream (Column I) were similar to percent similarity values observed in the annual comparison of unimpacted conditions in Column III. Overlap in percent similarity values also was found for ratio of Chironomidae to total number of individuals (% Chironomidae) and ratio of individuals in numerically dominant family to total number of individuals (% dominant family). Statistically significant differences in ratio of Chironomidae individuals to EPT individuals were detected only at unimpacted sites (Columns III and VI), whereas ratio of individuals in the numerically dominant taxa to total number of individuals (% dominant taxa) was significantly different at both impacted and unimpacted sites. Therefore, these enumerations are not consistently accurate enough to be used effectively in rapid assessment approaches.

Of the *functional measures* used (calculated as usually done using trophic designations in Merritt and Cummins, 1984, but remember Cummins' 1988 admonition mentioned above), ratio of scrapers to total number of individuals (% scrapers) showed reduced percent similarity and statistically significant differences when impact occurred (Table 6.2E, Columns I and II), but were not reduced in the spatial and temporal comparisons where impact did not occur (Columns III–VII). For this measure, impacted conditions had <60% similarity, whereas unimpacted conditions had >69% similarity. The other three functional measures showed overlap in percent similarity under impacted and unimpacted conditions. Although ratio of shredders to total number of individuals (% shredders) should be assessed by using a leaf pack or coarse particulate organic matter (CPOM) sample (which was not done for this analysis), it should be noted that the streams studied here have much higher amounts of periphyton than CPOM, which may explain why ratio of

Table 6.2. Numbers represent percent similarities calculated as either impact/control[(a)] or control/impact[(b)] × 100, depending on measure. A value could be >100% if the value for the analysis of impact year exceeded that of the control (or baseline) year. The asterisk (*) represents statistical significance of change (ANOVA or t-tests with $\alpha \le 0.003$ to maintain procedure-wise error = 0.05); see text for further explanation. EPT = Ephemeroptera, Plecoptera, Trichoptera.

	IMPACT		NO IMPACT				
	I Temporal comparison (annual): first-order acid-impacted stream	II Spatial comparison: second-order thermally impacted stream	III Temporal comparison (annual): second-order stream	IV Temporal comparison (annual): second-order stream (downstream of III)	V Temporal comparison (annual): first-order stream	VI Temporal comparison (seasonal): second-order stream	VII Spatial comparison: two second-order sites
A. *Richness measures*							
Number of taxa[a]	67*	33*	91	84	> 100	72*	> 100
Number of EPT taxa[a]	33*	14*	91	70	> 100	69	>100*
Number of combined taxa[a] (generic distinctions, Chironomidae at family level)	42*	30*	93	72	91	94	> 100
Number of families[a]	45*	37*	91	77	85	89	> 100
B. *Community indices*							
Margalef's Index[a]	61*	28*	83	83	> 100	61*	> 100
Menhinick's Index[a]	61	31*	73	83	> 100	47*	99
Simpson's Index[a]	> 100	12*	87	> 100	> 100	71*	96
Shannon's Index[a]	97	15*	76	88	> 100	63*	> 100*
C. *Biotic Index*							
Family Biotic Index[b]	79*	75*	93	93	> 100	> 100*	> 100*

D. *Enumerations*							
Number of individuals[a,b]	a: >100 b: 70	> 100 73	> 100 64	99 > 100	69 > 100	> 100* 38*	> 100 79
Number of EPT individuals[a]	63	1*	49	> 100	> 100	> 100*	> 100*
Ratio of EPT to total number of individuals[a]	59	1*	43	> 100	> 100	> 100*	> 100*
Ratio of Chironomidae individuals to total number of individuals[b]	68	50*	63*	99	> 100	> 100*	> 100*
Ratio of individuals in numerically dominant family to total number of individuals[b]	90	60*	70*	100	> 100	> 100	> 100*
Ratio of Chironomidae individuals to EPT individuals[b]	45	5	19*	71	> 100	> 100*	91
Ratio of individuals in numerically dominant taxa to total number of individuals[a,b]	a: >100* b: >100	> 100* > 100*	43* 88	59 > 100	97 > 100	> 100* > 100*	29* > 100*
E. *Functional measures*							
Ratio of scrapers to total number of individuals[a]	57*	8*	76	> 100	> 100	69	> 100*
Ratio of collector-filterers to total number of individuals[b]	31	> 100	> 100	> 100	> 100*	> 100*	31*
Ratio of shredders to total number of individuals[a]	> 100	2*	18	59	87	> 100*	46
Ratio of scrapers to collector-filterers[a]	> 100	3	65	> 100	> 100	> 100	88

scrapers to total number of individuals was the most accurate of the functional measures.

In summary, the responses of seven measures (all four richness measures, Margalef's Index, the Family Biotic Index, and ratio of scrapers to total number of individuals) to both acid (acute) and thermal (chronic) impacts were accurate and statistically significant (Table 6.2, Columns I and II). Percent similarity of these seven measures was high in the unimpacted comparisons (Table 6.2, Columns III–V, VII; see discussion below for seasonal comparisons in Column VI) and statistically significant differences were not detected. Thus, the responses of these measures appear to be accurate representations of conditions in these streams. It should be noted, however, that seasonal differences were found for almost all of these measures (Table 6.2, Column VI); obviously, caution should be used if data from different seasons are compared. The other measures in Table 6.2 were not accurate in these streams because they did not detect impact, the values obtained overlapped under impacted and/or unimpacted conditions, or they were only able to detect chronic thermal impact. The approach used above also may be applicable in choosing appropriate measures, and setting criteria for reductions in percent similarity with impact, in other geographic areas or ecoregions.

6.7. Characteristics of an Accurate Rapid Assessment Protocol

The goal of rapid assessment protocols is a cost-effective approach to biomonitoring. However, accuracy remains the primary consideration; therefore, the question is how to design a rapid assessment protocol that is both accurate and cost-effective. Two important considerations should be mentioned immediately: (1) seasonal differences confound accuracy (Table 6.2, Column VI), and (2) widespread geographic applicability is difficult. The limitations of widespread geographic applicability are evident in terms of expected values of taxa richness and diversity indices, tolerance values used in a biotic index (especially at levels below family), and the nature of habitats present (e.g., autochthonous systems, in which scrapers may be most useful as measures of impact, and allochthonous systems where shredders may be most useful). Given these limitations, what features would best comprise a rapid assessment protocol for the Mediterranean-climate, northern California streams used in this analysis (Table 6.2)?

6.7.1. Protocol Choices

The sampling device used in rapid assessment protocols should provide a representative collection of taxa from the habitat, or from the variety of habitats, sampled. The various dip and kick nets, and the other netted samplers listed in Appendix 6.1, would appear to fulfill this goal. However, depending on habitat chosen, the following samplers would be best: in shallow riffles, kick nets or a Hess or other box-type sampler; in deeper (>1 m) areas of rivers, an airlift sampler or Ponar grab. As will be mentioned below, the need for improved sampling devices is still a relevant topic of research.

Two alternatives are acceptable in choosing a sampling habitat: either limit sampling to riffles or sample subhabitats in a reach according to their relative abundances. Riffle-only (or other limited-habitat) sampling approaches are a problem because the subhabitat chosen may not be affected by an impact, and the sensitivity of organisms in different subhabitats can vary. Sampling each subhabitat proportionately provides better characterization of a site, but it makes comparison with reference sites more difficult because of intersite differences that will always exist. Choice of habitats in rapid assessments of lakes could follow the same principle: limit sampling to a specific habitat (e.g., wave-swept shorelines, littoral areas, or profundal zones) or use a proportional combination of representative subhabitats.

The proportion of a sample unit examined may be one way to achieve cost reduction. Early-spring macroinvertebrate densities in the coastal California streams examined here are typically about 400 individuals/Surber sample (0.93 m^2). Examination of the first 100 to 200 organisms encountered may be sufficient. Whatever proportion of the sample unit is selected for analysis, it is important that the proportion examined be *unbiased*.

The taxonomic level of identification used in rapid assessment protocols clearly depends on the measure chosen. This topic is discussed in detail by Resh and McElravy in Chapter 5. For richness measures, identification to family, identification to a combination of genus for some groups and family for others, and identification to species were all accurate in the examples presented. Species-level discriminations (as EPT richness, for example) require a great deal of taxonomic expertise (which is costly) and more time (which is also costly). Also, finer taxonomic resolution, such as to species level, can increase the chances for misidentification, which would be critical if species-based tolerance values were being used. Simple discrimination rather than identification of taxa will provide an estimate of richness, but the data obtained cannot be used for measures that require identifications (e.g., biotic indices). It may be useful to establish different identification levels for different groups (as is done in the Belgian Biotic Index and to a certain extent in the scoring for some applications of the Biotic Index).

6.7.2. *Choice of Measures to Be Examined*

The analysis of several northern California coastal streams revealed that a variety of measures were effective: all richness measures, Margalef's Index, the Family Biotic Index, and the ratio of scrapers to total number of individuals (Table 6.2). Conversely, the other three community indices, all enumerations, and the other three functional group analyses did not accurately depict either impacted or unimpacted conditions. So, what are the implications for choosing measures for rapid assessment protocols?

To begin with, the approach in which multiple measures are being used (individually or as combined indices) seems preferable to reliance on a single measure, such as a biotic index score or a single richness measure. Given this, what components would be included in an ideal protocol for monitoring acid and thermal impacts in California coastal streams? First, richness would be chosen as a measure of community structure; species-level taxa richness would be preferred but family richness would seem to suffice. Second, a measure weighted for pollution tolerances, such as the Biotic Index with regionally modified tolerance scores, would be chosen. An eventual goal would be to refine its tolerance values at lower taxonomic levels (hence species-level identifications obtained above would be useful). Third, the ratio of scrapers to total numbers of individuals would be chosen as a functional measure. Perhaps in streams where CPOM is more abundant, the ratio of shredders to total numbers of individuals would be an effective alternative measure; in any event, measures using ratios between different functional groups would be avoided. In choosing functional groups, remember that additional examination of specimens (if used *sensu* Cummins 1988) or generic-level identifications (if used traditionally) will be required in incorporating these measures.

A final consideration in this ideal protocol is that, in Table 6.2, differences were always more evident when two different sites (Column VII) or different seasons at a site (Column VI) were compared than when annual differences observed at a single site were compared (Columns III–V). In future multiple-site analyses within a region, some application resembling the Before-After-Control-Impact (BACI) approach (Stewart-Oaten et al., 1986; see discussion in Cooper and Barmuta, Chapter 11), in which temporal and spatial changes are incorporated into the analysis, also may be appropriate.

6.8. What Is Needed in Rapid Assessment Approaches?

First, in general, rapid assessment protocols have been developed to apply to small, shallow streams. Protocols for larger streams and lakes, and sampling devices for certain types of habitats (especially those containing large boulders) in these systems need to be developed.

Second, tolerance scores that have been based traditionally on organic pollution need to be obtained for other types of pollution, and especially for multiple inputs (see review of this latter topic by Metcalfe, 1989).

More sophisticated methods of assigning scores also are needed; saprobic valency (Sládeček, 1979) seems to be a useful approach to this problem. In addition, scoring systems should attempt to indicate impact on a linear scale. Currently, a score of two is not twice as "good" or "bad" as a score of four; this seems to be a major problem with most scoring systems. Geographical variations in tolerance also must be considered when developing scores for different regions of North America.

Third, both measures and protocols need to be calibrated. For example, the analysis presented here only addressed two types of impact in northern California coastal streams; the results may not be applicable to other types of impact or to streams in other regions, even in California. Continued international or interagency cooperation in developing and calibrating protocols, and establishing ecoregion-based tolerances and background data (Hughes, 1989) is essential.

Fourth, legislative regulations involving rapid assessment protocols and assessments of impacts using them should be delayed until the influence of factors such as natural variability is better known or measures and protocols are calibrated.

Fifth, incorporation of caveats in the protocols may be useful. For example, extremely low densities of organisms (which would not be apparent when analysis is based on counts of a preestablished number of specimens) should be readily identifiable as a sign of impact.

Finally, guidelines must be developed for determining whether quantitative or qualitative approaches are to be used. Ultimately, this choice will depend on the purpose of the study and the sensitivity required.

Acknowledgments

We thank our colleagues listed in Appendix 6.1 for providing us with rapid assessment protocols and explanations of their use. R. Norris, B. Statzner, and M. Barbour provided valuable comments on this manuscript.

References

Beck, W.M., Jr. 1954. Studies in stream pollution biology. I. A simplified ecological classification of organisms. *Quarterly Journal of the Florida Academy of Sciences* 17:211–27.

Beck, W.M., Jr. 1955. Suggested method for reporting biotic data. *Sewage and Industrial Wastes* 27:1193–7.

Bode, R.W. 1988. *Quality Assurance Work Plan for Biological Stream Monitoring in New York State*. Stream Biomonitoring Unit, Bureau of Monitoring and Assessment, Division of Water, New York State Department of Environmental Conservation, Albany, NY.

Chutter, F.M. 1972. An empirical biotic index of the quality of water in South African streams and rivers. *Water Research* 6:19–30.

Courtemanch, D.L. 1989. *Trophic Classification of Maine Lakes Using Benthic Chironomid Fauna*. Unpublished ms.

Courtemanch, D.L. and S.P. Davies. 1987. A coefficient of community loss to assess detrimental change in aquatic communities. *Water Research* 21:217–22.

Cummins, K.W. 1988. Rapid bioassessment using functional analysis of running water invertebrates. In *Proceedings of the First National Workshop on Biological Criteria,* eds. T.P. Simon, L.L. Holst, and L.J. Shepard, pp. 49–54. EPA–905/9–89/003. U.S. Environmental Protection Agency, Chicago, IL.

Cummins, K.W. and M.A. Wilzbach. 1985. *Field Procedures for Analysis of Functional Feeding Groups of Stream Macroinvertebrates*. Contribution 1611, Appalachian Environmental Laboratory, Univ. of Maryland, Frostburg, MD.

Day, W.C. 1956. Ephemeroptera. In *Aquatic Insects of California,* ed. R.L. Usinger, pp. 79–105. Univ. of California Press, Berkeley, CA.

De Pauw, N. and G. Vanhooren. 1983. Method for biological quality assessment of watercourses in Belgium. *Hydrobiologia* 100:153–68.

DIN (Deutsches Institut für Normung e.V.). 1989. Entwurf. Biologisch-ökologische Gewässeruntersuchung (Gruppe M). Bestimmung des Saprobienindex (M 2). In *Deutsche Einheitsverfahren zur Wasser-, Abwasser- und Schlammuntersuchung*. Verlag Chemie, Weinheim, F.R.G.

Drake, C.M. and J.M. Elliott. 1983. A new quantitative air-lift sampler for collecting macroinvertebrates on stony bottoms in deep rivers. *Freshwater Biology* 13:545–59.

Edmunds, G.F., S.L. Jensen, and L. Berner. 1976. *The Mayflies of North and Central America*. Univ. of Minnesota Press, Minneapolis, MN.

Gale, W.F. and J.D. Thompson. 1975. A suction sampler for quantitatively sampling benthos on rocky substrates in rivers. *Transactions of the American Fisheries Society* 104:398–405.

Hilsenhoff, W.L. 1977. *Use of Arthropods to Evaluate Water Quality of Streams*. Technical Bulletin No. 100. Wisconsin Department of Natural Resources, Madison, WI.

Hilsenhoff, W.L. 1982. *Using a Biotic Index to Evaluate Water Quality in Streams*. Technical Bulletin No. 132. Wisconsin Department of Natural Resources, Madison, WI.

Hilsenhoff, W.L. 1987. An improved biotic index of organic stream pollution. *Great Lakes Entomologist* 20:31–9.

Hilsenhoff, W.L. 1988a. Rapid field assessment of organic pollution with a family-level biotic index. *Journal of the North American Benthological Society* 7:65–8.

Hilsenhoff, W.L. 1988b. Seasonal correction factors for the Biotic Index. *Great Lakes Entomologist* 21:9–13.

Hughes, R.M. 1989. Ecoregional biological criteria. In *Water Quality Standards for the 21st Century,* pp. 147–51. Office of Water, U.S. Environmental Protection Agency, Washington, DC.

Hynes, H.B.N. 1960. *The Biology of Polluted Waters*. Liverpool Univ. Press, Liverpool, England.

ISO (International Organization for Standardization). 1984. *Water Quality—Assessment of the Water and Habital Quality of Rivers by a Micro-Invertebrate "score."* ISO/TC 147/SC 5/WG 6 N 40, Draft Proposal ISO/DP 8689.

Jaccard, P. 1912. The distribution of flora in the alpine zone. *New Phytologist* 11:37–50

Kelso, J.R.M. 1987. *Long-Term Monitoring of the Response by the Biota of Lakes and Rivers to Acidification: Constituents of a D.F.O. Monitoring Program.* Unpublished ms. Department of Fisheries and Oceans, Ottawa, ON.

Lamberti, G.A. and V.H. Resh. 1985. Distribution of benthic algae and macroinvertebrates along a thermal stream gradient. *Hydrobiologia* 128:13–21.

Lenat, D.R. 1988. Water quality assessment of streams using a qualitative collection method for benthic macroinvertebrates. *Journal of the North American Benthological Society* 7:222–33.

Maine Department of Environmental Protection. 1987. *Methods for Biological Sampling and Analysis of Maine's Waters.* Maine Department of Environmental Protection, Augusta, ME.

Mason, W.T., Jr. 1979. A rapid procedure for assessment of surface mining impacts to aquatic life. In *Coal Conference and Expo V. Proceedings of a Symposium, Louisville, KY, October 23–25, 1979,* pp. 310–23. McGraw-Hill, New York.

Merritt, R.W. and K.W. Cummins, eds. 1984. *An Introduction to the Aquatic Insects of North America,* 2nd ed. Kendall/Hunt Publishing, Dubuque, IA.

Metcalfe, J.L. 1989. Biological water quality assessment of running waters based on macroinvertebrate communities: history and present status in Europe. *Environmental Pollution* 60:101–39.

Ohio EPA (Environmental Protection Agency). 1987. *Biological Criteria for the Protection of Aquatic Life. Vols. I-III.* Surface Water Section, Division of Water Quality Monitoring and Assessment, Ohio Environmental Protection Agency, Columbus, OH.

Pinkham, C.F.A. and J.G. Pearson. 1976. Applications of a new coefficient of similarity to pollution surveys. *Journal of the Water Pollution Control Federation* 48:717–23.

Plafkin, J.L., M.T. Barbour, K.D. Porter, S.K. Gross, and R.M. Hughes. 1989. *Rapid Bioassessment Protocols for Use in Streams and Rivers. Benthic Macroinvertebrates and Fish.* EPA/444/4–89/001. Office of Water Regulations and Standards, U.S. Environmental Protection Agency, Washington, DC.

Raddum, G.G., A. Fjellheim, and T. Hesthagen. 1988. Monitoring of acidification by the use of aquatic organisms. *Internationale Vereinigung für Theoretische und Angewandte Limnologie Verhandlungen* 23:2291–7.

Resh, V.H. and G. Grodhaus. 1983. Aquatic insects in urban environments. In *Urban Entomology: Interdisciplinary Perspectives,* eds. G.W. Frankie and C.S. Koehler, pp. 247–76. Praeger Pubs., New York.

Resh, V.H., G.A. Lamberti, E.P. McElravy, J.R. Wood, and J.W. Feminella. 1984. *Quantitative Methods for Evaluating the Effects of Geothermal Energy Development on Stream Benthic Communities at The Geysers, California.* California Water Resources Center Contribution No. 190. California Water Resources Center, Univ. of California, Davis, CA.

Resh, V.H. and D.M. Rosenberg. 1989. Spatial-temporal variability and the study of aquatic insects. *Canadian Entomologist* 121:941–63.

Rosenberg, D.M. and V.H. Resh. 1982. The use of artificial substrates in the study of freshwater benthic macroinvertebrates. In *Artificial Substrates,* ed. J. Cairns, Jr., pp. 175–235. Ann Arbor Science Pubs., Ann Arbor, MI.

Ross, L.T. and D.A. Jones, eds. 1979. *Biological Aspects of Water Quality in Florida.* Technical Series Vol. 4, No. 3. Department of Environmental Regulation, State of Florida, Tallahassee, FL.

Shackleford, B. 1988. *Rapid Bioassessments of Lotic Macroinvertebrate Communities: Biocriteria Development.* Biomonitoring Section, Arkansas Department of Pollution Control and Ecology, Little Rock, AR.

Sládeček, V. 1979. Continental systems for the assessment of river water quality. In *Biological Indicators of Water Quality,* eds. A. James and L. Evison, Chap. 3. John Wiley, Chichester, England.

Statzner, B. and R. Müller. 1989. Standard hemispheres as indicators of flow characteristics in lotic benthos research. *Freshwater Biology* 21:445–59.

Stewart, K.W. and B.P. Stark. 1988. *Nymphs of North American Stonefly Genera (Plecoptera).* Vol. 12. The Thomas Say Foundation, Entomological Society of America, Lanham, MD.

Stewart-Oaten, A., W.W. Murdoch, and K.R. Parker. 1986. Environmental impact assessment: "pseudoreplication" in time? *Ecology* 67:929–40.

Tuffery, G. and J. Verneaux. 1968. *Méthode de Détermination de la Qualité Biologique des Eaux Courantes. Exploitation Codifiée des Inventaires de la Faune du Fond.* Section Pêche et Pisciculture, Centre National d'Etude Techniques et de Recherches Technologiques pour l'Agriculture, les Forêts et l'Equipment Rural, Ministère de l'Agriculture, Paris, France.

USEPA (U.S. Environmental Protection Agency). 1982. *Standard Operating Procedures.* Vol. II. Environmental Biology Section, Environmental Service Division, U.S. Environmental Protection Agency, Athens, GA.

Vermont Department of Environmental Conservation. 1988a. Indirect discharge rules. In *Environmental Protection Rules,* Appendix pp. D–1 to D–45. Agency of Natural Resources, Department of Environmental Conservation, Waterbury, VT.

Vermont Department of Environmental Conservation. 1988b. *Field and Laboratory Standard Operating Procedures Manuals.* Agency of Natural Resources, Department of Environmental Conservation, Waterbury, VT.

Voshell, J.R., Jr., R.J. Layton, and S.W. Hiner. 1989. Field techniques for determining the effects of toxic substances on benthic macroinvertebrates in rocky-bottomed streams. In *Aquatic Toxicology and Hazard Assessment: 12th Volume,* eds. U.M. Cowgill and L.R. Williams, pp. 134–55. American Society for Testing and Materials Special Technical Publication 1027. American Society for Testing and Materials, Philadelphia, PA.

Washington, H.G. 1984. Diversity, biotic and similarity indices. A review with special relevance to aquatic ecosystems. *Water Research* 18:653–94.

Weber, C.I., ed. 1973. *Biological Field and Laboratory Methods for Measuring the Quality of Surface Waters and Effluents.* EPA-670/4–73–001. U.S. Environmental Protection Agency, Cincinnati, OH.

Wiggins, G.B. 1977. *Larvae of the North American Caddisfly Genera (Trichoptera)*. Univ. of Toronto Press, Toronto.
Winget, R.N. and F.A. Mangum. 1979. *Biotic Condition Index: Integrated Biological, Physical, and Chemical Stream Parameters for Management*. U.S. Forest Service Intermountain Region, U.S. Department of Agriculture, Ogden, UT.
Woodiwiss, F.S. 1964. The biological system of stream classification used by the Trent River Board. *Chemistry and Industry* 11:443–7.
Wright, J.F., P.D. Armitage, M.T. Furse, and D. Moss. 1988. A new approach to the biological surveillance of river quality using macroinvertebrates. *Internationale Vereinigung für Theoretische und Angewandte Limnologie Verhandlungen* 23:1548–52.
Zelinka, M. and P. Marvan. 1961. Zur Präzisierung der biologischen Klassifikation der Reinheit fliessender Gewässer. *Archiv für Hydrobiologie* 57:389–407.

Appendix 6.1. Selected examples of qualitative assessment protocols (arranged alphabetically). EPT = Ephemeroptera, Plecoptera, Trichoptera; CPOM = coarse particulate organic matter.

Study	System	Device	Habitat	Proportion of Sample Examined	Level of Taxonomy	Measures Used
Bode (1988)	Streams	Quantitative approach: 1 of 2 artificial substrate sample units; tissue analysis also done	Deeper runs and pools	1/2, or 1/4 if >1000 individuals	Genus/species	For both quantitative and qualitative approaches: number of individuals or biomass; Shannon's Index; number of taxa; number of EPT taxa; Biotic Index; % individuals in numerically dominant taxa; % individuals in major groups. Impact assessment is made by assignment to categories developed from background data collected regionally
		Qualitative approach: traveling kick sample	Riffle	100 organisms	Genus/species	
De Pauw and Vanhooren (1983)	Streams	D-net over a 10–20 m stretch of stream; 3 min of collecting for streams <2 m wide; ≤5 min for larger streams	All habitats	Whole	Family or genus for most orders	Belgian Biotic Index. Impact assessment is made by assignment to categories developed with background data collected from sites throughout Belgium
DIN (1989)	Streams	Sampler appropriate to habitat, such as dip net, grab, dredge, or suction	All major habitats typical of whole reach being studied	Whole	For selected taxa: species for most orders; some to genus	Saprobien Index (i.e., Saprobic Index of Zelinka and Marvan, 1961), with a

		sampler; specific devices are recommended for different habitats; sample unit is standardized for a given time or spatial area				variance test of applicability. Impact assessment is made by assignment of indicator species to pollution categories developed with background data collected from sites throughout Germany
Hilsenhoff (1987, 1988a)	Streams	D-frame net	Riffle or shallow run	100-organism representative subsample	Family (for Family Biotic Index), genus/species (for Biotic Index)	Assessment uses Family Biotic Index and Biotic Index; categories are developed from background data
ISO (1984) Score	Streams	Several sample units collected by a hand net or a Surber/cylinder sampler	All microhabitats of eroding zone of riffle/run are sampled	Whole	Family	A list of families is used to calculate a Total ISO Score and Mean ISO Score Impact assessed as [(New Score–Previous Score)/(Previous Score)] × 100 Sample units are scored either separately or together
Kelso (1987) and I.J. Davies (Freshwater Institute, Winnipeg, MB, personal communication)	Streams	One Surber sample unit with 100–μm mesh and one Surber sample unit with 450–μm mesh in Habitat 1; handpick 10 boulders in Habitat 2; hand net in Habitat 3	Habitat 1 is a riffle with velocity 100–400 cm/sec; depth 10–30 cm; stones 2–10 cm diam.; Habitat 2: across-stream transect; Habitat 3: stream margins	Whole	EPT genus/species	Community structure based on species richness, presence/absence, relative abundance, multivariate analysis and multidimensional scaling

(continued)

Appendix 6.1. (Continued)

Study	System	Device	Habitat	Proportion of Sample Examined	Level of Taxonomy	Measures Used
	Lakes	(1) 1 to 5 sweep net sample units in Habitat 1 (2) 10 traps for mobile species in Habitat 1 (3) 3 grab sample units in Habitat 2 (4) 3 vertical net tows at night (1 m × 1 m × column) in Habitat 2	Habitat 1: 5 sites in littoral areas, 0–1 m deep; Habitat 2: 5 sites in sublittoral/profundal zones	Whole	(1) Species if possible (2) Species; also size and age measures (3) Species (4) Species	Community structure based on species richness, presence/absence, relative abundance; multivariate analysis and multidimensional scaling
Lenat (1988 and personal communication)	Streams	Kick net: 2 sample units in Habitat 1; dip net: 3 sample units in Habitat 2; wash bucket: 1 sample unit in Habitat 3; PVC cylinder: 3 sample units in Habitat 4; Nitex bag: 1 sample unit in Habitat 5; visual observations in Habitat 6	Habitat 1: riffles and snags; Habitat 2: stream banks; Habitat 3: leaf packs; Habitat 4: rocks and logs; Habitat 5: sand; Habitat 6: large rocks and logs	400–800 organisms, roughly proportional to abundance	Genus/species	Number of EPT taxa and relative abundance; number of taxa; also uses tolerance data (indicator assemblages, biotic indices), number of unique species per site, functional feeding group type, approximate estimates of relative abundance Assessment is made on categories determined from background data in different regions, including modifications for seasonal and stream-size differences
Lenat (personal communication)	Streams	Single kick net, sweep net, and leaf pack samples; visual observations	As above	Whole but only for EPT individuals	EPT species	Number of EPT taxa

Maine Department of Environmental Protection (1987)	Streams/ rivers (primarily for assessing point source impacts)	Rock-filled artificial substrates	< 90 cm deep, flowing, with mineral substrate	Whole or subsampled to yield >100 organisms per sample	Genus/species	Total number of individuals; number of taxa; indicator organism presence/absence; number of EPT taxa; % non-dipteran insects; functional feeding group measures; ratio of specialist to generalist; Shannon's Index; similarity index; Coefficient of Community Loss; Biotic Index
		Rock-filled, remote, retrievable artificial substrates	>90 cm deep, riverine or impounded with compacted substrate			
		Ekman grab	> 90 cm deep, soft dredgable substrate			Impact is assessed by use of a trichotomous decision key with impact criteria derived from reference sites and statewide baseline database
	Streams (primarily for assessing non-point source impacts)	Kick net and hand collections (as per Lenat, 1988)	All available habitats in wadable streams			Assessment as above
	Lakes (based on Courtemanch, 1989)	5 or more Ekman grab sample units are taken, until at least 50 chironomid individuals are collected	Profundal zone of lake	Whole	Genus/species	Indicator organisms presence/absence; relative abundance of taxa Assessment is made by assignment to one of 14 lake types defined by chlorophyll-*a*, phosphorus, and mean depth characteristics

(*continued*)

Appendix 6.1. (Continued)

Study	System	Device	Habitat	Proportion of Sample Examined	Level of Taxonomy	Measures Used
Mason (1979)	Stream or lake shoreline	Kick net or dip net using a shovel; 2 sample units/site or 75% of total diversity (in a trial sample of 10 sample units)	All habitats	Whole; only Niche Occupant Forms (roughly equivalent to number of taxa, see Table 6.1) identified	Niche Occupant Forms identified	Number of Niche Occupant Forms (roughly equivalent to number of taxa)
						Assess impact as proportional change: # observed/# expected = index value; # expected is from a reference site or previous studies of watershed; unity implies perfect agreement with reference site
Ohio EPA (1987)	Streams	Quantitative approach: 5 Hester-Dendy artificial substrate sample units; can use Surber sampler or grabs as appropriate or necessary	Runs preferred over pools or riffles; uniformity between sites required	Whole; Folsom sample splitter used to subsample some invertebrate groups if unmanageable numbers	Genus/species	Invertebrate Community Index (ICI) is calculated (see Table 6.1F)
						Categories of ICI are from Ohio background data
		Qualitative approach: A-frame net and hand collections; grab sampler where necessary	All available habitats	Whole	Genus/species	Incorporated as one metric in the ICI; a combination index for qualitative data is being developed

Plafkin et al. (1989)	Streams Protocol 1	Dip or kick net, or hand collections	Sample all available habitats in stream	Whole; relative abundance then assessed	Order/family	Impairment = absence of EPT taxa; dominance of tolerant groups; low number of individuals; low number of taxa
	Protocol 2	Kick net, CPOM collection	Riffle/run habitat sample (sample is of two 1-m^2 sample units); CPOM sample	100-organism representative subsample; 20–60 organisms from CPOM sample	Order/family	Biological Condition Score is calculated (see Table 6.1F), but with Family Biotic Index and percent dominant family Assess change by comparison with reference site or background data
	Protocol 3	Kick net, CPOM collection	Riffle/run habitat sample (sample is of two 1-m^2 sample units); CPOM sample	100-organism representative subsample; 20–60 organisms from CPOM sample	Lowest level possible	Biological Condition Score is calculated (see Table 6.1F) Assess change by comparison with reference site or background data
Raddum et al. (1988)	Streams	Surber sampler and kick net	Not stated; presumably all habitats	Whole, which was generally >200 individuals	Genus/species	An acidification number, based on the presence/absence of organisms indicating different levels of acidity, is calculated. This is used to compare acidification in different regions and for comparing trends in acidification over time

(*continued*)

Appendix 6.1. (Continued)

Study	System	Device	Habitat	Proportion of Sample Examined	Level of Taxonomy	Measures Used
Ross and Jones (1979)	Streams	Quantitative approach: 3 Hester-Dendy sample units; if >200 organisms in the first sample unit, remaining 2 sample units are not examined; otherwise, all 3 sample units are examined	Quantitative: at middepth of stream or 1 m, whichever is less	Quantitative: whole	Genus/species	Quantitative: Shannon's Index; assessments based on values assigned to categories determined from Florida surveys
	Lakes	1 sample unit (with Ekman or petite-Ponar grab) is collected per site (1 sample unit consists of the number of grabs required to obtain a total of 15 individuals)	3–4 stations in small lakes (<50 acres), up to 10 stations in large lakes (>500 acres)	Whole	Genus/species	Assessment as above
	Streams and rivers	Qualitative approach: D-frame net	Qualitative: all habitats	Qualitative: field sorting to collect a number of specimens of each taxon; sampling is considered complete when no different taxa are found after 10 min of sampling	Genus/species	Qualitative: Florida Index; assessments based on values assigned to categories determined from Florida surveys

Shackleford (1988)	Streams	A-frame net	Riffles and pools	100 organisms in representative proportion	Family: Chironomidae; genus: others	Mean Biometric Score is calculated (see Table 6.1F); assessment based on scores determined by comparing study site with reference site
USEPA (1982)	Streams	Quantitative approach: three sample units using drift net, artificial substrate samplers, or other appropriate devices	All available habitats	Whole	Genus/species	Quantitative: biomass or number of individuals in benthos and drift collections
		Qualitative approach: hand net or handpicked sample	All available habitats	Whole	Genus/species	Qualitative: assessment of impact is based on classifying taxa using tolerances; comparative analyses of associations also used
Vermont Department of Environmental Conservation (1988a,b) and S. Fiske (personal communication)	Lakes and streams Protocol 1	5 Ekman-grab sample units in lakes; 5 rock-filled basket or Surber-sample units in streams	Profundal zone in lakes; riffles in streams	Whole, or subsample and check whole sample for EPT	Genus/species	Pinkham-Pearson Similarity Index applied only to the dominant taxa in the communities (i.e., ≥3.5%); mean number of EPT taxa; number of individuals; Biotic Index Significant alteration is assessed by comparison with reference site

(continued)

Appendix 6.1. (Continued)

Study	System	Device	Habitat	Proportion of Sample Examined	Level of Taxonomy	Measures Used
	Protocol 2	3 or 4 D-net sample units	Littoral zone in lakes; midstream riffle in streams	Subsample with a gridded tray to examine 300 organisms or 1/4 of sample unit; entire sample unit examined for EPT taxa	Genus/species	Mean number of EPT taxa; mean Biotic Index; mean Shannon's Index; % dominant genera; EPT abundance/chironomid abundance These are assigned to categories for different stream types; lake impact assessment not yet developed
Winget and Mangum (1979)	Streams	3 or 4 1-ft^2 sample units with a modified Surber sampler	Riffle or rubble	3/8 of sample unit sorted or 200 to 300 individuals using Water's-type rotating subsampler with fine screen in bottom of eight pans	Family for Chironomidae; genus for most taxa; species for some taxa	Biotic Condition Index (BCI); Community Tolerance Quotient (CTQ); number of taxa; number of individuals; dry-weight biomass; Dominance and Taxa Diversity Index; functional feeding group balance

						Assessment is made using comparison with CTQ values predicted from stream gradient, substrate, total alkalinity, and sulfate concentrations; BCI and CTQ classified by categories
Wright et al. (1988)	Streams	Combination of 3-min sample units in each of 3 seasons with a pond net	All major habitats	Whole	Family	Biological Monitoring Working Party (BMWP) Score: total score and average score per taxon (ASPT); impact assessed as a proportional change: score observed/score expected; score expected is predicted from multivariate models of physical, chemical, and benthic macroinvertebrate data

7

Analysis and Interpretation of Benthic Macroinvertebrate Surveys

Richard H. Norris and Arthur Georges

7.1. Introduction

Biologists often find themselves in the position of having to use patterns of distribution and abundance of organisms to detect environmental change and to infer the cause of the change by associating changes in biological variables with corresponding changes in physicochemical variables (Norris and Georges, 1986; Barmuta, 1987). Inevitably, biological and environmental correlates of water quality are compared across sites and times, or against set standards, to assess impacts of disturbances or management initiatives, to develop models useful for prediction, or to establish cause and effect. Study designs should facilitate the making of these comparisons through the collection of relevant data, elimination of confounding effects, and selection of appropriate analyses.

Biologists working with benthic macroinvertebrates have long been aware of the problems of variability in what they measure (e.g., Needham and Usinger, 1956) and have emphasized the need to account for variability of benthic macroinvertebrate data in their sampling designs (Downing, 1979; Resh, 1979; Allan, 1984; Morin, 1985; Norris and Georges, 1986; Canton and Chadwick, 1988; Resh and McElravy, Chapter 5). These sampling designs usually include some level of replication to enable subsequent analyses to be performed (see Resh and McElravy, Chapter 5).

Several approaches have been developed to cope with the need for sometimes large numbers of replicate collections of macroinvertebrates (Needham and Usinger, 1956; Chutter and Noble, 1966; Downing, 1979; Resh, 1979; Allan, 1984) and the need for rapid return of results (Cairns and Van Der Schalie, 1980). Many indices have been created and used in benthic monitoring studies (see reviews by Cairns et al., 1972; Fager, 1972; Poole, 1974; Washington, 1984; Hellawell, 1986), but many workers have avoided the problem altogether by relying on direct measurement or by presenting data on number of individuals or taxa with little or no analysis (lotic environments: Frost and Sinniah, 1982; Whiting and Clifford, 1983; Chadwick and Canton, 1984; Pistrang and Burger, 1984; lentic environments: Cherry et al., 1984; Voshell and Simmons, 1984; Hynes and Yadav, 1985). Fryer (1987) made an impassioned plea for balance and sense in the use of quantitative and qualitative information in the study of living organisms. Although this chapter will lean strongly toward the analysis of quantitative data (and mostly of counts of macroinvertebrates), Fryer's view of the need to observe and understand the animals themselves should be heeded.

Several steps, given in the order of their implementation in a sampling program, will affect the analysis of data:

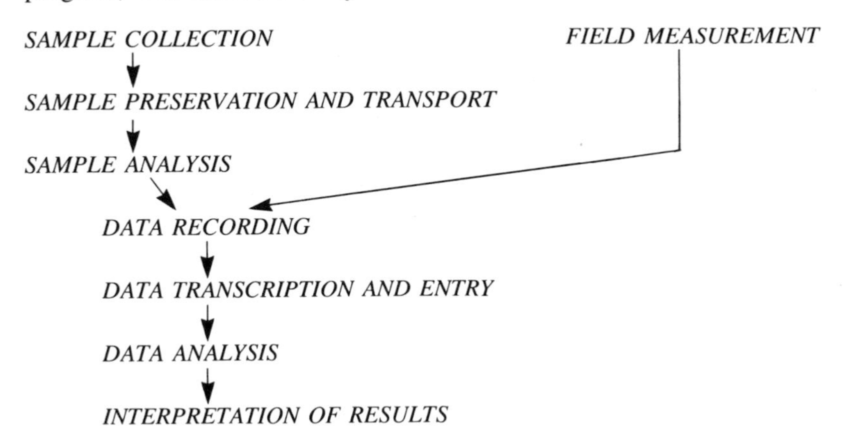

Sample collection, field measurement, and sample preservation and transport are considered by Resh and McElravy in Chapter 5. Errors in data that occur in these first stages may be serious and usually are irretrievable. Procedures are required to minimize errors during sample analysis, data recording, and data transcription and entry, and these procedures are presented below in the section on exploratory analysis.

Some studies will need only the simplest presentation of data by graphical or tabular means. Others will require statistical analysis so that generalities may be drawn with some measure of confidence, or will require complex numerical analysis to detect and highlight patterns or trends in complex data sets comprised of many variables. This chapter intends to provide a balanced but critical view of the options available for analysis of benthic biomonitoring data based on the need for comparisons and for inferring causal relationships. Clearly, the treatment of all possible approaches for analyzing data cannot be included. However, detailed statistical information can be found in biostatistical texts such as Sokal and Rohlf (1981) and the excellent book on sampling design and approaches to analysis by Green (1979). This chapter will consider common analytical approaches used and those that show great potential.

7.2. Elucidating Mechanisms

A problem often encountered in assessing water quality concerns the effects of a range of environmental variables on species that comprise a biological community. As our knowledge and understanding of aquatic systems improve, an increased need also exists for assessment of the effects of interrelationships of factors having minor or subtle outcomes on such communities. Studies assessing water quality often provide large data sets with many variables. The relationships of cause and effect between the variables may be complex and difficult to analyze, often needing multivariate methods (Green, 1979). Approaches used for the collection of physicochemical data have been different from those used for the collection of biological data, and this has resulted in difficulties in associating the two types of data during analysis. These broad divisions in data types need not be considered separately (Norris and Georges, 1986).

When interpreting the relationships between biological and physicochemical data, it is important to base the interpretations on real biological properties that relate to the environment rather than just on statistical interpretations (Taylor, 1980; Anderson et al., 1982; Fryer, 1987). Strictly speaking, causal relationships only can be determined through direct experimental work in which the features of concern are tested under controlled conditions (see Cooper and Barmuta, Chapter 11).

7.3. Data Quality Assurance and Exploratory Analysis

7.3.1. Precision and Analysis

Assurance of the quality of data from sampling involves sound measurement procedures and an understanding of sampling theory. For data to be useful, knowledge of the variability of repeated measurements is required. If single samples of benthic macroinvertebrates, or a single biological indicator such as a diversity index, are highly variable and differ at two stations, the researcher is faced with an ambiguity: do they differ because values of the indicator actually differ at the two stations, or do they differ solely because any two samples would be expected to differ as much, even if collected at the same time and from the same station? Environmental variability is a fundamental problem facing those interested in assessing changes in water quality through space and time. High environmental variability and logistical constraints on sample collection and analysis often may result in data that are too variable to demonstrate the impact of a disturbance, or management initiatives, on water quality (see also Resh and Jackson, Chapter 6).

Conclusions that water quality at particular sites or times actually differs only can be made when observed differences in means between sites or times are greater than would be expected based on observed variation within sites or times. This reasoning forms the basis of statistical procedures such as analysis of variance (ANOVA), and demonstration of statistically significant differences in measurements of physicochemical or biological characteristics between sites or times necessarily will need replicate collections.

7.3.2. Exploratory Data Analysis

Study design in the statistical sense involves decisions on the spatial distribution of sampling stations and decisions on the frequency of sampling at those stations (see Resh and McElravy, Chapter 5). Poorly designed programs, no matter how rigorously and energetically implemented, fundamentally may lack the potential to address the objectives of the program. The destructive influence of confounding variables on interpretation is well-known to statisticians. Unfortunately, the converse is not true: well-designed studies do not necessarily yield data that can be used in addressing project objectives. The additive effects of failure to implement laboratory procedures leading to reproducible results, equipment malfunction unnoticed by inexperienced staff, poorly designed data sheets that promote transcription errors, inappropriate analysis, and so on, may act to frustrate achievement of project objectives.

In most data sets, some errors are inevitable, and even gross errors may go undetected, profoundly affecting the outcome of analysis. Data verification and procedures designed to minimize errors can be performed at each of the several steps in the implementation of a monitoring program.

When screening data for outliers that may represent erroneous data, the outliers should only be deleted from the data when an *a priori* explanation of the aberrance is possible, when detailed scrutiny of equipment or records reveals the source of the error, or when the value (e.g., pH 14 in natural waters) is impossible (Cochran, 1947). Automatic data elimination or correction without close scrutiny by the data analyst is undesirable. The computer should be used to flag potential errors, which then can come under greater scrutiny. In addition to visual checks by people experienced with the nature of the data, initial checks on data as they are entered should include the following:

1. Are the known ranges in numbers of individuals and species per replicate for the sites exceeded? Does the recorded station exist in a list of stations? Do numbers allocated to species names coincide with the proper species? It is still possible to have numbers outside these ranges that are correct, say because of an unexpected organic load, and it is also possible to have incorrect values within the

ranges (e.g., a species with the incorrect number allocated to it during sorting or data entry). Further checks will be necessary.

2. Calculation of means and standard deviations for subsets (possibly by date or site) of the data is useful because aberrantly high or low means, or aberrantly high standard deviations, indicate errors.
3. Univariate checks for outliers can be performed by testing for values more than three standard deviations from overall means. Since counts rarely are distributed normally, often being highly skewed, this procedure is bound to produce a high number of suspicious observations. For skewed data, calculating percentiles and scrutinizing values that fall above the ninety-fifth or ninety-ninth percentiles may be preferable.
4. The frequency of occurrence of particular species in a set of replicates from particular sites can be cross-checked independently. These occurrences should fall within specified acceptable limits based on previous work. The calculation of cumulative numbers for the different species collected can be compared with the total numbers recorded for each replicate.
5. Calculation of ratios of the numbers of selected species, which will be site- or subcatchment-specific, will help to locate which counts of particular species, or identifications, are in error.
6. Bivariate checks for outliers, using regressions of the variable under scrutiny with other related variables in the data set, may enable identification of erroneous data. For example, numbers of animals collected may be related linearly to distance from a point source of organic pollution. If sources are relatively constant, then departures of three or four standard deviations from the regression will highlight procedural or real environmental problems.
7. Plots of variables against each other or against time or distance from a point of impact are useful for visual checks on whether the data follow logical sequences.
8. A check on total counts is useful for verifying that counts of individual species have been entered correctly. Data sheets may consist of a list of macroinvertebrate species, and the data may be counts of individuals in each species. The total number of animals counted, although redundant because it can be calculated easily by statistical computer packages, enables a valuable check against mispunching of counts of individual species that are otherwise very difficult to verify.

7.3.3. Missing Data

Exploratory data analysis will reveal the number of missing values for different variables measured. Missing values often can seriously complicate further analysis. For example, although ANOVA can handle unequal sample sizes because of missing values, it is not nearly as robust with regard to violations of homogeneity of variances as it would be if care had been taken to balance the design. Ordination, classification, multiple regression, Discriminant Function Analysis, and time-series analysis all require complete sets of measurements for each site and time. Several options are available to deal with missing data:

1. If values are missing from one site or one time, all the records from that site or time can be discarded. This is the default option for most statistical packages.
2. If values are missing from one variable (e.g., one species), all values for that variable can be disregarded.
3. Sample means can be substituted for the missing values.
4. The missing value can be estimated using spatial or temporal relationships or relationships established with other variables.

The first two options may be unacceptably wasteful of data. The last two options, although presenting no conceptual difficulties for descriptive statistics, need adjustment of degrees of freedom if significance testing is required. This adjustment will not necessarily be a feature of statistical packages because the modern approach to unbalanced designs is to develop linear models based on available data (Freund et al., 1986). The choice of which approach to take will depend on the nature of the data and the analysis contemplated, but in all cases it is preferable to avoid missing values at the time of data collection.

7.4. Analysis Options

7.4.1. Univariate Approaches

Studies of spatial and temporal changes in benthic communities have required quantification of the "state" of the communities, often in the form of a single, summary measure. This measure may be simply biomass or total abundance of benthic macroinvertebrates (Table 5.6 in Resh and McElravy, Chapter 5), species richness (de March, 1976; Hynes and Yadav, 1985), or it may be an index of community structure combining elements of both abundance and species number (e.g., species diversity or evenness). The measure also may be based on data selected from one or more indicator species (Harman, 1972), a so-called biotic index (Washington, 1984). All biological measures are designed to capture some aspect of the community at a particular site or time.

Nearly 40% of all lake and river studies surveyed in Chapter 5 (Table 5.6) used some type of index. The most common one was the Shannon (1948) index, followed by percent similarity (Whittaker, 1952) in lentic studies and the Simpson (1949) diversity index in lotic studies. Reviews of indices relative to aquatic systems can be found in Cairns et al. (1972), Washington (1984), Hellawell (1986, p. 430), Abel (1989), and Johnson et al. (Chapter 4). According to Green (1979), Auclair and Goff's (1971) field data on upland forests provide an excellent example of spatial and temporal patterns of variation, and covariation, of all commonly used diversity indices.

Indices may be used for the following reasons: (1) they are seen as a useful way to condense complex data and thus aid interpretation (Wilhm, 1972; Hellawell, 1986); (2) people with little biological expertise can understand them easily (Cairns et al., 1968; Wilhm and Dorris, 1968) and can gather the data to create some of them (e.g., Cairns et al., 1968); (3) they are of more general value than physical and chemical measures (Hellawell, 1986); (4) they allow comparisons of sites or times where collections have been made using different sample sizes, methods, or habitats; and (5) their data needs are seen as relatively less expensive than other more traditional statistically based approaches. Clearly, the use of indices is popular in studies of benthic macroinvertebrates. Therefore, it is appropriate to consider data needs of these indices and the assumptions on which they are based, because these will affect data analysis and interpretation.

In his extensive review of indices, Washington (1984) identified three groups: diversity or community structure indices, biotic indices (which includes most of the rapid assessment indices discussed in Resh and Jackson, Chapter 6), and similarity indices.

7.4.1.1. Diversity Indices

Diversity indices usually require a count of the total number of individuals and a total count for each of the taxa. The taxa need to be separated but not necessarily identified. Separation is often at the species level, but it is sometimes at the generic or family level (Hughes, 1978).

The combination of abundance and richness in a diversity index supposedly indicates the state of the community. It seems to be generally accepted that values of most indices decrease with decreasing water quality. Also, low diversity supposedly indicates a stressed community that tends to be unstable (Goodman, 1975).

Washington (1984) divided the diversity indices he reviewed into eight groups, which he claimed are based on different assumptions. A brief discussion of the main groups follows.

Simpson's (1949) index is based on the assumption that in more diverse communities, a lower probability exists that individuals chosen at random will belong to the same taxon. However, this assumption disregards the possibility that members of the same taxon will be clumped for reasons of microhabitat, breeding, or behavior. The index is independent of any theory about the form of the frequency distribution of abundance of different taxa, but it is not independent of sample size. Thus, comparing values of the index will be difficult if different collection methods are used or different areas are sampled.

Species per 1,000 individuals (Odum et al., 1960), like other indices that "standardize" sample size, are based on the idea (supported empirically) that the number of species is related to the number of individuals sampled. Therefore, to assess properly the number of species that exist at different sites, the same number of individuals should be analyzed. Another similar approach is rarefaction (Hurlbert, 1971), which has been discussed in detail by Simberloff (1979), who recommended it as a distribution-free method of expressing and estimating diversity. However, the validity of methods such as these will depend on the manner in which the data are gathered. Sampling methods must be consistent, and samples must be collected from similar habitats (Simberloff, 1979). The approach assumes that the individuals are randomly dispersed. Clumping will cause an overestimate in the number of species, a problem that is reduced if larger subsample sizes are used (Simberloff, 1979). Sampling of different areas in sites that are to be compared may invalidate comparisons because the number of species collected also will be determined by the area sampled (Preston, 1948). Extrapolation of a rarefaction curve, which is distribution-free beyond the number of individuals collected, is not permissible because doing so assumes an underlying statistical distribution of the data (Simberloff, 1979). By assuming an underlying statistical distribution, a species-number relationship can be extrapolated to estimate the number of species at a site, but this also is subject to errors and is affected by sample size (Bechtel and Copeland, 1970). Workers will need to consider the relative importance of the number of taxa/number of individuals by comparison with the number of taxa/area of habitat. Both ratios may be important, but they represent different ecological relationships.

The diversity indices of Gleason (1922) and Margalef (1958) are similar (Washington, 1984), and they are based on "guesses" at fitting curves to species abundance distributions. The dubious assumption is made that the number of individuals is *directly* proportional to the area sampled. Of course, these indices are likely to be highly dependent on sample size (Murphy, 1978). More rigorous approaches are available for fitting curves to frequency distributions of species abundance, including the "characteristic" of Fisher et al. (1943) and Yule (1944) and the log-normal curve of Preston (1948) (Washington, 1984). The log-normal distribution is probably the most widely used of these approaches. Some doubt exists as to the biological meaning of frequency distributions, and no consideration seems to have been given to how environmental stress (including pollution) will affect the relationship. The indices developed by this procedure only should be used when the curves are a good fit to the data (Pielou, 1975), which may be difficult to satisfy because the relationship is likely to change with environmental stress. Krebs (1985) and Goodman (1975) have discussed the log-normal distribution particularly in relation to estimating the total number of taxa, or the number of rare taxa, at a site. This suggests the existence of a true biological relationship (i.e., niche subdivision; Sugihara, 1980) represented by the log-normal distribution of species frequency curves.

The most widely used diversity indices are those derived from information theory (e.g., Shannon, 1948). Washington (1984) provided a full discussion of this type of index and pointed out the rather tenuous biological links that have been attributed to them. Indices from information theory purport to measure "uncertainty" in the data, which may be considered to be the same as "information content," and consequently "diversity" (Washington, 1984). However, a direct link of biological relevance between these factors is doubtful (Goodman, 1975). The Shannon Index (H′) reaches its maximum value when all species are distributed evenly. Biologically, this is assumed to be the most desirable situation, although it contradicts the evidence provided by the log-normal distribution for many different communities (Goodman, 1975; Krebs, 1985). Hurlbert (1971) showed that many indices derived from information theory are correlated because they use the same variables in their calculation, and Krebs (1985, p. 523) concluded that "in practice it seems to matter very little which of these different measures of diversity we use." Goodman (1975) pointed out that these indices may be affected by the degree of clumping, problems of different body size in the organisms collected, and different habitat needs. Hughes (1978) listed six factors, other than pollution, that affected these diversity indices, including sampling method, sample size, depth of sampling, duration of sampling, time of year, and taxonomic level used. Thus, diversity indices based on information theory should be interpreted and compared with caution because their values will depend on study design.

Diversity indices often are based on ecological theories, such as the diversity/stability hypothesis (Goodman, 1975) or competitive interaction (Hurlbert, 1971). As such, authors promoting their use argue that real ecological properties are being measured. For example, the meaning of diversity and how it might be measured has been the subject of considerable debate (Hurlbert, 1971; Goodman, 1975), which remains inconclusive (Washington, 1984). Different pollution types may have different impacts in different habitats; this subject hardly seems to be debated at all. Therefore, the measurement of diversity and the impact of pollution on a diversity index both need to be resolved before such indices can be used and the results interpreted effectively. When such indices are applied to pollution studies, the predicted impacts of pollution on the ecological attributes supposedly measured by the index are rarely stated.

7.4.1.2. Biotic Indices

Biotic indices usually have been developed empirically as a means for assessing pollution impact, mostly in rivers (see Johnson et al., Chapter 4). Many are specific to a site and pollution type (usually organic). Rapid assessment techniques often include biotic indices (Resh and Jackson, Chapter 6). Calculation of biotic indices requires: (1) a total count of individuals or total counts of taxa; (2) counts (or biomass measurements) of specific groups such as all insects and tubificid worms, or the number of mayflies, stoneflies, and caddisflies; (3) detailed lists of the responses of different taxa to pollution; or (4) division of invertebrates into groups with different feeding strategies (see Chapter 6).

Biotic indices clearly are related to the conditions that led to their development. Because many are recommended for general use, the assumptions on which they are based need to be considered.

Taxonomic richness (as measured from the species to the order level) generally decreases with decreasing water quality. Number of individuals and biomass may increase, or decrease, depending on the type of pollution and the organisms involved. Many biotic indices first classify taxa according to their responses to certain pollution types (e.g., Patrick 1950; Beck, 1955; Woodiwiss, 1964; Beak, 1965). Most of these take little account of abundance (except see Chandler, 1970). Considerably more research is needed to classify the responses of different taxonomic groups to pollution; this is an increasingly difficult task because new pollutants continue to appear and effluents (including sewage) are becoming more complex. The difficulties of working above the species level have been demonstrated clearly (Resh and Unzicker, 1975).

Chutter (1972) based his index on the assumptions that the faunal communities of rivers and streams are definable, that these communities will change in a predictable way as organic matter is added, and that the addition of more oxidizable organic matter will cause greater faunal change. It has not been until the recent work of Wright et al. (1984), Armitage et al. (1987), and Ormerod and Edwards (1987) that macroinvertebrate communities expected in particular habitats could be defined with confidence. The work of these authors has been based on extensive surveys over large areas of Britain and is yet to be repeated in other areas of the world. The approach has considerable merit and may greatly strengthen the basis of some biotic indices.

Ephemeroptera, Trichoptera, and Plecoptera are sensitive to most types of pollution, so the numbers of individuals in these orders will decrease with a decrease in water quality. The numbers of some Diptera and tubificid worms may increase in response to organic pollution. These responses have been used as indices (e.g., Balloch et al., 1976), as the ratios between tubificids and other organisms (e.g., King and Ball, 1964), or just as counts of the number of taxa belonging to the sensitive groups (e.g., Plafkin et al., 1989) (see Johnson et al., Chapter 4). Virtually all of the indices or other measurements using these assumptions have been developed in relation to organic pollution of rivers. However, some species of Trichoptera and Ephemeroptera are highly tolerant of trace metal pollution (Norris et al., 1982; Norris, 1986), so caution is advised in the general application of indices based on the assumptions just discussed. Other difficulties include: the large amount of initial work that may be needed to define pollution tolerances and "clean" freshwater communities, and the limited number of taxonomic keys to many species (Resh and Jackson, Chapter 6).

Some biotic indices are based on the assumption that the ratios of organisms with different feeding strategies will change with pollution (e.g., collectors will be more abundant than shredders under polluted conditions) or that trophic generalists will be more pollution-tolerant than trophic specialists. Some doubt exists as to whether these general rules hold true and even whether it is possible to assign taxa to different feeding strategies (see Chapter 6).

7.4.1.3. Statistical Needs of Indices

Indices sometimes are applied on the assumption that their calculation in some way replaces the need for hypothesis testing (Resh and Jackson, Chapter 6) or statistical calculations (Beck, 1955). Clear questions must be asked before studies are designed and before the data from them can be analyzed in any sensible way (Green, 1979). Often, temporal trends in a diversity or biotic index are presented, or values of a diversity index are compared with little or no statistical analysis (e.g., Gupta and Pant, 1983; Chadwick and Canton, 1984), as if they were absolute measures characteristic of the community in question and not subject to sampling error. Clearly, this assumption is no truer than for any other finite set of measurements made, and replicated determinations of a diversity or biotic index can be expected to vary by chance alone. Many variance formulae for diversity indices are for the sample variance and not the variance of the sampling distribution of the diversity index. This latter value is needed for statistical inference. Without replication and a knowledge of the sampling distribution of the index, statistical procedures cannot be applied to determine if observed trends or differences result because of sampling error, or if they are a reflection of true trends or differences in the community under study. The uncritical interpretation of trends in an index of community structure or condition is a major shortcoming in their current application.

The case for measuring biota to assess the state of the environment has been argued adequately by many authors, possibly the most notable of which is Hynes (1960). The proliferation of various indices, especially for pollution assessment, has occurred in recognition of the value of using the biota. Unfortunately, the assumptions on which many indices are based are rather tenuous, not generally applicable, or are in need of substantiation. For these reasons, it is recommended that their features be considered and that they be used with caution. The condensation of data that indices are designed to achieve will always make it difficult to associate them with physical and chemical properties of water. In fact, no good reason exists for such correlation (Washington, 1984). Thus, the factors responsible for observed changes will be difficult to identify if a lone number, resulting from an index, is associated somehow with other water quality data.

Few of the reasons for using indices listed at the beginning of this section can be regarded as supporting their use in favor of more traditional approaches. In fact, the data needed for calculating many indices often are the same (or more in the case of pollution tolerances) as those needed for more traditional statistical methods. Some general recommendations on the use of indices follow:

1. Indices should not be seen as a way to avoid deciding on an appropriate study design or to avoid properly analyzing the data.
2. Many indices rely on detailed biological information on tolerances or feeding strategies. Such information may be specific to species, sites, and type of pollution. Therefore, caution is recommended when applying established indices to new situations.
3. Indices that result in a single number should be interpreted in light of other biological evidence, which will require biological expertise.
4. The index chosen should have biological significance and should measure actual changes because of pollution in the environment.
5. The quality of the data from which the index is calculated, and their limitations, must be known.
6. Where comparisons are being made, errors in measurement and calculation of the index need to be known.

7.4.2. Analysis of Variance

Analysis of variance (ANOVA) is a well-established parametric technique for comparing means of a single variable (Sokal and Rohlf, 1981). Its major uses in the analysis of data from benthic surveys include: assessing the statistical significance of differences in a biological measure among various water bodies, among different locations in the same water body, or among samples taken from one location at different times; and partitioning total variability in values of a biological measure into components attributable to variation among sites or times and components attributable to each of one or more levels of replication. This latter analysis is usually a prelude to optimizing sampling design.

ANOVA arose from the study of replicated samples and is founded in particular on a linear relationship between the variance of means of replicated samples and the average variance of the values that make up each sample. The observed variance among the means for different sites or times can be compared with that expected to arise from this relationship; the ratio of the two variances can be compared using an F-test.

An example of this kind of analysis is provided by Tiller (1988) who studied the effects of human disturbance on the benthic macroinvertebrate fauna of the Thredbo River in Kosciusko National Park, New South Wales, Australia. The river passes by a ski resort, Thredbo Village, which discharges treated sewage 1.5 km downstream. To assess the effects of this potential source of pollution on the fauna, Tiller chose sampling stations above and below the village itself (sites I, II), below a refuse dump and above the sewage outflow (site III), and at various distances downstream of the outflow (sites IV-VIII) (Fig. 7.1). Ten replicate collections of benthic macroinvertebrates were taken at each sampling station and total numbers were used as an appropriate biological measure likely to be affected by the sewage outflow. A separate analysis also was performed on species richness. Both of these variables generally are accepted as having skewed distributions (Elliott, 1977), so they were log-transformed before analysis. The results of the ANOVA for logged abundance are shown in Table 7.1. The significance of the F-value in the ANOVA table indicates that variation of macroinvertebrate abundance among sites was unlikely to have occurred by chance alone.

The next step in the analysis was to determine which sites were significantly different from the others, and Tiller chose from a range of possible procedures (reviewed by Keppel, 1973; Sokal and Rohlf, 1981; Day and Quinn, 1989) to do pair-wise multiple comparisons using the Student-Newman-Kuels procedure. Site II, below the village, and site IV, immediately below the sewage outflow, each had significantly greater numbers of benthic macroinvertebrates than any of the other sites, demonstrating an impact on

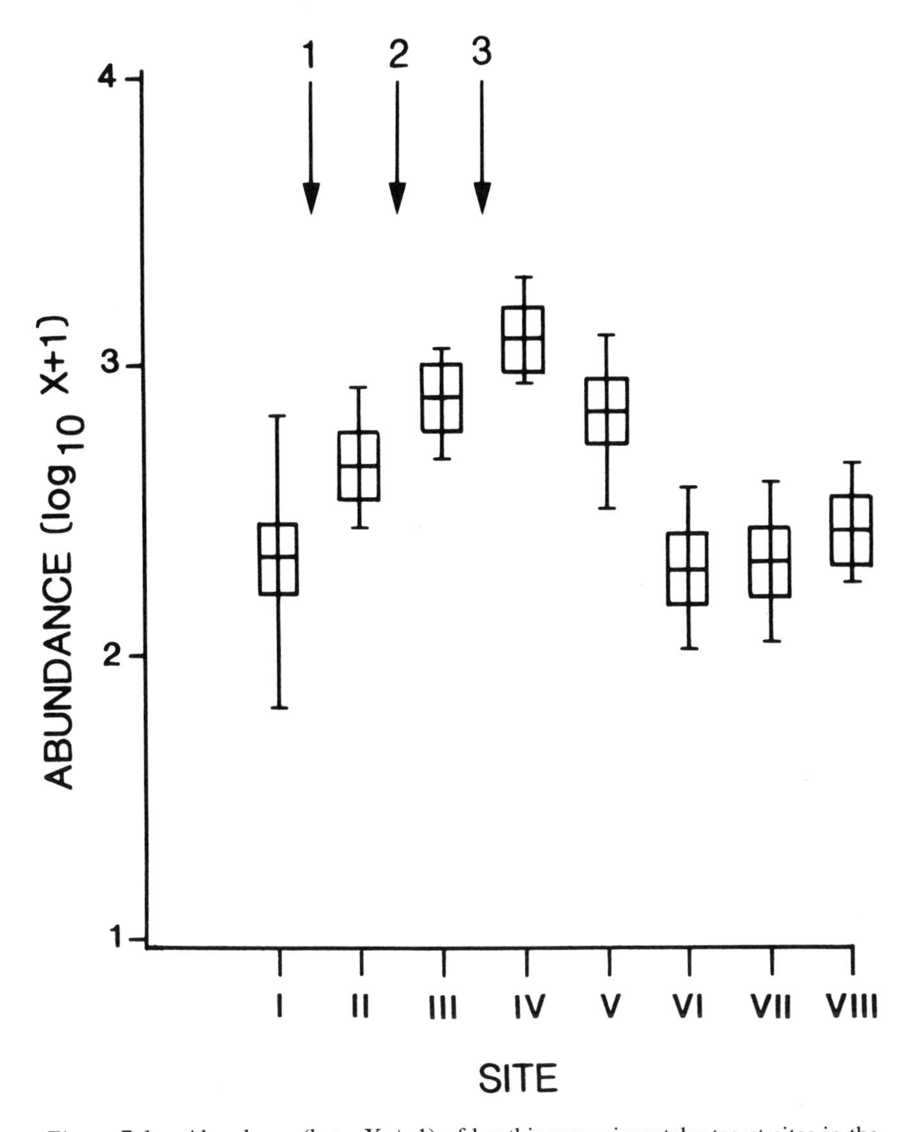

Figure 7.1. Abundance ($\log_{10}$ X + 1) of benthic macroinvertebrates at sites in the Thredbo River, N.S.W., Australia, subjected to effluent inputs from: (1) Thredbo Township; (2) a domestic refuse dump; and (3) secondary treated sewage. The box represents the mean (middle line) and 95% confidence limits (upper and lower lines); the vertical lines represent the range. Confidence limits are based on the pooled estimates of the common population variance. Results are for February 1983. From Tiller (1988).

Table 7.1. Analysis of variance table of results from the effects of sewage effluent on abundance ($\log_{10}$ X + 1) of macroinvertebrates in the Thredbo River, Australia.

Source	DF	Sum of Squares	Mean Square	F	Probability
Site	7	6.7255	0.9608	28.03	0.0001
Error	72	2.4679	0.0343		
Total	79	9.1934			

the stream by both the village itself and the effluent outflow. The site upstream of the village and sites some distance downstream of the outflow were not significantly different, so a persistent effect on the stream fauna could not be demonstrated.

The advantage of ANOVA over more qualitative approaches lies in its ability to distinguish between true trends that occur in the river (significant results) and those trends likely to have arisen in the sample through chance alone (nonsignificant results), because of sampling error. However, the technique has a number of limitations, which are described next.

7.4.2.1. *The Need to Replicate*

Replication of the biological measure used as a summary of community structure or community conditions is a necessary prerequisite to ANOVA. Only by replication can the magnitude of differences between sites or times be compared against the magnitude of differences that would be expected to occur by chance.

The number of replicates needed in a study are decided at the design stage (see Resh and McElravy, Chapter 5), and are related to the interpretations to be made with the data, the magnitude of differences to be detected, and the type of analyses to be performed (see Section 7.3.1, "Precision and Analysis," above). Replication occurs at two levels: within a site and time and among sites and times.

The design of many benthic macroinvertebrate studies in rivers, and to a lesser extent in lakes, is difficult because of problems measuring changes before and after the beginning of an impact and because sites cannot be replicated easily. Consider the case where sites are located upstream and downstream of a harmful discharge and replicated collections are made on each sampling occasion before and after the discharge is released. This would appear to be a simple two-way ANOVA with two factors, each with two levels: area (control and impact) by time (before and after) (Barmuta, 1987). However, treatments within the design are not properly replicated. Hurlbert (1984) referred to this design as "pseudoreplicated" because it is possible that some factor other than that being tested may affect the downstream impact area but not the control (see Cooper and Barmuta, Chapter 11). Thus, the results of tests of significance need to be qualified because they depend on the validity of using upstream sites as a control for downstream sites.

When samples at each site and time have not been replicated, it is possible to use samples through time as replicates (Stewart-Oaten et al., 1986). However, this practice should be used only as a last resort. Although it is valid in the sense that the significant results obtained can be believed, statistical testing in ANOVA using samples through time as replicates will be inefficient unless true trends in time are identical in magnitude and direction for all sites. Normally, this would not be expected in most studies, so to ensure the best opportunity for detecting true differences between sites, it is necessary to replicate sample collection on each visit to each site.

7.4.2.2. *Assumptions and the Need to Transform*

Randomness in sampling and independence among measurements within and across samples are necessary if sample statistics, such as abundance, richness, and diversity and biotic indices, are to be representative of the community that is sampled. These assumptions are fundamental to all qualitative and quantitative analyses where characteristics of the entire community are inferred from characteristics of finite samples. They are achieved generally by sound sampling design (Green, 1979). The assumptions that replicated values of a biological measure are distributed normally and that variances are equal (in order for the samples to be compared) often are beyond the control of the investigator at the time of data collection and must be dealt with during preliminary analysis. Violations of these assumptions generally are expressed as a relationship between the variance and the mean, which should be independent for normally distributed data. Such a relationship is common for counts of benthic macroinvertebrates (Downing, 1979; Morin, 1985; Norris and Georges, 1986). If replicate collections have been taken at each site or time, a plot of the sample variances against the sample means, followed by a test of the correlation between the two, will indicate whether a transformation is necessary (Fig. 7.2).

Traditionally, the square-root transformation ($Y' = \sqrt{Y + 1/2}$) is applied if the variance and mean are approximately equal (corresponding to a random distribution of benthic macroinvertebrates) and a log transformation [$Y' = \log_{10}(Y + 1)$] is applied when the variance is consistently greater than the mean (corresponding to a clumped distribution) (Elliott, 1977). The variances of the transformed data then are plotted against their corresponding means to see if the transformation has been successful (Fig 7.2 and 7.3).

A data set may be skewed (e.g., when it conforms to a Poisson or negative-binomial distribution), which violates the assumption of normality. In

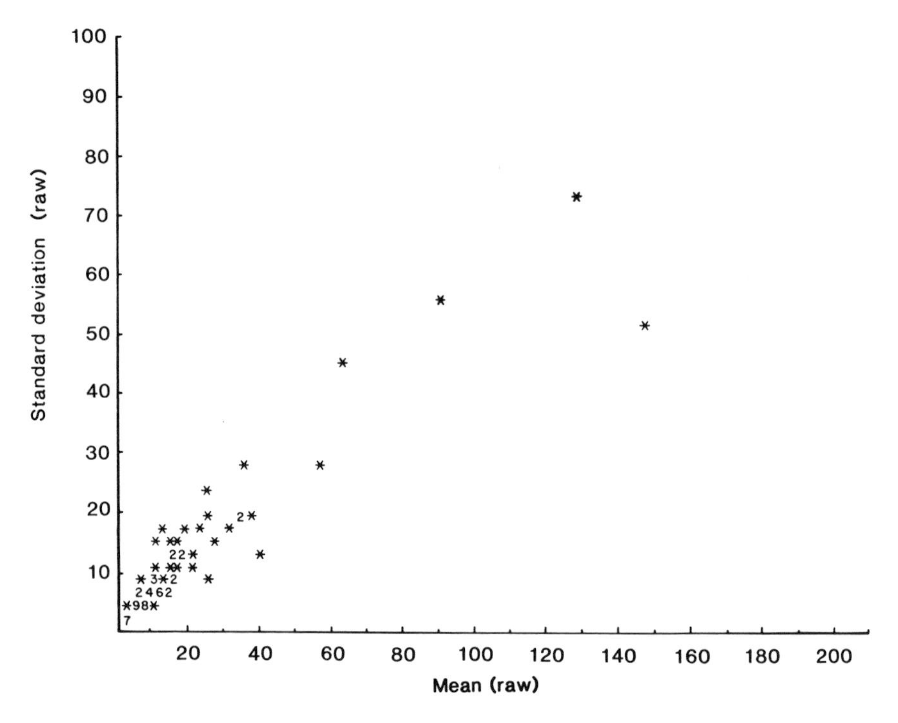

Figure 7.2. Relationship between standard deviations and means for counts of benthic macroinvertebrates from the South Esk River, northeastern Tasmania, Australia. $n = 80$, $r = 0.96$, $P < 0.001$.

extreme cases, no amount of transforming will render such data normal, and the researcher may need to resort to ANOVA models based on alternatives to the normal distribution (e.g., the GLIM package; Numerical Algorithms Group, 1986) or to nonparametric alternatives to ANOVA (Siegel, 1956; Conover, 1980). This situation might occur in water bodies that are highly polluted by some toxic waste, but fortunately the interpretation of biological data such as these usually is a trivial matter hardly needing the use of statistics (Norris and Georges, 1986).

More sophisticated approaches to transformation include those of Taylor (1961, 1980) and Box and Cox (1964). Box and Cox recommended the following transformations, which are attuned to characteristics of the data:

$$Y' = (Y^k - 1)/k \qquad \text{if } k \neq 0$$
$$Y' = \log_e(Y) \qquad \text{if } k = 0$$

where k is the maximum value of the expression:

$$(-v/2)\log_e S^2 + (k - 1)\,(v/n)\log_e Y$$

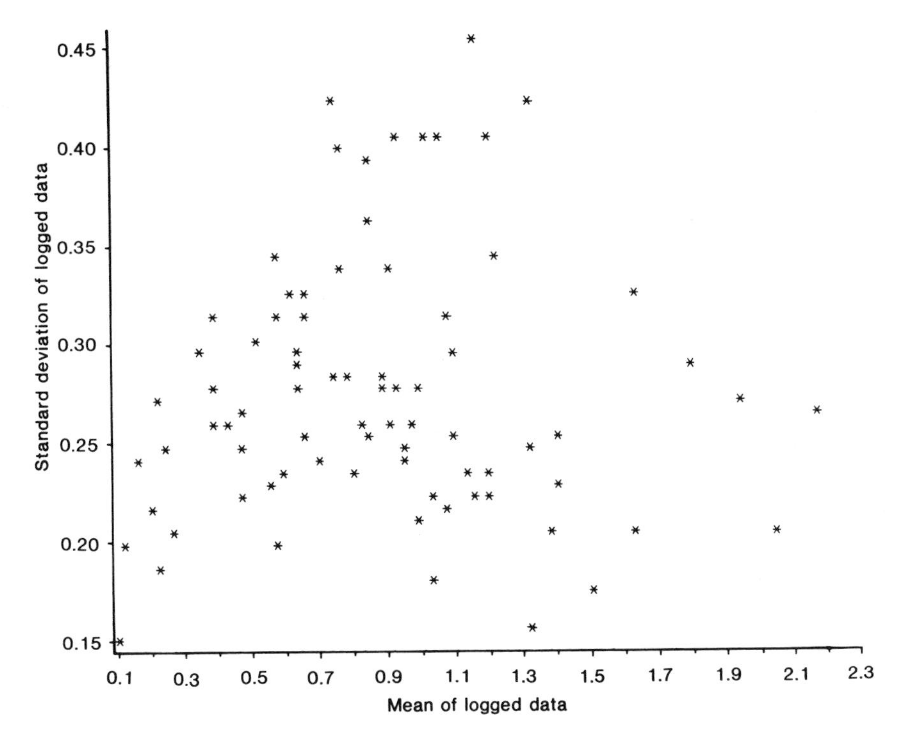

Figure 7.3. Effectiveness of $\log_{10}$ (X + 1) transformation of counts of benthic macroinvertebrates from the South Esk River, northeastern Tasmania, Australia. n = 80, r = 0.37, P > 0.05.

which is substituted in the equations above for the best transformation to normality (S^2 is the variance of the transformed values, v is the degrees of freedom associated with this variance, and Y represents the original untransformed data values.) The value of k must be solved iteratively. This transformation has been used infrequently in benthic studies, possibly because the computational procedure is complicated.

Taylor (1961, 1980) found that, for abundance data, the following power relationship exists between the variance and the mean:

$$S^2 = a\bar{Y}^b$$

and that $Y' = Y^{1-0.5b}$ or $Y' = (Y + c)^{1-0.5b}$ is the appropriate transformation for breaking the relationship between the mean and the variance. The parameter b is obtained by linearizing the above equation with a log-log transformation ($Y' = \log_{10} Y$; $X' = \log_{10} X$) and applying linear regression techniques. When $b = 2$, a log transformation is appropriate (Green, 1979).

When $b = 0$, the distribution is considered normal and no transformation is necessary.

Downing (1979) examined the relationship between mean and variance for benthic macroinvertebrates collected with several types of samplers and from various substrata. He found that b was surprisingly constant at 1.5, suggesting that a fourth-root transformation was appropriate for all samples examined. Subsequent studies have failed to confirm Downing's suggested universality of the fourth-root transformation. Allan (1982) collected 12 replicate samples at each of several sites on 31 occasions over two years. Although these data showed a tendency for b to fall near 1.5 for seven out of 21 taxa, estimates of b were significantly different from 1.5 for the other taxa. Taylor (1980) found that, in some instances, square-root and log transformations were superior to the fourth-root transformation. Morin (1985) showed that, because the relationship between the logarithm of the variance and the logarithm of the mean is quadratic, the best transformation will depend upon the range of the means being compared.

Caution also is required when using Taylor's procedure because a log-log transformation may yield a grossly biased estimate of b (Zar, 1968; Sprugel, 1983), and the degree of bias will depend, in part, on the amount of scatter about Taylor's power relationship. A more appropriate solution for b can be obtained by nonlinear least squares regression (NLIN procedure; SAS Institute, 1987) applied directly to the means and variances suspected of following a power curve. Caution also should be exercised with the procedures of Box and Cox (1964) and Taylor (1961, 1980) where the transformation is optimized for the sample at hand, when what really is required is a transformation that will correct the population from which the sample was drawn. Sampling error may lead to selection of an inappropriate transformation, so the traditional log and square-root transformations should not be rejected too readily.

Finally, transformations alter the statistical properties of the data, and they also may alter the way in which underlying biological-environmental relationships are expressed. If a habitat had all but one of the characteristics that were necessary to support a species, this habitat would not support the species. A logarithmic transformation of habitat characteristics would render them multiplicative, rather than additive; a zero value would correctly define the site as uninhabitable, whereas untransformed data would remain additive and possibly obscure the true situation (Meffe and Sheldon, 1988). The consequences of transformation on the interpretations also may need to be considered.

7.4.2.3. Suitability of Diversity and Biotic Indices for ANOVA

Most of what has been said above about transformations applies to simple biological measures, such as total abundance or species richness. The behavior of other biological measures, such as diversity or biotic indices, is less well-known, principally because of a lack of suitable data sets in which these measures have been replicated. However, some examples of replication of indices, followed by tests of significance, are provided by Mayack and Waterhouse (1983), Cushman and Goyert (1984), and Pinder and Farr (1987). Statistical analyses of such measures occasionally have been undertaken or recommended (Bowman et al., 1971; Smith and Grassle, 1977; Narf et al., 1984), but in general the sampling distribution of many indices is unknown, and choice of an appropriate transformation is a matter of faith. For such data, nonparametric alternatives to ANOVA, such as the Kruskal-Wallis procedure (Siegel, 1956; Conover, 1980), should be considered; however, such options are limited for multifactor ANOVA.

7.4.3. Multiple Regression

Having selected an appropriate biological measure of community structure and used it to compare several sites or times of interest, one might choose to examine relationships between the biological measure and various environmental factors considered to be important determinants of the state of the community. A commonly used statistical approach to this broad area of enquiry is multiple regression. Its major uses in the analysis of benthic surveys include, first, the development of models that can be used to predict the state of the community from one or more environmental variables. Generally, one is interested in selecting a subset of environmental variables that provides adequate prediction of the biological measure. Second, multiple regression can be used as an exploratory tool to investigate relationships between a biological index and selected environmental variables with a view toward generating hypotheses that later may be tested by environmental manipulation and experimentation (see Cooper and Barmuta, Chapter 11).

Multiple regression is a univariate procedure in the sense that only one variable, the dependent variable, need be outside the control of the experiment. The analysis assumes that environmental variables are independent, but this assumption is seldom satisfied because many environmental variables are highly correlated. Disregard for correlation among variables is one of the most common causes of misinterpretation of the results of a multiple regression analysis. A stepwise approach to the analysis, whereby independent variables are included in the analysis one at a time, while excluding from consideration variance in the biological index that is explained by variables already in the model, often is considered a partial solution to correlation among the independent variables. However, it is crucial to note that the standardized regression coefficients that describe the final model do not indicate the strength of the functional relationships between environmental variables and the biological index. Both the judgment of the significance of each environmental variable and the evaluation of their rank order of importance are meaningless when the "independent" variables are correlated (Green, 1979). Application of Principal Components Analysis to environmental variables before multiple regression will overcome violations of assuming independence among variables but will frustrate the major objective of providing a minimum subset of variables for prediction.

Contrary to common perception, multiple regression cannot provide insight into causal relationships between environmental variables and the biological measure of interest. Causal relationships only can be established by experimentation, where environmental factors can be manipulated (see Cooper and Barmuta, Chapter 11). At the very best, multiple regression may provide a subset of environmental variables that can be considered to be the most parsimonious explanation for variation in the biological measure, but then parsimony serves only as a foundation for future experimentation and testing. On its own, a multiple regression seldom provides great insight into the true state of affairs. However, where true controls are unavailable and true replication of treatments (sites or times) often is impossible (as in most field situations), meeting the requirements for manipulative experiments may not be achievable. Under these circumstances, many take the view that, as an exploratory tool for gaining insight into causal relationships, only multiple regression and related procedures are available. Although multiple regression is unable to demonstrate conclusively a causal link between variables, it is superior to just staring at a table of data or at a neat representation of an ordination.

The strength of multiple regression lies in its ability to yield predictive models. For example, Downing (1986) successfully used regression techniques to establish predictive relationships between the number of organisms for each of several epiphytic invertebrate taxa and the biomass of each macrophyte species. In this example, the relative significance of environmental variables in the model, or whether or not they were causally related to the dependent variable, were not of concern. An empirical tool useful for prediction was of interest, based on some minimum set of environmental variables.

If used in stepwise fashion, multiple regression will yield a subset of available environmental variables that is best able to predict the value of the biological index of interest. Only these variables need be measured to obtain the prediction, often with considerable savings in time and cost. The resulting model may be used to predict the value of the biological index at sites where only measurements of environmental variables are available or to predict changes in the biological index on the basis of postulated changes in environmental variables. The model may be used to predict the value of a biological index in an impacted area, which then can be compared with the value observed, to assess the impact.

7.4.4. Univariate vs. Multivariate Statistics

The insight that can be obtained from univariate analyses, such as ANOVA and regression, is limited. Nonsignificant differences or trends in each of several variables may prove highly significant if analyzed by a multivariate procedure. For example, a composite variable like total macroinvertebrate abundance, if analyzed by a univariate procedure such as ANOVA, may hide different but compensating trends in the abundance of individual species (Green, 1979; Gauch, 1982). Derived biological measures (e.g., diversity indices) can be ambiguous indicators of change (Bernstein and Smith, 1986), because they cannot always be expected to change monotonically along pollution gradients. Indicator species often are superior in this regard, but the natural variability in abundance of individual species can make it difficult to quantify change consistently and precisely (Bernstein and Smith, 1986).

In contrast to univariate analyses, multivariate procedures consider each species to be a variable and the presence/absence or abundance of each species to be an attribute of a site or time. Subtle changes in the species composition across sites or in the abundance of particular species across sites are not inherently masked by the need to summarize the combined characteristics of the site as a single value. Multivariate techniques, therefore, show greater promise than univariate comparisons for detecting and understanding spatial and temporal trends in benthic fauna.

Empirical relationships between the abundance of an indicator species and various physical and chemical aspects of its environment have provided a basis for drawing inferences about the history, quality, successional status, etc., of a variety of environments (e.g., Barnes, 1983; Sheldon, 1985; Ormerod, 1987; Osborne and Davies, 1987). The same is true for relationships between levels of physical or chemical measurements and types of environment. An alternative to basing such inferences on knowledge of a few key indicator variables is to use information from groups of variables. It can be argued that patterns of variability in a group of variables provide more information than does variation in a single variable (Sprules, 1977). Many environmental problems involve multiple variables and, therefore, should be analyzed using multivariate statistics (Green, 1979). The multivariate procedures that will be discussed in the following sections include the various clustering procedures, ordination, Discriminant Function Analysis, and time-series analysis.

7.4.5. Classification

Classification involves grouping species or samples that are similar. The main grouping procedures include: (1) table arrangement, in which compositionally similar samples/sites and taxa with similar distributions are located together in a table; (2) nonhierarchical classification procedures that group the most homogeneous samples or taxa (the investigator usually predetermines the number of groups that are formed); (3) hierarchical classification, which forms groups but arranges them using a hierarchy (dendrogram); and (4) ordination.

Groups generally are formed by convenient management of information. The groups so formed usually are informationally somewhat homogeneous, a feature in which grouping offers a simplification of the complexities of the natural world (Gauch, 1982). The relationships among several group-forming procedures are illustrated in Table 7.2.

Table arrangement, which is used for classifying a "taxa by sample" matrix, was suggested first by Braun-Blanquet (1932) for work with plant communities. The approach subsequently has been used widely in Europe by plant ecologists (Gauch, 1982). The data originally were compiled manually for display in a table of compositionally similar samples and distributionally similar species. The table indicates relationships among groups of sites and/or species along a continuum, rather than in distinct groups as is the case with classification. This is the basis for the Saprobien system (Kolkwitz and Marsson, 1909; see Cairns and Pratt, Chapter 2, and Johnson et al., Chapter 4). The method has been used by Norris et al. (1982) in their "total numbers classification," which grouped taxa based on abundance and distribution relative to effluent from an abandoned metal mine (Table 7.3). Here, sites 1–3 are upstream controls, whereas sites 5–8 indicate a gradient of recovery downstream. Group 1 taxa were tolerant of trace metals, Group 2 were sensitive, and Group 3 were taxa that reached maximum numbers downstream of the effluent inflow (Norris et al., 1982). Allocation of taxa to groups is subjective (e.g., helminthid beetles in Group 1), so table arrangements by others may differ, usually at group boundaries (Gauch, 1982). Nevertheless, the method is useful for showing gradients in abundance of taxa, in groups or community types, and in sites.

Numerical techniques generally begin with the calculation of some measure of association between pairs of samples. The indices then are grouped using a mathematical technique. Many different indices are available (see reviews by Goodman and Kruskal, 1954, 1959; Sokal and Sneath, 1963; Southwood, 1978; Washington, 1984; Hruby, 1987; Johannsson and Minns, 1987). Many of the indices presented have different theoretical justifications (Grassle and Smith, 1976; Southwood, 1978; Hruby, 1987; Johannsson and Minns, 1987), so it is important for anyone approaching the field to decide which index is best suited to their particular needs. The methods used to group the indices are basically of two types (Table 7.2): classification procedures and ordination (e.g., Principal Components Analysis, Principal Coordinates Analysis, and Multidimensional Scaling).

Several features of classification procedures need to be considered when

Table 7.2. Relationships among some group-forming procedures.

Approach for Forming Groups	Methods on Which Procedure Is Based	Mathematical Procedures	Sorting Strategies
I: Subjective (visual)	I: A. Graphic and/or tabular		
II: Numerical (statistical, mathematical)	II: A. Graphic and/or tabular		
	II: B. Index of association or distance[1]	II: B.1. Ordination	
		II: B.2. Classification	II: B.2.a) Hierarchical
			II: B.2.b) Nonhierarchical

[1]Association is used to cover proximity, distance, dissimilarity, similarity, affinity, and correlation (Belbin 1987).

Table 7.3. An example of a table arrangement: the "total numbers classification" of Norris et al. (1982). Mine waste pollution enters between sites 3 and 4. Numbers are totals collected in 30 replicates on 10 occasions over two years.

	Site							
	1	2	3	4	5	6	7	8
Group 1								
Ephemeroptera								
Baetidae *Baetis baddamsae*	252	616	681	366	28	299	416	63
Coleoptera Helminthidae								
Austrolimnius sp. (adult)	27	70	40	15	8	3	2	1
Austrolimnius sp. (larvae)	314	285	148	20	2	1	6	0
Trichoptera								
Rhyacophilidae *Taschorema ferulum*	65	100	56	65	14	35	13	5
Leptoceridae *Oecetis* sp. 1	75	2,302	1,404	1,030	1,681	672	203	2
Hydropsychidae *Asmicridea* sp.	504	23	24	231	1	127	1	33
Group 2								
Basomatophora								
Hydrobiidae *Rivisessor gunnii*	597	13,511	4,491	0	0	0	0	0
Heterodonta								
Sphaeriidae *Sphaerium tasmanicum*	7	643	107	0	0	1	0	1
Ephemeroptera								
Leptophlebiidae								
Atalophlebia australis	0	127	65	0	0	0	0	0
Atalophlebioides sp. 1	773	342	205	28	2	0	0	0
Atalophlebioides sp. 2	75	724	480	25	1	0	12	0
Atalonella sp.	356	485	337	27	2	1	2	0
Caenidae *Tasmanocaenis* sp.	4	27	99	0	0	0	0	1
Diptera								
Chironomidae Orthocladiinae sp.	104	3	3	8	4	0	12	1

Trichoptera								
Helicopsychidae								
Helicopsyche murrumba	11	150	186	3	0	0	0	0
Glossosomatidae *Agapetus* sp.	376	63	261	2	0	0	0	0
Conoesucidae								
Unidentified genus sp. 1	21	23	81	10	8	3	0	1
Unidentified genus sp. 2	152	160	189	1	0	3	6	0
Group 3								
Basomatophora								
Planorbidae *Isidorella hainesii*	0	4	6	0	0	3	21	189
Amphipoda								
Ceinidae								
Austrochiltonia australis	0	9	14	0	0	4	0	174
Diptera								
Simuliidae								
Austrosimulium sp. (larvae)	32	88	133	5	5	122	5	361
Austrosimulium sp. (pupae)	0	9	11	17	0	43	14	231
Chironomidae								
Cricotopus albitibia	12	36	46	65	36	35	809	74
Eukiefferiella sp.	0	0	0	0	0	20	1,063	0
Coelopynia pruinosa	0	0	2	1	0	8	127	2
Trichoptera								
Ecnomidae *Ecnomus* sp. 1	5	18	18	46	21	59	293	3
Hydroptilidae								
Unidentified genus sp. 1	0	5	14	6	1	6	23	71
Hydracarina								
Unidentified genus sp. 1	2	25	23	2	31	2	35	40

choosing an approach (Gauch, 1982). Nonhierarchical methods will produce groups of sites or samples that are as similar as possible, but they do not provide information on the relationships among groups. Hierarchical methods, the most commonly used procedures, seek to find the most efficient step at each stage either in the progressive synthesis (agglomerative) of the population or in its subdivision to individuals (divisive), but the route may sacrifice homogeneity of the groups through which it passes. It is uncertain whether any method simultaneously can maximize hierarchical efficiency and cluster homogeneity (Lance and Williams, 1966). Classification methods can use either qualitative (presence or absence data: e.g., Crossman et al., 1974; Wright et al., 1984; Ormerod, 1987; Ormerod and Edwards, 1987) or quantitative data (abundance: e.g., Norris et al., 1982; Barnes, 1983; Osborne and Davies, 1987).

Divisive classification strategies use mathematical techniques that begin with all entities together and divide them into successively smaller groups until each one contains a single member or until a limit is reached that is determined by the researcher. Predetermined limits are useful because they save computing time and because individual entities are difficult to interpret (otherwise a classification procedure would not have been used in the first place). Agglomerative techniques begin with individual entities and form successive groups until all are included.

Monothetic approaches divide the sets of entities (usually sites or times) according to presence or absence of a single species. Such an approach would prove useful for determining individual species that are most indicative of particular pollution or habitat conditions (e.g., Murphy and Edwards, 1982). Polythetic methods use the entire taxonomic composition of samples when deriving clusters. When single indicator taxa are used to split groups, such monothetic methods may be only divisive, whereas polythetic techniques may be divisive or agglomerative (Gower, 1967). Polythetic, agglomerative approaches are the most commonly used methods (Gauch, 1982).

An important initial decision is whether or not the *relationships* among groups are needed (hierarchical vs. nonhierarchical clustering). Hierarchical methods will yield groups that are in some order. A hierarchy is the most efficient pathway for obtaining a number of groups, but not necessarily the most efficient means of obtaining final subdivisions. For example, if a group of taxa does not appear near a closely related group early in the hierarchy, it will be more dissimilar from succeeding groups, and its final representation in the hierarchy may be most dissimilar from groups to which it is, in reality, closely related. Nonhierarchical classification methods such as the REMUL program (Lance and Williams, 1975) may form groups, the members of which are as similar to each other as possible.

For example, several clustering methods were used by Norris et al. (1982) to create groups of taxa, the distribution and abundance of which responded in similar ways to trace metal pollution. In Fig. 7.4, an agglomerative polythetic clustering method with average linkage sorting (Option CM = 4 of the GENSTAT statistical package; Anonymous, 1977) is compared to a nonhierarchical clustering program (REMUL, Taxon Package of the Division of Computing Research, CSIRO; Dale et al., 1980). Comparison of the results yielded a good example of the splitting of similar groups referred to above.

The taxa of Subgroup 2A (Fig. 7.4) occurred at two of the three sites upstream of the inflow of trace metals to the river, but not at sites downstream. As the groups were formed in the hierarchy, Subgroup 2A taxa initially were not grouped with other sensitive taxa (Group 2). Thus, they became more dissimilar with each successive step in the hierarchy until they were joined at the end with taxa that all had relatively low abundance at upstream sites. The problem was resolved through observation of the clear gradient between sites along the river (Table 7.3) and by comparison with the groups of taxa produced by the nonhierarchical method (Norris et al., 1982).

If the procedure had been performed on taxa collected from a mosaic of habitats, or geographic areas, or from a water body with multiple impacts, then the misclassification noted by Norris et al. (1982) may not have been determined as easily. Additionally, the comparison with a second grouping method was beneficial (see also Murphy and Edwards, 1982).

The groups that finally are accepted from one or a number of clustering strategies (e.g., Norris et al., 1982) may be somewhat arbitrary. A difficulty with classification procedures has been the lack of objective criteria for determining the biological significance of the groups formed. Initially, no matter which of the many group-forming procedures might have been used, the final groupings essentially were dependent on the user (Lance and Williams, 1966; Hall, 1969). However, numerical grouping methods should be used principally to generate, but not to test, hypotheses.

More recently, methods have been developed for testing the success of explaining the input data, the robustness of the groups formed by the various procedures, or for testing the perceived ecological structure represented by groupings with respect to extrinsic environmental variables. "Bootstrapping" (Efron, 1979; Efron and Gong, 1983; Felsenstein, 1985) involves resampling the data set and randomizing the selection of, for example, species with replacement, but leaves the same pattern of distribution among sites. As a result, an original species may be represented a number of times in the new data set. The covariance structure between variables is not broken down (i.e., the relative abundance of species that occur together at the same site does not change). If two species are selected, then the abundance at which they were sampled will remain the same. The new randomized data set then is reanalyzed to test the robustness of the original groups. Nemec

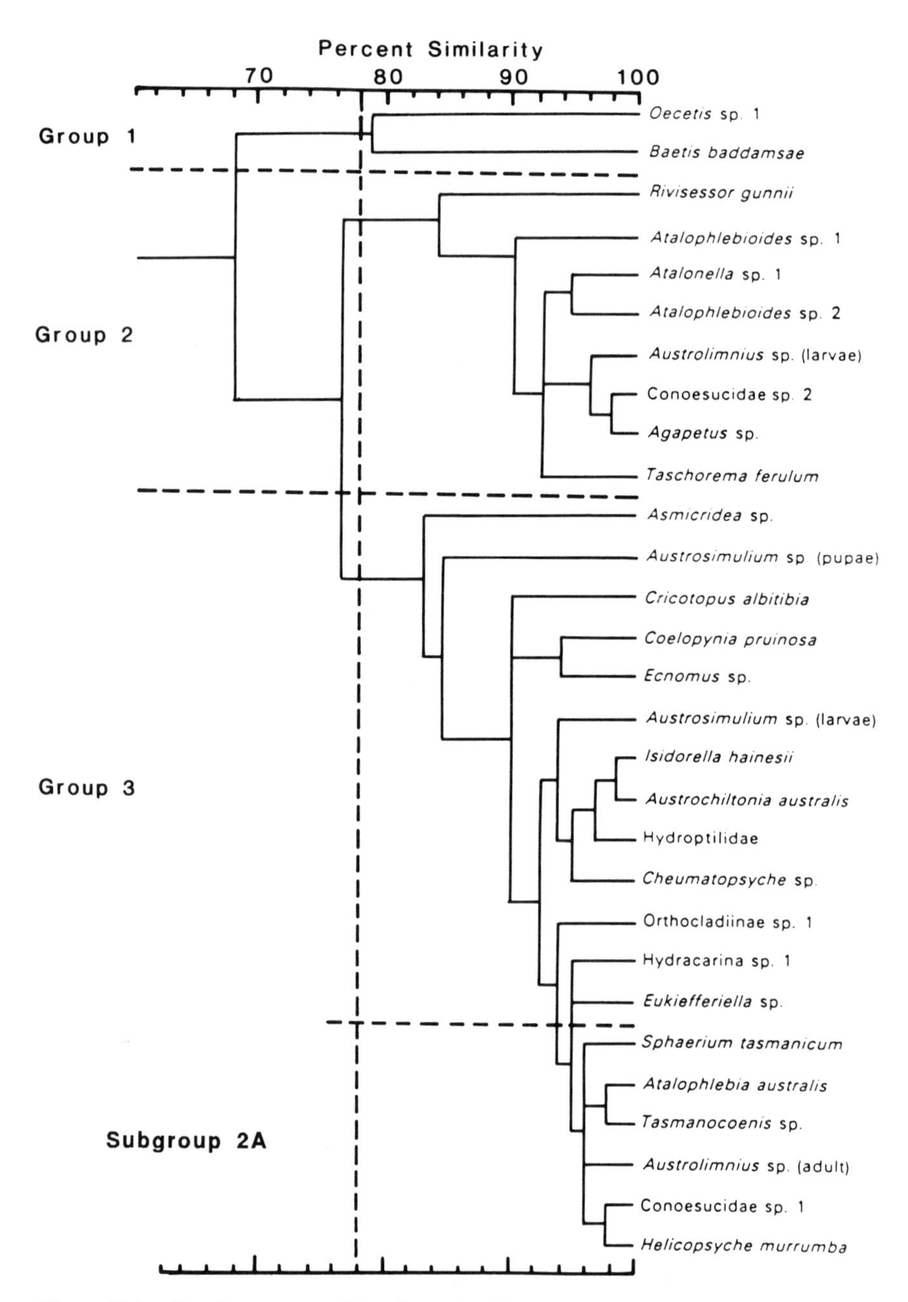

Figure 7.4. Dendrogram resulting from the hierarchical classification using an agglomerative polythetic clustering method with average linkage sorting. Subgroup 2A is composed of taxa whose distribution and abundance were close to Group 2, but which joined at the end of the hierarchy through a chaining effect. From Norris et al. (1982), reprinted by permission of the CSIRO Editorial and Publishing Unit.

and Brinkhurst (1988a, b) discussed this approach and provided examples using benthic macroinvertebrate data. Another major approach is that of "Monte Carlo simulation" (Hope, 1968), which is another randomization procedure, but one that includes each species only once and randomly changes the order of abundance values attributed to them. Because this procedure breaks down the manner in which the species covary, it tests whether the pattern evident in the data results from this covariance. The method was used to test the perceived ecological structure of data, with respect to extrinsic variables, in a study of benthic macroinvertebrates in which Hybrid Multi-Dimensional Scaling was used (Faith and Norris, 1989) (see Section 7.4.8., "Ordination," below).

These approaches test how well input data are explained by the analysis and whether or not the initial data set could have yielded the same result by using random data. The methods have not been used widely in freshwater studies, but they are seen as profitable approaches for more rigorous analysis and interpretation of data following multivariate analysis.

Often, final groups of taxa or sites, rather than the relationships between them, are used for further analysis. For example, Discriminant Function Analysis (see below) commonly is used for relating environmental variables to the groups created (Green, 1979; Marchant et al., 1984; Armitage et al., 1987; Ormerod, 1987; Osborne and Davies, 1987).

7.4.6. *Indices of Association*

The mathematical characteristics of various indices of association are described by several authors (e.g., Sneath and Sokal, 1973; Marriott, 1974; Legendre and Legendre, 1983; Belbin, 1987). Another group of authors have assessed the biological applicability of these indices (e.g., Brock, 1977; Washington, 1984; Hruby, 1987; Johannsson and Minns, 1987).

The number of alternative indices available is large (Sneath and Sokal, 1973; Legendre and Legendre, 1983), and the ones that are best to use need to be identified. Criteria for selecting indices of association are lacking but need to be developed on the basis of intrinsic and ecological properties (Hruby, 1987).

Types of data available and the nature of the sample need to be considered. Abundance data often are important for comparing communities, and an appropriate index should be used, if available (Clifford and Stephenson, 1975; Hruby, 1987). Nevertheless, presence/absence data have been used to group river sites (Murphy and Edwards, 1982; Moss et al., 1987). Murphy and Edwards (1982) considered that their samples provided unreliable abundance data, so these authors used the Jaccard Coefficient, which is specifically designed for presence/absence data (Dyer, 1978).

If large variations in abundance exist between sites or samples, then the data should be transformed (Noy-Meir, 1973; Gauch, 1982; Hruby, 1987), or an index should be chosen that accounts for these differences (e.g., the Canberra Metric Index, Clifford and Stephenson, 1975; or the Morisita Index, Hruby, 1987). The need for transformation was discussed by Gauch (1982) who suggested that ranges of 0 to 10 are best, but that ranges of 0 up to 300 can be tolerated. When community samples are relatively homogeneous, and variation is manifested in small differences in abundance, a transformation may compress the values, thereby destroying important information (Gauch, 1982). Additionally, rare taxa, which usually are defined by some arbitrary limit (e.g., Gauch, 1982; Norris et al., 1982; Marchant et al., 1984), often are deleted from the data matrix before analysis. Such taxa should not be disregarded totally because they may provide useful data particularly in relation to conservation (e.g., reanalysis by Faith and Norris, 1989, of Metzeling et al., 1984, from the La Trobe River, Victoria).

Indices of association that are based on the total number of taxa found in all samples and that treat the abundance of a taxon from samples being compared as a point of similarity, should be disregarded (Hruby, 1987). Coabsence is considered to be of little ecological significance (Clifford and Stephenson, 1975; Boesch, 1977), especially in data sets that may have many zero entries (Hruby, 1987). Additionally, Morisita's Index has abundance in the denominator, so its use can be rejected for any data sets that include sample records with no animals, a situation that may be common in polluted areas.

Euclidean distance gives more weight to abundant taxa (Clifford and Stephenson, 1975; Washington, 1984; Hruby, 1987). Pinkham and Pearson's Coefficient is sensitive to changes in rare taxa, which also may make it sensitive to normal sampling error.

The Bray-Curtis Index is favored by some (e.g., Boesch, 1977) because it varies linearly to changes in species numbers and abundance; the Canberra Metric Index does not (Bloom, 1981). Percent similarity also is useful (Johannsson and Minns, 1987) and also has been shown to respond linearly to community overlap (Bloom, 1981; Gauch, 1982).

Whittaker (1952) noted that percent similarity failed when relative proportions of taxa remained similar but overall abundance varied, because it only can be sensitive to changes in relative abundance.

Indices should be chosen relative to the type of data being analyzed, which may be transformed accordingly. Investigators should use several indices, rather than relying on a single, all-purpose one (Brock, 1977).

7.4.7. Clustering Strategies

A procedure must be selected to build up the hierarchy; such procedures are described by Sneath and Sokal (1973), Gauch (1982), and Belbin (1987).

Techniques such as Complete-Linkage Clustering, also called the Maximum Neighbor Method, produce very tight clusters (Sneath and Sokal, 1973) that may be difficult to interpret. Average Linkage Clustering uses the average dissimilarity between taxa or groups. Several methods can be used to calculate an average, so several different Average Linkage techniques are available (Gauch, 1982; Belbin, 1987). The method most commonly used is Unweighted Pair-Groups using Arithmetic Averages (UPGMA), which uses a simple unweighted arithmetic average (Gauch, 1982). Sneath and Sokal (1973) recommend this technique when no specific reason exists for choosing another.

Polythetic divisive techniques include Detrended Correspondence Analysis (DECORANA; Hill, 1979a) and Two-Way Indicator Species Analysis (TWINSPAN; Hill, 1979b). Both of these programs have been used successfully to analyze benthic macroinvertebrate data (Wright et al., 1984; Moss et al., 1987; Ormerod, 1987; Ormerod and Edwards, 1987). The TWINSPAN program arranges the clusters of sites and individuals in a two-way table similar to the Braun-Blanquet approach already discussed.

7.4.8. *Ordination*

When two environmental or biological variables are correlated, knowledge of one provides information on the other. This redundancy of information inherent in most multivariable data sets provides scope for analysis using ordination. Ordination has one or both of the following objectives: (1) to reduce the dimensionality of a complex multivariate data set with minimal loss of information; and (2) to extract a set of uncorrelated variables from a set of correlated variables (called components or factors), perhaps as a prelude to multiple regression.

Ordination represents the entities that are the subject of study (usually the sites or samples) in a multidimensional space that is defined by taking each of the attributes (usually species or environmental variables) as an axis. A simplified version of the basis for ordination is shown in Fig. 7.5. The distance between any two samples plotted in this space corresponds in some way to the value of the distance measure used to represent their dissimilarity. New axes, or factors, then are defined as linear combinations of the measurement variables such that they are orthogonal (i.e., independent), and these new axes are ordered so that the first factor accounts for most of the variation among samples and the last factor explains the least. In the hypothetical example of Fig. 7.5, the first factor explains 93% of the total variation among samples and the second factor explains the remaining 7%. A subspace of preferably two or three dimensions then is chosen as a summary of the structure in the data. In the hypothetical example, Factor II

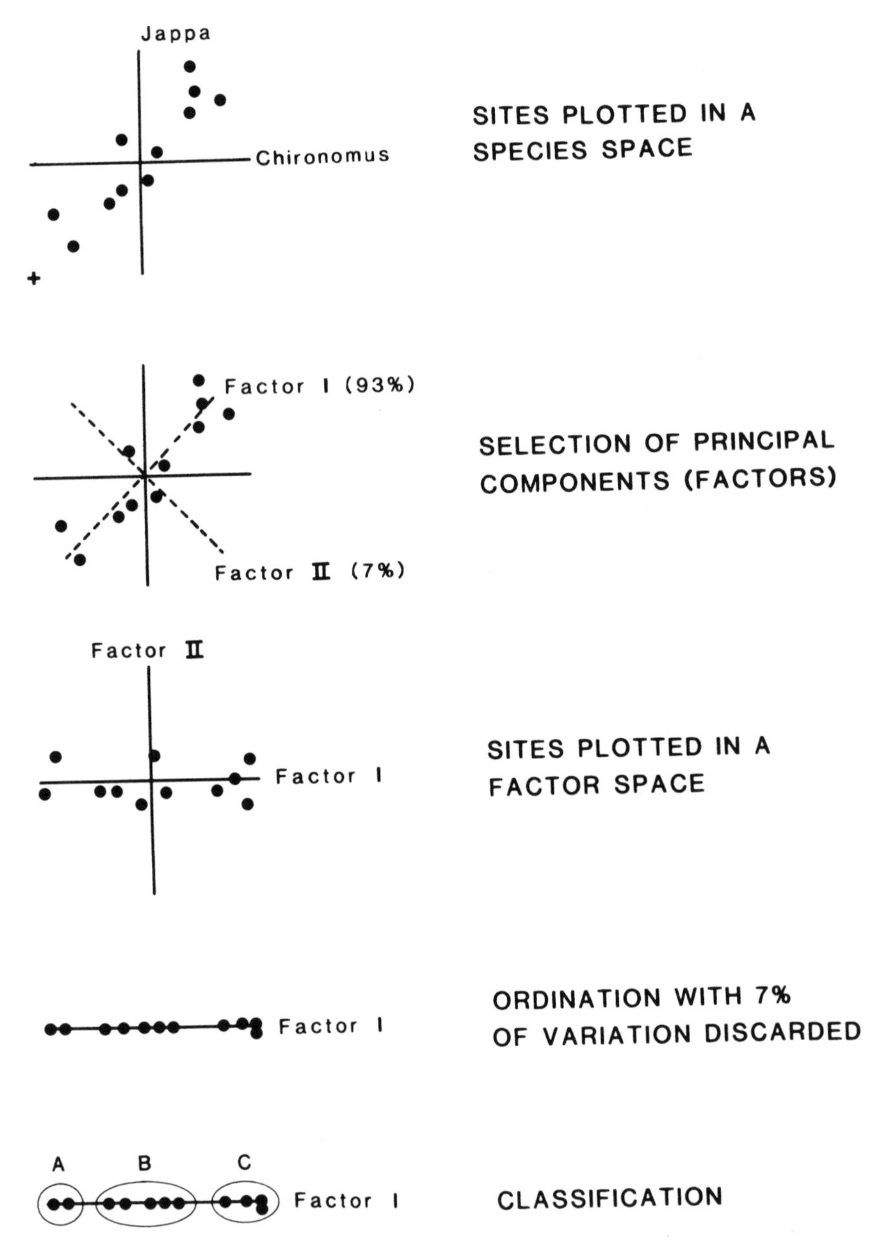

Figure 7.5. A visual rationale for ordination by the method of Principal Components. The first step is to plot sites or samples in a space defined by taking each species as an axis and letting abundance represent the position of the sample along each axis. The coordinate axes then are centered on the multivariate mean (+ is the origin before centering) and new orthogonal factors are defined progressively, each in the direction of maximal remaining variation. Sites or samples then are visualized in the new factor space, and least informative dimensions are discarded, with some loss of information. Patterns in the reduced space then are identified and trends or clusters defined.

could be disregarded with a loss of only 7% of variation. Thus, rejection of lower dimensions is justified, provided that a substantial proportion of the total variation among samples is explained by the first two or three axes that define the subspace. The relationships among samples, whether they are clusters or continuous trends, then may be evident in the resulting two- or three-dimensional plots (Fig. 7.5).

Ordination is used extensively in studies of benthic macroinvertebrates (Table 5.6 in Resh and McElravy, Chapter 5). For example, in assessing the impact of a point discharge of industrial effluents on benthic communities in central Canada, the relationships among sampling stations above and below the discharge were summarized in two dimensions defined by the first and second principal components (Clarke, 1977). A clear impact of the discharge was demonstrated, and the analysis provided indicator species that were responsible for the separation of stations along each of the two axes. Similarly, Descy (1973) used ordination to identify species of algae that were sensitive to, or tolerant of, organic pollution. Ormerod and Edwards (1987) used ordination to summarize relationships between environmental variables and benthic macroinvertebrate taxa among 45 sites on the River Wye catchment in the United Kingdom. The attributes were 13 physicochemical variables, and the ordination yielded two axes that explained 72% of total variation. Natural groupings of sites were identified from a plot of the two axes, and correlations between the original physicochemical variables and the two new axes yielded a basis for interpreting which physicochemical variables were associated with separation of groups. Leland et al. (1986) used Detrended Correspondence Analysis, a form of ordination, to detect underlying environmental or temporal gradients that were useful in explaining the spatial distribution of benthic macroinvertebrates. Clearly, a wide range of applications of ordination exist in the study of benthic macroinvertebrates. The reader is referred to Gauch and Whittaker (1972), Green (1979), Culp and Davies (1980), Leland et al. (1986), and Minchin (1987) for detailed evaluations.

The many approaches to ordination are introduced in the readable text by Manly (1986) and are described in some detail by Gauch (1982), Legendre and Legendre (1983), and Pielou (1984). The first procedure developed was Principal Components Analysis (PCA) (Hottelling, 1933), which is based essentially on the correlation coefficient as a measure of similarity between samples. This measure often is considered inappropriate for ecological data, so a wide range of similarity/dissimilarity measures has been developed and used instead for studies of benthic macroinvertebrates (see above). A generalization of PCA that allows a wide range of dissimilarity measures is Principal Coordinates Analysis (PCoA) (Gower, 1966), which requires only that the dissimilarity measure be a metric (essentially that it satisfy the triangle inequality). In PCoA, the dissimilarity between each pair of samples is measured, using some chosen measure of dissimilarity, and the points then are plotted to make the squared distance between every pair of points correspond to the squared dissimilarity. An initial "sample by species" matrix is not required by the analysis. Any metric distance matrix summarizing the relationships between samples can be used so that the researcher is not limited to similarities based on the correlation coefficient.

Should the dissimilarity measure be nonmetric, the procedure will produce satisfactory results provided that care is taken not to include axes in the final solution for which the cumulative percentage variation explained equals or exceeds 1.0, or for which the associated eigenvalues are negative. The often-quoted assumption of normality of the measurements that form the data set also may be disregarded, provided no statistical tests or inferences are to be made and that the technique is to be used only in the search for pattern (Marriott, 1974).

One difficulty with PCA and PCoA is that the solution is optimized in terms of squared distances, so that larger distances between samples are given disproportionate weight. This may be an advantage if one is seeking clusters, because the first few axes will emphasize distances between natural groupings at the expense of distances between samples within the groupings. However, less information will be summarized in the same number of axes compared with a technique such as Multidimensional Scaling (MDS) (Kruskal, 1964a,b). MDS endeavors to find, using an iterative procedure, the best fit between the input dissimilarities and the distances between samples in the resulting ordinated space. MDS now is chosen by many over PCA and PCoA when dealing with ecological data. MDS also is less likely to produce the distorted representations of underlying gradients that affect PCA (Noy-Meir and Austin, 1970; Austin and Noy-Meir, 1971). PCA and PCoA assume a linear response in the abundance of species along environmental gradients, which is a poor reflection of reality. Nonmetric Multidimensional Scaling (NMDS), so-called because only the rank order of the dissimilarities between samples is preserved in the geometric representation, can accommodate a much wider range of response functions (Minchin, 1987) but still assumes monotonicity in the response of species abundances to environmental gradients. A disadvantage of MDS in comparison with PCA and PCoA is that the analyst must provide the dimension of the solution in advance, and the most appropriate dimension may not be very evident. One must repeat the computationally expensive analysis for each of several dimensions and look for a significant decline in "stress" (Manly, 1986) with increasing dimension. Stress is a measure of the fit of the dissimilarity between samples and the distance between them in the ordinated space. In PCA and PCoA, one needs only to peruse the list of eigenvalues to decide on the most appropriate dimension for the solution.

7.4.9. Discriminant Function Analysis

It is often desirable to develop predictive relationships between groups of sites or taxa, which are established using clustering methods applied to macroinvertebrate abundance, and physicochemical characteristics. Discriminant Function Analysis (DFA) is appropriate. The value of a multivariate technique such as DFA over univariate approaches is shown in Fig. 7.6. Here, the original variables, filterable reactive phosphorus (FRP) and total Kjeldahl nitrogen (TKN), when taken alone, each discriminate poorly between the site groups. When used in combination to define a discriminant function in the direction of maximal variation between groups, the discrimination is absolute. Discriminant functions are chosen as independent linear combinations of the original physicochemical variables that best discriminate between previously established groups of sites.

DFA differs from multiple regression in that the variable to be predicted, site group, is discrete rather than continuous, and the independent variables in DFA are assumed to be normally distributed. It differs from PCA because groups of sites are defined before the analysis is undertaken, and variation between sites within groups is disregarded in the process of calculating the factors. These factors are called discriminant functions.

One of the major uses of DFA in the analysis of benthic surveys includes the development of models that can be used to predict the membership of a site or group of taxa in a previously established classification of sites or taxa. For example, the detection of pollution through differences between predicted and actual faunal assemblages might be an objective (Ormerod and Edwards, 1987). Selection of a subset of environmental variables that provides adequate prediction of the biological measure is usually of interest. DFA also is used as an exploratory tool to investigate relationships between the macroinvertebrate groups established and physicochemical variables, with a view toward generating hypotheses that later may be tested by environmental manipulation and experimentation (see Cooper and Barmuta, Chapter 11).

An example of the application of DFA to benthic surveys is provided by Marchant et al. (1984) from a study of the La Trobe River in southeastern Australia. The river was subjected to a number of different impacts caused by agricultural runoff, sewage effluent, a dam, a weir, heated water from a power station, saline drainage from a coal mine, and industrial effluents. The authors were concerned with establishing the thermal effects of the power station, which involved separating the effects of heated water from those of other impacts. Sampling of benthic macroinvertebrates was stratified, with 15 replicates from each of the main channel and the river margins. Twenty-three variables commonly used to assess water quality were measured on each sampling occasion over two years.

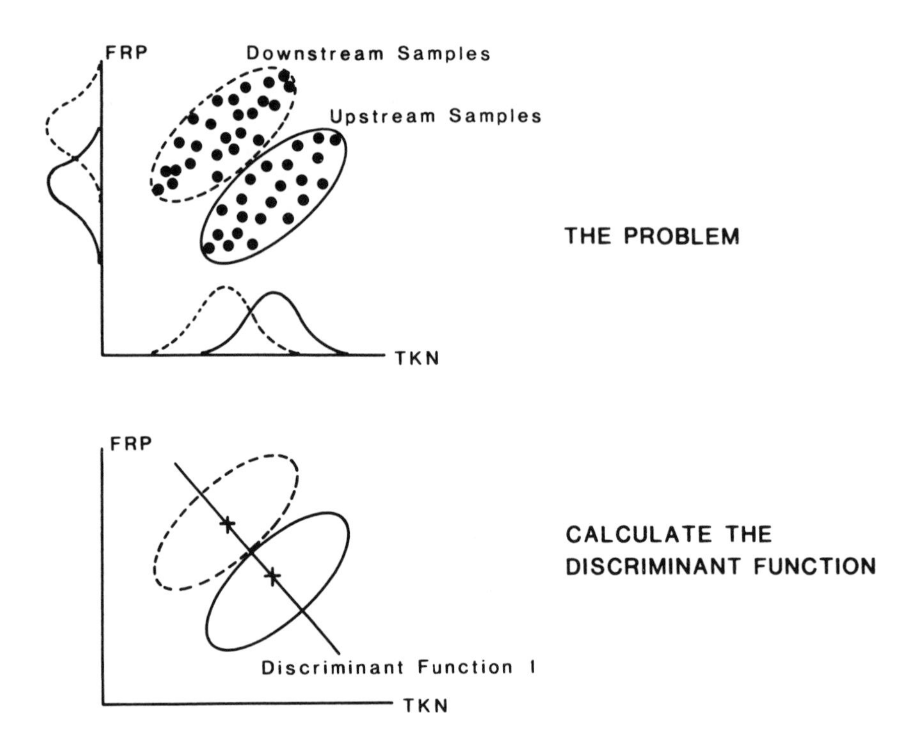

Figure 7.6. A visual rationale for Discriminant Function Analysis. In this hypothetical example, samples are classified before the analysis as being either upstream or downstream of a point impact. When the environmental variable, filterable reactive phosphorus (FRP), is considered alone, discrimination is poor between upstream and downstream sites. Another variable, total Kjeldahl nitrogen (TKN), is equally poor for discrimination. When taken together and used to define a linear discriminant function, these variables are able to discriminate absolutely between upstream and downstream samples.

Sites were grouped based on composition and abundance of the fauna using the Czekanowski and the Canberra Metric indices, and hierarchical classifications were formed using Average Linkage Clustering. DFA then was used to determine the extent to which the water quality data could distinguish groups of sites based on their faunal similarity. Physical and chemical variables that clearly discriminated between site groups were considered to be associated closely with changes in the fauna (Marchant et al., 1984).

Two DFAs were performed: one for the site groups based on the main channel samples and one for the groups based on the river margin samples. Separation of the groups of main channel sites occurred largely with conductivity (70% of variation explained) in the first discriminant function. The second discriminant function accounted for the remaining variation and was related mainly to changes in suspended solids, conductivity, and potassium concentrations. The groups of river margin samples also were discriminated mostly by conductivity (57% of variation explained) in the first discriminant function. A further 28% of the variation in the second discriminant function was related to FRP, and 12% was related to suspended solids.

The discriminant functions were related easily to impacts on the river, thus implicating saline mine drainage, sewage effluent, and the dam as affecting the fauna at particular sites. Heated water from the power plant and effluents from a pulp mill had relatively little impact on the fauna (Marchant et al., 1984).

Like multiple regression analysis, DFA assumes that the physicochemical variables are independent, and a stepwise approach is recommended for finding a minimum subset of such variables that provides adequate prediction of group membership. Care should be taken to use a procedure that allows for the removal of variables from the analysis should they become redundant following the addition of subsequent variables. Forced retention of redundant variables actually may reduce discriminatory power (Kleinbaum and Kupper, 1978). Also, like multiple regression, it is important to realize that the standardized coefficients that describe the final model do not provide an indication of the strength of functional relationships between physicochemical variables and the hierarchical structure of sites revealed in the cluster analysis. Nor can DFA provide clear insight into causal relationships between physicochemical variables and the group membership of particular sites. When performing DFA, concern is neither with the relative significance of environmental variables in the model, nor whether or not they are related causally to the dependent variable. Interest is only in an empirical tool useful for prediction, but based on some minimum set of environmental variables.

DFA has a major advantage over other multivariate techniques because the outcome of the analysis is relatively easy to validate. The most common approach is to resubstitute the physicochemical data for each site into the discriminant functions to determine the proportion of sites that are misclassified (e.g. Marchant et al., 1984). This approach is circular because the sites used to validate the predictive model are those used to formulate it initially, but it provides a rough indication of the power of discrimination. A more rigorous approach is to validate on the basis of a set of sites of known group membership, but which were not used in developing the discriminant functions, or to use bootstrapping (where DFA is calculated from a subset of the data and validated using the remainder, and the process is repeated for an estimate of the reliability of discrimination).

7.4.10. Time-Series and Trend Analyses

7.4.10.1. Time Series

Many study designs for the collection of benthic macroinvertebrates involve sampling through time. The numbers of animals collected at successive samplings in time and space may be correlated highly with the numbers collected in previous samples. Such autocorrelation (Hurlbert, 1984; Stewart-Oaten et al., 1986; Barmuta, 1987) may invalidate the use of many parametric statistical tests because the assumption of independence is violated. A family of statistical methods called "time-series analysis" may be appropriate for analyzing these types of data, but the methods seem to have been overlooked in benthic macroinvertebrate studies. The approach has been reviewed by Green (1979) and Millard et al. (1985), and Van Latesteijn and Lambeck (1986) have applied time-series analysis to the effects of estuary closure on oyster catchers in Holland. An easily read introductory text on the subject is provided by Chatfield (1984). Examples of the application of time-series analysis to benthic biomonitoring studies are lacking, but because it is a potentially useful approach with which few seem to be familiar, a description of the procedure follows.

Time-series analysis involves the calculation of autocorrelations and cross-correlations. The autocorrelation of a series of observations taken over time is a measure of the extent to which a variable is dependent on its own past values. Calculation of the autocorrelation function (ACF) of a time series involves determining the correlation of each value with values 1, 2, 3, . . . k intervals (lags) preceding it. "k" of these autocorrelations will exist for a series, together making up the autocorrelation function (Box and Jenkins, 1976). The ACF of a time series with a seasonal pattern will exhibit a peak at a lag that corresponds to the length of the cycle. A time series with no serial dependence will produce an ACF with scatter around zero. The ACF of a random series can be shown to have a mean of zero, a variance of $1/N$ and a normal distribution, so values that lie outside $\pm 1.96 / \sqrt{N}$ are significant at the 5% probability level (Chatfield, 1984). Calculation of the

5% probability limits on the ACF allows identification of significant structure in the series.

Calculation of the ACF is a necessary preliminary step to calculation of the cross-correlation function (CCF). The CCF is a measure of the dependence of one variable upon another, not only at a particular time but also with past values of the other variable. This function comprises a set of ordinary correlation coefficients between pairs of observations in the two series at increasing lag. The cross-correlation coefficient at lag zero corresponds to the normal correlation coefficient. In a plot of the CCF against lag, the lag at which significant peaks occur will indicate how far out of step the two series are. As with the ACF, it can be shown that for two uncorrelated random series the CCF has an expected mean of zero and variance of $1/N$ (Chatfield, 1984), so values outside the interval $\pm 1.96 / \sqrt{N}$ are significantly different from zero ($P < 0.05$). If the two series being cross-correlated are unrelated, 5% or less of the cross-correlations in the CCF should be significant ($P < 0.05$). The presence of more significant cross-correlations in the CCF is an indication of a significant association between the series. Because cross-correlations in the CCF may be positive at some lags and negative at others, it may be difficult to categorize significant associations between series as positive or negative.

Estimates of the cross-correlations can be shown to be asymptotically unbiased and consistent. However, variance of the CCF depends on the ACFs of the two component series (Chatfield, 1984). Consequently, the CCF of two totally unrelated series, each with high autocorrelation, will show large, spurious cross-correlations solely because of their high autocorrelation. To eliminate this effect, the series first can be filtered to remove autocorrelation, by fitting Autoregressive Integrated Moving Average (ARIMA) models (Box and Jenkins, 1976), before computing cross-correlations (Jenkins and Watts, 1968). The residuals from the ARIMA models are the filtered series.

The problem of "aliasing" may arise if the sampling interval is longer than the frequency of change of a variable. Aliasing is the false representation of high-frequency variation in the ACF as lower-frequency variation (Chatfield, 1984). Aliasing will bias the ACF, but it will not produce a significant ACF from a random series. Filtering the two series before cross-correlation will remove serial dependence of observations whether it is real or produced by aliasing, so that the subsequent cross-correlations will not be influenced by aliasing. However, relationships between variables with lags shorter than the sampling interval cannot be identified by the cross-correlation analysis.

In practice, time-series analysis requires observations at regular intervals with no missing values, but missing values can be interpolated using cubic spline-fitting procedures. Spline functions have the advantage of producing a closer fit to the data than a global polynomial because the fit of the function for each region is determined by local, not global, conditions (Wold, 1974).

Time-series analysis techniques require that variables be distributed normally and that no relationship exist between the mean and variance (Chatfield, 1984). Many limnological variables are distributed approximately lognormally, so a logarithmic transformation is appropriate (Platt et al., 1970; Rutherford, 1984). This has a further benefit: if variables have a seasonal component, it will convert any multiplicative relationship between the mean, the seasonal term, and the error term to an additive one, which allows the series to be fitted to a linear equation (Chatfield, 1984).

Time-series analysis further requires that any long-term trend in variables be removed before analysis (Chatfield, 1984). This can be done by "differencing" the data, that is, producing a new series made up of the differences between successive pairs of observations in the original series. A linear trend will be removed by differencing once, a quadratic trend by differencing the already differenced series, and so on. Seasonal variation also may be removed by seasonal differencing, that is, subtracting pairs of observations that are a season apart. This allows resolution of features that may have been concealed by the magnitude of seasonal variation.

The first steps in analysis are interpolation of missed observations and, where necessary, transformation of variables (pH excepted). ACFs of each variable then may be calculated and, using these functions, ARIMA models of the series can be fitted. The residual series from these models, the filtered series, are used to calculate the CCFs between variables.

7.4.10.2. Trends

Trends in water quality data may be detected by a variety of methods, including nonparametric, parametric, spectral, and time-series analysis. Nonparametric methods have been suggested for water quality data by Hirsch et al. (1982) and Van Belle and Hughes (1984) because of the problems of nonnormality, nonlinearity, nonindependence, missing values, censored data, and periodic cycles. The seasonal Kendall test can be recommended (Hirsch et al., 1982) because, under realistic stochastic processes (exhibiting seasonality, skewness, and serial correlation), it is robust in comparison to parametric alternatives, although it is not an exact test in the presence of serial correlation (autocorrelation). A second test recommended by Hirsch et al. (1982) is the Kendall slope estimator, which estimates trend magnitude and is an unbiased estimator of the slope of a linear trend. It has considerably higher precision than a regression estimator in which data are highly skewed, but it has lower precision when the data are distributed normally. The third procedure recommended provides a means of testing for change over time in the relationship between constituent concentrations and flow, thus avoiding the problem of identifying trends in water quality that are artifacts of the particular sequence of discharges observed (e.g., drought effects). In this last method, a flow-adjusted concentration is defined as the residual (actual minus conditional expectation) based on a regression of concentration on some function of discharge. These flow-adjusted concentrations, which also may be seasonal and nonnormal, then can be tested for trend by using the seasonal Kendall test.

The simplest parametric method is to calculate means over some interval (e.g., monthly), plot these against time, and then fit a linear regression through the points (Ellis and Lacey, 1980). The line is tested for significance of fit and slope, and if they are significantly different from zero, either positively or negatively, a trend will be identified.

Spectral analysis is presented in detail by Chatfield (1984). Basically, it is concerned with approximating trends or periods in the data by a sum of sine and cosine terms. A detailed description of this approach is outside the scope of this chapter, but Chatfield (1984) is a useful text for this procedure.

Many ongoing monitoring programs are yielding long-term data sets for which analysis of trends is often a primary objective. Time-series analysis may avoid autocorrelation problems that can occur when measurements are repeated through time; autocorrelation violates the assumption of independence required for parametric tests. Analysis using time-series methods potentially is a powerful tool that has been used infrequently in studies of benthic macroinvertebrates.

7.5. Summary and Recommendations

1. Most studies of benthic macroinvertebrates need comparisons of data among different sites or times, or association with physical and chemical measurements to infer causal relationships. Interpretation of causal relationships should be based on real biological properties relating to the environment rather than just on statistical interpretations. However, causal relationships only can be determined fully through direct experimental work.
2. In making any kind of comparison, environmental variability and its effects on sampling will need to be accounted for, first, in the study design and, second, in the analysis of data.
3. The quality of data and the assurance that they are free from recording, measurement, transcription, and other errors is an important first step in data analysis. Apart from strict field and laboratory protocols to ensure data quality, many subsequent checks can be made for ranges, means, standard deviations, frequency

of occurrences, ratios, regressions, plotting, and subcounts vs. total counts. Missing data also need to be identified and a strategy to deal with them adopted (either by deletion of records or by some means of estimating the missing value).

4. Diversity and biotic indices are used commonly to condense presence/absence and abundance data for further analysis and interpretation. Many of the arguments in favor of using indices are not theoretically valid, many shortcomings are evident in their application, and the biological or ecological meaning of what they purport to measure is poorly understood. Most biotic indices are specific to the conditions and sites of their development, and much background information on responses of taxa to pollution usually is needed before they can be implemented. They are as subject to error as any other community measures would be, and the level of uncertainty in their use needs to be established before they can be interpreted properly. It is concluded that indices should be used with caution.
5. ANOVA is a major method of analysis used to make comparisons and to partition total variability into components of the study. It relies on replicated sampling and requires that certain assumptions be met. Normally distributed data is one assumption that may be attended to after data collection; this is done by transformation. Caution is advised in the use of diversity or biotic indices with ANOVA, unless the statistical distribution of the data is known and appropriate transformations (if needed) can be performed.
6. Multiple regression may be a useful technique for developing predictive models and for hypothesis generation. When a subset of environmental variables is required that is best able to predict the biological measure of interest, a stepwise application of the method should be used.
7. Many benthic studies will be multiple-variable problems and, therefore, are amenable to the use of multivariate statistical techniques. Multivariate procedures consider each species to be a variable and the presence/absence or abundance of each species to be an attribute of a site or time. Subtle changes in the species composition or in the abundance of particular species across sites are not inherently masked by the need to summarize the combined characteristics of the site as a single value. Multivariate techniques, therefore, show greater promise than univariate comparisons for detecting and understanding spatial and temporal trends in the benthic macroinvertebrate fauna.
8. Several methods are available for the classification or grouping

of data, including table arrangement and hierarchical and nonhierarchical procedures. The most commonly used procedures are polythetic, agglomerative ones, which group attributes of the data set through a hierarchy. If the relationships among groups are needed, then hierarchical methods are recommended, but if the final groups alone are important, then nonhierarchical or ordination procedures might be more appropriate. The indices of association that are used in classification procedures need to be considered with respect to the type of data being analyzed and features of the index. Recent advances have been made in methods for testing the significance of groups formed using multivariate procedures, including bootstrapping and Monte Carlo simulation. Introduction of these methods is seen as a major advance in the use of multivariate analysis.

9. Ordination may be used to reduce the dimensionality of complex multivariate data sets with minimal loss of information or to extract a set of uncorrelated variables from a set of correlated ones, perhaps as a prelude to multiple regression. In benthic studies, ordination has been used to define indicator species that best represent differences among sites that may have been impacted by pollution. Several ordination procedures may be used, depending on objectives of the study from which the data are derived.

10. Difficulties often arise in associating physical and chemical data with biological data so that the change in environmental variables most closely associated with the change in biological ones can be defined. To this end, Discriminant Function Analysis (DFA) can be used to develop a predictive relationship between groups of sites or taxa that have been established using clustering or ordination techniques. DFA also may be used as an exploratory tool to investigate relationships between groups established from the macroinvertebrate fauna and physical and chemical variables, with a view to generating hypotheses that later may be tested by environmental manipulation and experimentation. The method has been used successfully in benthic studies to separate the effects of multiple impacts on a river.

11. Time-series analysis is a powerful tool that virtually has been ignored in benthic studies. It may be used to develop predictive models based on variation in past time series. Time-series analysis also may overcome the problem of autocorrelation in data between sites or times, which may invalidate the use of many parametric statistical tests. The method also can determine the occurrence of trends, often a primary aim of monitoring studies. Time-

series analysis relies on data that are collected regularly at intervals more frequent than the period of variation among the variables of interest. These requirements, although quite stringent, are fulfilled by most monitoring programs.

Acknowledgments

We thank the Faculty of Applied Science, University of Canberra, for supplying facilities during the preparation of this manuscript, D. Tiller for providing unpublished data on the Thredbo River, and P. Liston for his extensive input on time-series analysis.

References

Abel, P.D. 1989. *Water Pollution Biology*. Ellis Horwood, Chichester, England.

Allan, J.D. 1982. The effects of reduction in trout density on the invertebrate community of a mountain stream. *Ecology* 63:1444–55.

Allan, J.D. 1984. Hypothesis testing in ecological studies of aquatic insects. In *The Ecology of Aquatic Insects,* eds. V.H. Resh and D.M. Rosenberg, pp. 484–507. Praeger Pubs., New York.

Anderson, R.M., D.M. Gordon, M.J. Crawley, and M.P. Hassell. 1982. Variability in the abundance of animal and plant species. *Nature* (London) 296:245–8.

Anonymous. 1977. *General Statistical Programme*. Rothamsted Experimental Station, Harpendon, Hertfordshire, England.

Armitage, P.D., R.J.M. Gunn, M.T. Furse, J.F. Wright, and D. Moss. 1987. The use of prediction to assess macroinvertebrate response to river regulation. *Hydrobiologia* 144:25–32.

Auclair, A.N. and F.G. Goff. 1971. Diversity relations of upland forests in the western Great Lakes area. *American Naturalist* 105:499–528.

Austin, M.P. and I. Noy-Meir. 1971. The problem of non-linearity in ordination: experiments with two-gradient models. *Journal of Ecology* 59:763–73.

Balloch, D., C.E. Davies, and F.H. Jones. 1976. Biological assessment of water quality in three British Rivers: the North Esk (Scotland), the Ivel (England) and the Taf (Wales). *Water Pollution Control* 76:92–110.

Barmuta, L.A. 1987. Polemics, aquatic insects and biomonitoring: an appraisal. In *The Role of Invertebrates in Conservation and Biological Survey,* ed. J.D. Majer, pp. 65–72. Western Australian Department of Land Management Report. Western Australian Department of Land Management, Perth, Western Australia.

Barnes, L.E. 1983. The colonization of ball-clay ponds by macroinvertebrates and macrophytes. *Freshwater Biology* 13:561–78.

Beak, T.W. 1965. A biotic index of polluted streams and its relationship to fisheries. In *Advances in Water Pollution Research. Proceedings of the Second International Conference, Tokyo, August, 1964, Vol. 10*, ed. O. Jaag, pp. 191–210. Pergamon Press, Oxford, England.

Bechtel, T.J. and B.J. Copeland. 1970. Fish species diversity indices as indicators of pollution in Galveston Bay, Texas. *Contributions in Marine Science* 15:103–32.

Beck, W.M., Jr. 1955. Suggested method for reporting biotic data. *Sewage and Industrial Wastes* 27:1193–7.

Belbin, L. 1987. *PATN: Pattern Analysis Package Reference Manual*. Division of Wildlife and Rangelands Research, Commonwealth Scientific and Industrial Research Organization, Canberra, Australia.

Bernstein, B.B. and R.W. Smith. 1986. Community approaches to monitoring. In *Oceans '86 Conference Record: Science-Engineering-Adventure. Proceedings of a Conference Hosted by the Washington D.C. Section of the Marine Technology Society, Washington, D.C., September 23–25, 1986*, pp. 934–9. Institute of Electrical and Electronic Engineering, Piscataway, NJ.

Bloom, S.A. 1981. Similarity indices in community studies: potential pitfalls. *Marine Ecology Progress Series* 5:125–38.

Boesch, D.F. 1977. *Application of Numerical Classification in Ecological Investigations of Water Pollution*. Special Scientific Report No. 77. Virginia Institute of Marine Science, Gloucester Point, VA. (Also issued as U.S. Environmental Protection Agency Report EPA-600/3–77–033. Available from National Technical Information Service, Springfield, VA as PB-269 604/5BE.)

Bowman, K.O., K. Hutcheson, E.P. Odum, and L.R. Shenton. 1971. Comments on the distribution of indices of diversity. In *Statistical Ecology. Vol. 3. Populations, Ecosystems and Analysis*, eds. G.P. Patil, E.C. Pielou, and W.E. Waters, pp. 315–66. Pennsylvania State Univ. Press, State College, PA.

Box, G.E.P. and D.R. Cox. 1964. An analysis of transformations. *Journal of the Royal Statistical Society* (Series B) 26:211–52.

Box, G.E.P. and G.M. Jenkins. 1976. *Time Series Analysis, Forecasting and Control*. Holden-Day, San Francisco.

Braun-Blanquet, J. 1932. *Plant Sociology: The Study of Plant Communities*. Authorized English translation of *Pflanzensoziologie*, 1st ed. Translated, edited, and revised by C.D. Fuller and H.S. Conard. McGraw-Hill, New York.

Brock, D.A. 1977. Comparison of community similarity indexes. *Journal of the Water Pollution Control Federation* 49:2488–94.

Cairns, J., Jr., D.W. Albaugh, F. Busey, and M.D. Chanay. 1968. The sequential comparison index—a simplified method for non-biologists to estimate relative differences in biological diversity in stream pollution studies. *Journal of the Water Pollution Control Federation* 40:1607–13.

Cairns, J., Jr., G.R. Lanza, and B.C. Parker. 1972. Pollution related structural and functional changes in aquatic communities with emphasis on freshwater algae and protozoa. *Proceedings of the Academy of Natural Sciences of Philadelphia* 124:79–127.

Cairns, J., Jr. and W.H. Van Der Schalie. 1980. Biological monitoring. Part I—Early warning systems. *Water Research* 14:1179–96.

Canton, S.P. and J.W. Chadwick. 1988. Variability in benthic invertebrate density estimates from stream samples. *Journal of Freshwater Ecology* 4:291–7.

Chadwick, J.W. and S.P. Canton. 1984. Inadequacy of diversity indices in discerning metal mine drainage effects on a stream invertebrate community. *Water, Air, and Soil Pollution* 22:217–23.

Chandler, J.R. 1970. A biological approach to water quality management. *Water Pollution Control* 69:791–2.

Chatfield, C. 1984. *The Analysis of Time Series: Theory and Practice,* 3rd ed. Chapman and Hall, London.

Cherry, D.S., R.K. Guthrie, E.M. Davis, and R.S. Harvey. 1984. Coal ash basin effects (particulates, metals, acidic pH) upon aquatic biota: an eight-year evaluation. *Water Resources Bulletin* 20:535–44.

Chutter, F.M. 1972. An empirical biotic index of the quality of water in South African streams and rivers. *Water Research* 6:19–30.

Chutter, F.M. and R.G. Noble. 1966. The reliability of a method of sampling stream invertebrates. *Archiv für Hydrobiologie* 62:95–103.

Clarke, R.McV. 1977. The use of multivariate techniques in analysing effects of industrial effluents on benthic communities in central Canada. In *Biological Monitoring of Inland Fisheries,* ed. J.S. Alabaster, pp. 133–41. Applied Science Pubs., London.

Clifford, H.T. and W. Stephenson. 1975. *An Introduction to Numerical Classification.* Academic Press, New York.

Cochran, W.G. 1947. Some consequences when the assumptions for the analysis of variance are not satisfied. *Biometrics* 3:22–38.

Conover, W.J. 1980. *Practical Nonparametric Statistics,* 2nd ed. John Wiley, New York.

Crossman, J.S., R.L. Kaesler, and J. Cairns, Jr. 1974. The use of cluster analysis in the assessment of spills of hazardous materials. *American Midland Naturalist* 92:94–114.

Culp, J.M. and R.W. Davies. 1980. Reciprocal averaging and polar ordination as techniques for analysing lotic macroinvertebrate communities. *Canadian Journal of Fisheries and Aquatic Sciences* 37:1358–64.

Cushman, R.M. and J.C. Goyert. 1984. Effects of a synthetic crude oil on pond benthic insects. *Environmental Pollution* (Series A) 33:163–86.

Dale, M., D. Hain, G. Lance, P. Milne, D. Ross, M. Thomas, and W.T. Williams. 1980. *Taxon Users Manual,* 2nd ed. Division of Computing Research, Commonwealth Scientific and Industrial Research Organization, St. Lucia, Qld., Australia.

Day, R.W. and G.P. Quinn. 1989. Comparisons of treatments after an analysis of variance in ecology. *Ecological Monographs* 59:433–63.

de March, B.G.E. 1976. Spatial and temporal patterns in macrobenthic stream diversity. *Journal of the Fisheries Research Board of Canada* 33:1261–70.

Descy, J.-P. 1973. La végétation algale benthique de la Meuse Belge et ses relations avec la pollution des eaux. *Lejeunia* (nouveau Series) 66:1–62.

Downing, J.A. 1979. Aggregation, transformation, and the design of benthos sampling programs. *Journal of the Fisheries Research Board of Canada* 36:1454–63.

Downing, J.A. 1986. A regression technique for the estimation of epiphytic invertebrate populations. *Freshwater Biology* 16:161–73.

Dyer, D.P. 1978. An analysis of species dissimilarity using multiple environmental variables. *Ecology* 59:117–25.

Efron, B. 1979. Bootstrap methods: another look at the jackknife. *Annals of Statistics* 7:1–26.

Efron, B. and G. Gong. 1983. A leisurely look at the bootstrap, the jackknife, and cross-validation. *American Statistician* 37:36–48.

Elliott, J.M. 1977. Some methods for the statistical analysis of samples of benthic invertebrates, 2nd ed. *Freshwater Biological Association Scientific Publication* No. 25:1–156.

Ellis, J.C. and R.F. Lacey. 1980. Sampling: defining the task and planning the scheme. *Water Pollution Control* 79:452–67.

Fager, E.W. 1972. Diversity: a sampling study. *American Naturalist* 106:293–310.

Faith, D.P. and R.H. Norris. 1989. Correlation of environmental variables with patterns of distribution and abundance of common and rare freshwater macroinvertebrates. *Biological Conservation* 50:77–98.

Felsenstein, J. 1985. Confidence limits on phylogenies: an approach using the bootstrap. *Evolution* 39:783–91.

Fisher, R.A., A.S. Corbet, and C.B. Williams. 1943. The relation between the number of species and the number of individuals in a random sample of an animal population. *Journal of Animal Ecology* 12:42–58.

Freund, R.J., R.C. Little, and P.C. Spector. 1986. *SAS System for Linear Models,* 1986 ed. SAS Institute, Cary, NC.

Frost, S. and L.B. Sinniah. 1982. Effect of particulate Abate insecticide on invertebrate stream drift communities in Newfoundland. *International Journal of Environmental Studies* 19:231–43.

Fryer, G. 1987. Quantitative and qualitative: numbers and reality in the study of living organisms. *Freshwater Biology* 17:177–89.

Gauch, H.G., Jr. 1982. *Multivariate Analysis in Community Ecology*. Cambridge Univ. Press, Cambridge, England.

Gauch, H.G., Jr. and R.H. Whittaker. 1972. Comparison of ordination techniques. *Ecology* 53:868–75.

Gleason, H.A. 1922. On the relation between species and area. *Ecology* 3:158–62.

Goodman, D. 1975. The theory of diversity-stability relationships in ecology. *Quarterly Review of Biology* 50:237–66.

Goodman, L.A. and W.H. Kruskal. 1954. Measures of association for cross classifications. *Journal of the American Statistical Association* 49:732–64.

Goodman, L.A. and W.H. Kruskal. 1959. Measures of association for cross classifications. II: Further discussion and references. *Journal of the American Statistical Association* 54:123–63.

Gower, J.C. 1966. Some distance properties of latent root and vector methods used in multivariate analysis. *Biometrika* 53:325–38.

Gower, J.C. 1967. A comparison of some methods of cluster analysis. *Biometrics* 23:623–37.

Grassle, J.F. and W. Smith. 1976. A similarity measure sensitive to the contribution of rare species and its use in investigation of variation in marine benthic communities. *Oecologia* (Berlin) 25:13–22.

Green, R.H. 1979. *Sampling Design and Statistical Methods for Environmental Biologists*. John Wiley, New York.

Gupta, P.K. and M.C. Pant. 1983. Macrobenthos of Lake Naini Tal (U.P., India) with particular reference to pollution. *Water, Air, and Soil Pollution* 19:397–405.

Hall, A.V. 1969. Avoiding informational distortion in automatic grouping programs. *Systematic Zoology* 18:318–29.

Harman, W.N. 1972. Benthic substrates: their effect on fresh-water Mollusca. *Ecology* 53:271–7.

Hellawell, J.M. 1986. *Biological Indicators of Freshwater Pollution and Environmental Management*. Elsevier, London.

Hill, M.O. 1979a. *DECORANA—A FORTRAN Program for Detrended Correspondence Analysis and Reciprocal Averaging*. Cornell Univ., Ithaca, NY.

Hill, M.O. 1979b. *TWINSPAN—A FORTRAN Program for Arranging Multivariate Data in an Ordered Two-Way Table by Classification of the Individuals and Attributes*. Cornell Univ., Ithaca, NY.

Hirsch, R.M., J.R. Slack, and R.A. Smith. 1982. Techniques of trend analysis for monthly water quality data. *Water Resources Research* 18:107–21.

Hope, A.C.A. 1968. A simplified Monte Carlo significance test procedure. *Journal of the Royal Statistical Society* (Series B) 30:582–98.

Hottelling, H. 1933. Analysis of a complex of statistical variables into principal components. *Journal of Educational Psychology* 24:417–41.

Hruby, T. 1987. Using similarity measures in benthic impact assessments. *Environmental Monitoring and Assessment* 8:163–80.

Hughes, B.D. 1978. The influence of factors other than pollution on the value of Shannon's diversity index for benthic macro-invertebrates in streams. *Water Research* 12:359–64.

Hurlbert, S.H. 1971. The nonconcept of species diversity: a critique and alternative parameters. *Ecology* 52:577–86.

Hurlbert, S.H. 1984. Pseudoreplication and the design of ecological field experiments. *Ecological Monographs* 54:187–211.

Hynes, H.B.N. 1960. *The Biology of Polluted Waters*. Liverpool Univ. Press, Liverpool, England.

Hynes, H.B.N. and U.R. Yadav. 1985. Three decades of post-impoundment data on the littoral fauna of Llyn Tegid, North Wales. *Archiv für Hydrobiologie* 104:39–48.

Jenkins, G.M. and D.G. Watts. 1968. *Spectral Analysis and Its Applications*. Holden-Day, London.

Johannsson, O.E. and C.K. Minns. 1987. Examination of association indices and formulation of a composite seasonal dissimilarity index. *Hydrobiologia* 150:109–21.

Keppel, G. 1973. *An Introduction to the Design and Analysis of Experiments*. Prentice-Hall, Englewood Cliffs, NJ.

King, D.L. and R.C. Ball. 1964. A quantitative biological measure of stream pollution. *Journal of the Water Pollution Control Federation* 36:650–3.

Kleinbaum, D.G. and L.L. Kupper. 1978. *Applied Regression Analysis and Other Multivariable Methods*. Duxbury Press, North Scituate, MA.

Kolkwitz, R. and M. Marsson. 1909. Ökologie der tierischen Saprobien. Beiträge zur Lehre von der biologischen Gewässerbeurteilung. *Internationale Revue der gesamten Hydrobiologie und Hydrographie* 2:126–52.

Krebs, C.J. 1985. *Ecology. The Experimental Analysis of Distribution and Abundance*, 3rd ed. Harper and Row, New York.

Kruskal, J.B. 1964a. Multidimensional scaling by optimizing goodness of fit to a non-metric hypothesis. *Psychometrika* 29:1–27.

Kruskal, J.B. 1964b. Non-metric multidimensional scaling: a numerical method. *Psychometrika* 29:115–29.

Lance, G.N. and W.T. Williams. 1966. A generalized sorting strategy for computer classifications. *Nature* (London) 212:218.

Lance, G.N. and W.T. Williams. 1975. REMUL: a new divisive polythetic classificatory program. *Australian Computer Journal* 7:109–12.

Legendre, L. and P. Legendre. 1983. *Numerical Ecology. Developments in Environmental Modelling, Vol. 3.* Elsevier, New York.

Leland, H.V., J.L. Carter, and S.V. Fend. 1986. Use of detrended correspondence analysis to evaluate factors controlling spatial distribution of benthic insects. *Hydrobiologia* 131:113–23.

Manly, B.F.J. 1986. *Multivariate Statistical Methods. A Primer.* Chapman and Hall, London.

Marchant, R., P. Mitchell, and R. Norris. 1984. Distribution of benthic invertebrates along a disturbed section of the La Trobe River, Victoria: an analysis based on numerical classification. *Australian Journal of Marine and Freshwater Research* 35:355–74.

Margalef, R. 1958. Information theory in ecology. *General Systems* 3:36–71.

Marriott, F.H.C. 1974. *The Interpretation of Multiple Observations.* Academic Press, London.

Mayack, D.T. and J.S. Waterhouse. 1983. The effects of low concentrations of particulates from paper mill effluent on the macroinvertebrate community of a fast-flowing stream. *Hydrobiologia* 107:271–82.

Meffe, G.K. and A.L. Sheldon. 1988. The influence of habitat structure on fish assemblage composition in southeastern blackwater streams. *American Midland Naturalist* 120:225–40.

Metzeling, L., A. Grasser, P. Suter, and R. Marchant. 1984. The distribution of aquatic macroinvertebrates in the upper catchment of the La Trobe River, Victoria. *Occasional Papers from the Museum of Victoria, Australia* 1:1–62.

Millard, S.P., J.R. Yearsley, and D.P. Lettenmaier. 1985. Space-time correlation and its effects on methods for detecting aquatic ecological change. *Canadian Journal of Fisheries and Aquatic Sciences* 42:1391–1400.

Minchin, P.R. 1987. An evaluation of the relative robustness of techniques for ecological ordination. *Vegetatio* 69:89–107.

Morin, A. 1985. Variability of density estimates and the optimization of sampling programs for stream benthos. *Canadian Journal of Fisheries and Aquatic Sciences* 42:1530–40.

Moss, D., M.T. Furse, J.F. Wright, and P.D. Armitage. 1987. The prediction of the macro-invertebrate fauna of unpolluted running-water sites in Great Britain using environmental data. *Freshwater Biology* 17:41–52.

Murphy, P.M. 1978. The temporal variability in biotic indices. *Environmental Pollution* 17:227–36.

Murphy, P.M. and R.W. Edwards. 1982. The spatial distribution of the freshwater macroinvertebrate fauna of the River Ely, South Wales, in relation to pollutional discharges. *Environmental Pollution* (Series A) 29:111–24.

Narf, R.P., E.L. Lange, and R.C. Wildman. 1984. Statistical procedures for applying Hilsenhoff's biotic index. *Journal of Freshwater Ecology* 2:441–8.

Needham, P.R. and R.L. Usinger. 1956. Variability in the macrofauna of a single riffle in Prosser Creek, California, as indicated by the Surber sampler. *Hilgardia* 24:383–409.

Nemec, A.F.L. and R.O. Brinkhurst. 1988a. Using the bootstrap to assess statistical significance in the cluster analysis of species abundance data. *Canadian Journal of Fisheries and Aquatic Sciences* 45:965–70.

Nemec, A.F.L. and R.O. Brinkhurst. 1988b. The Fowlkes-Mallows statistic and the comparison of two independently determined dendrograms. *Canadian Journal of Fisheries and Aquatic Sciences* 45:971–5.

Norris, R.H. 1986. Mine waste pollution of the Molonglo River, New South Wales and the Australian Capital Territory: effectiveness of remedial works at Captains Flat mining area. *Australian Journal of Marine and Freshwater Research* 37:147–57.

Norris, R.H. and A. Georges. 1986. Design and analysis for assessment of water quality. In *Limnology in Australia,* eds. P. De Deckker and W.D. Williams, pp. 555–72. Commonwealth Scientific and Industrial Research Organization, Melbourne, Australia, and Junk Publs., Dordrecht, The Netherlands.

Norris, R.H., P.S. Lake, and R. Swain. 1982. Ecological effects of mine effluents on the South Esk River, north-eastern Tasmania. III. Benthic macroinvertebrates. *Australian Journal of Marine and Freshwater Research* 33:789–809.

Noy-Meir, I. 1973. Data transformations in ecological ordination. I. Some advantages of non-centering. *Journal of Ecology* 61:329–41.

Noy-Meir, I. and M.P. Austin. 1970. Principal component ordination and simulated vegetational data. *Ecology* 51:551–2.

Numerical Algorithms Group. 1986. *The Generalized Linear Interactive Modelling System. Release 3.77. Reference guide.* Numerical Algorithms Group Central Office, Oxford, England.

Odum, H.T., J.E. Cantlon, and L.S. Kornicker. 1960. An organizational hierarchy postulate for the interpretation of species-individual distributions, species entropy, ecosystem evolution, and the meaning of a species-variety index. *Ecology* 41:395–9.

Ormerod, S.J. 1987. The influences of habitat and seasonal sampling regimes on the ordination and classification of macroinvertebrate assemblages in the catchment of the River Wye, Wales. *Hydrobiologia* 150:143–51.

Ormerod, S.J. and R.W. Edwards. 1987. The ordination and classification of macroinvertebrate assemblages in the catchment of the River Wye in relation to environmental factors. *Freshwater Biology* 17:533–46.

Osborne, L.L. and R.W. Davies. 1987. The effects of a chlorinated discharge and a thermal outfall on the structure and composition of the aquatic macroinvertebrate communities in the Sheep River, Alberta, Canada. *Water Research* 21:913–21.

Patrick, R. 1950. Biological measure of stream conditions. *Sewage and Industrial Wastes* 22:926–38.

Pielou, E.C. 1975. *Ecological Diversity.* John Wiley, New York.

Pielou, E.C. 1984. *The Interpretation of Ecological Data.* John Wiley, New York.

Pinder, L.C.V. and I.S. Farr. 1987. Biological surveillance of water quality—2. Temporal and spatial variation in the macroinvertebrate fauna of the River Frome, a Dorset chalk stream. *Archiv für Hydrobiologie* 109:321–31.

Pistrang, L.A. and J.F. Burger. 1984. Effect of *Bacillus thuringiensis* var. *israelensis* on a genetically-defined population of black flies (Diptera: Simuliidae) and associated insects in a montane New Hampshire stream. *Canadian Entomologist* 116:975–81.

Plafkin, J.L., M.T. Barbour, K.D. Porter, S.K. Gross, and R.M. Hughes. 1989. *Rapid Bioassessment Protocols for Use in Streams and Rivers. Benthic Macroinvertebrates and Fish.* EPA/444/4–89/001. Office of Water Regulations and Standards, U.S. Environmental Protection Agency, Washington, DC.

Platt, T., L.M. Dickie, and R.W. Trites. 1970. Spatial heterogeneity of phytoplankton in a near-shore environment. *Journal of the Fisheries Research Board of Canada* 27:1453–73.

Poole, R.W. 1974. *An Introduction to Quantitative Ecology*. McGraw-Hill, New York.

Preston, F.W. 1948. The commonness, and rarity, of species. *Ecology* 29:254–83.

Resh, V.H. 1979. Sampling variability and life history features: basic considerations in the design of aquatic insect studies. *Journal of the Fisheries Research Board of Canada*. 36:290–311.

Resh, V.H. and J.D. Unzicker. 1975. Water quality monitoring and aquatic organisms: the importance of species identification. *Journal of the Water Pollution Control Federation* 47:9–19.

Rutherford, J.C. 1984. Trends in Lake Rotorua water quality. *New Zealand Journal of Marine and Freshwater Research* 18:355–65.

SAS Institute. 1987. *SAS/STAT Guide for Personal Computers, Version 6*. SAS Institute, Cary, NC.

Shannon, C.E. 1948. A mathematical theory of communication. *Bell System Technical Journal* 27:379–423, 623–56.

Sheldon, A.L. 1985. Perlid stoneflies (Plecoptera) in an Appalachian drainage: a multivariate approach to mapping stream communities. *American Midland Naturalist* 113:334–42.

Siegel, S. 1956. *Nonparametric Statistics for the Behavioral Sciences*. McGraw-Hill, New York.

Simberloff, D. 1979. Rarefaction as a distribution-free method of expressing and estimating diversity. In *Ecological Diversity in Theory and Practice*, eds. J.F. Grassle, G.P. Patil, W. Smith, and C. Taillie, pp. 159–76. International Co-operative Publishing House, Fairland, MD.

Simpson, E.H. 1949. Measurement of diversity. *Nature* (London) 163:688.

Smith, W. and J.F. Grassle. 1977. Sampling properties of a family of diversity measures. *Biometrics* 33:283–92.

Sneath, P.H.A. and R.R. Sokal. 1973. *Numerical Taxonomy*. W.H. Freeman, San Francisco.

Sokal, R.R. and F.J. Rohlf. 1981. *Biometry*, 2nd ed. W.H. Freeman, London.

Sokal, R.R. and P.H.A. Sneath. 1963. *Principals of Numerical Taxonomy*. W.H. Freeman, San Francisco.

Southwood, T.R.E. 1978. *Ecological Methods with Particular Reference to the Study of Insect Populations,* 2nd ed. Chapman and Hall, London.
Sprugel, D.G. 1983. Correcting for bias in log-transformed allometric equations. *Ecology* 64:209–10.
Sprules, W.G. 1977. Crustacean zooplankton communities as indicators of limnological conditions: an approach using principal component analysis. *Journal of the Fisheries Research Board of Canada* 34:962–75.
Stewart-Oaten, A., W.W. Murdoch, and K.R. Parker. 1986. Environmental impact assessment: "pseudoreplication" in time? *Ecology* 67:929–40.
Sugihara, G. 1980. Minimal community structure: an explanation of species abundance patterns. *American Naturalist* 116:770–87.
Taylor, L.R. 1961. Aggregation, variance and the mean. *Nature* (London) 189:732–5.
Taylor, L.R. 1980. New light on the variance/mean view of aggregation and transformation: comment. *Canadian Journal of Fisheries and Aquatic Sciences* 37:1330–2.
Tiller, D. 1988. The impact of sewage effluent on the benthic macroinvertebrate community of the upper Thredbo River. M.Sc. dissertation, Canberra College of Advanced Education, Canberra, Australia.
Van Belle, G. and J.P. Hughes. 1984. Nonparametric tests for trend in water quality. *Water Resources Research* 20:127–36.
Van Latesteijn, H.C. and R.H.D. Lambeck. 1986. The analysis of monitoring data with the aid of time-series analysis. *Environmental Monitoring and Assessment* 7:287–97.
Voshell, J.R., Jr. and G.M. Simmons, Jr. 1984. Colonization and succession of benthic macroinvertebrates in a new reservoir. *Hydrobiologia* 112:27–39.
Washington, H.G. 1984. Diversity, biotic and similarity indices. A review with special relevance to aquatic ecosystems. *Water Research* 18:653–94.
Whiting, E.R. and H.F. Clifford. 1983. Invertebrates and urban runoff in a small northern stream, Edmonton, Alberta, Canada. *Hydrobiologia* 102:73–80.
Whittaker, R.H. 1952. A study of summer foliage insect communities in the Great Smoky Mountains. *Ecological Monographs* 22:1–44.
Wilhm, J.L. 1972. Graphic and mathematical analyses of biotic communities in polluted streams. *Annual Review of Entomology* 17:223–52.
Wilhm, J.L. and T.C. Dorris. 1968. Biological parameters for water quality criteria. *BioScience* 18:477–81.
Wold, S. 1974. Spline functions in data analysis. *Technometrics* 16:1–11.
Woodiwiss, F.S. 1964. The biological system of stream classification used by the Trent River Board. *Chemistry and Industry* 11:443–7.
Wright, J.F., D. Moss, P.D. Armitage, and M.T. Furse. 1984. A preliminary classification of running-water sites in Great Britain based on macro-invertebrate species and the prediction of community type using environmental data. *Freshwater Biology* 14:221–56.
Yule, G.U. 1944. *The Statistical Study of Literary Vocabulary*. Cambridge Univ. Press, London.
Zar, J.H. 1968. Calculation and miscalculation of the allometric equation as a model in biological data. *BioScience* 18:1118–20.

8

Monitoring Freshwater Benthic Macroinvertebrates and Benthic Processes: Measures for Assessment of Ecosystem Health

Seth R. Reice and Margaret Wohlenberg

8.1. Introduction

An ecosystem perspective is essential for understanding and detecting the effects of disturbance in natural systems. Since an ecosystem is the interacting biotic community together with the abiotic environment, the ecosystem perspective is necessarily holistic and process-oriented. Determination of the structure and dynamics of the benthic community is a key to understanding the state of a freshwater ecosystem and how it works. This chapter will focus on the advantages that benthic macroinvertebrates offer for assessment of the state of the ecosystem.

The distinction between community (structure-oriented) and ecosystem (process-oriented) is often obscure. Here, changes in benthic community structure will be used to make inferences about consequent changes in ecosystem-level processes such as decomposition and productivity. Important insights into the dynamics of freshwater ecosystems can be gleaned by concentrating on the benthos as a whole community, interacting with and reflecting changes in the ecosystem.

The benthos is an excellent biotic indicator of ecosystem change for several reasons. First, the life history of benthic macroinvertebrates, especially length of their life cycles, provides for long-term exposure to toxic substances, relative to other system constituents such as zooplankton. Second, benthic macroinvertebrates live in intimate contact with the sediments, which enhances their contact with many pollutants. Thus, body burdens of toxins build up to easily detectable levels in benthic macroinvertebrates. Third, decomposition, the fundamental process wherein dead organic matter is broken down into CO_2 and simple inorganic molecules, takes place principally in the benthos; studying rates of decomposition, therefore, provides an effective vehicle for assessment of ecosystem functioning.

The field of ecotoxicology recently has recognized that classical monitoring of chemical and physical aspects of water quality and single-species toxicity testing are insufficient to detect and interpret ecosystem responses to pollution (Cairns, 1981; Kimball and Levin, 1985; Steines and Wharfe, 1987; Perry and Troelstrup, 1988; Courtemanch et al., 1989). Although chemical data provide a snapshot indicator of ambient conditions, biotic data provide a cumulative indication of conditions over time, the scale of which depends on the life span of the organism(s) being monitored (Steines and Wharfe, 1987). Single species testing in *ex situ* conditions often fails to detect effects on community interactions that whole-ecosystem manipulations have shown to be crucial. This recognition has led Perry and Troelstrup (1988) and La Point and Perry (1989) to advocate the use of controlled, replicated whole-ecosystem manipulations as a valuable tool in ecotoxicology (see also Cooper and Barmuta, Chapter 11).

Whole-ecosystem experiments are uncommon in aquatic systems, and few of those done have had a strong benthic component. However, controlled manipulations of natural systems have proven quite successful in elucidating processes and predicting the consequences of anthropogenic alterations.

Such studies also have shown the limits of classical monitoring approaches that often depend on single organisms (e.g., fish) or single processes (e.g., primary productivity). Consider the common monitoring strategy of periodically sampling adult fish populations in lakes and streams; this approach reflects interest in the state of the game fish populations in these systems. Such populations are inadequate indicators of stress because of significant time lags (up to five or six years) between the time recruitment has ceased and when decline in the number of adult fish becomes apparent. Likewise, primary and secondary productivity are amazingly stable under all but the most extreme conditions of environmental deterioration. In contrast, benthic macroinvertebrate populations and the benthic community display far greater sensitivity to several types of environmental disturbance.

The review of several ecosystem-level manipulations that follows will demonstrate that the zoobenthos is an especially sensitive community and therefore is useful for monitoring ecosystem deterioration.

8.2. Case Histories of Ecosystem Studies with Benthic Components

8.2.1. Lakes

Lake ecosystems often are perturbed by agricultural runoff, fish stocking programs, and acid precipitation. Several experimental studies have been performed to identify systematically the effects of these perturbations on the benthos and the system as a whole.

8.2.1.1. Cornell Pond Studies

One of the first aquatic whole-ecosystem manipulations was the pond study of Hall et al. (1970) at Cornell University in New York, which was a large-scale experiment designed to test the roles of nutrients and fish in ponds. Two fish densities were cross-classified with three nutrient fertilization regimes, and three replicate ponds were used per treatment combination. Fish, zooplankton, and zoobenthos were followed for three years. This study was a model of whole-ecosystem manipulations in pond systems for other investigators to follow. Benthic macroinvertebrates displayed both population and community changes in relation to manipulations of nutrients and fish. The most revealing changes were in mean body size of the benthic macroinvertebrates, across taxa and ages. With high fish predation, mean body size of macroinvertebrates was reduced. Compensatory changes in body size resulted: as large invertebrate predators were reduced by fish feeding, smaller-sized benthic taxa increased in abundance. Nutrient enrichment had its greatest effect on zoobenthic productivity in the first year. Changes in the benthic macroinvertebrates were extremely valuable for interpreting whole-pond dynamics and, particularly, fish responses to the experimental treatments.

8.2.1.2. Experimental Lakes Area (ELA) Studies

Many long-term and comprehensive manipulations have been performed in several lake basins at ELA in northwestern Ontario, Canada. Over the past two decades, ecosystem-level responses to eutrophication, acidification, and contamination with various chemical toxicants have been studied by making whole-lake additions of these pollutants to some lakes and comparing their responses at all levels of ecosystem function to the functioning of unmanipulated reference lakes nearby.

Among the most successful of the ELA experiments was the gradual acidification of Lake 223 from pH 6.8 to 5.0 between 1976 and 1983 (Schindler et al., 1985). The actual manipulation was undertaken after a two-year period of baseline studies. Although overall primary productivity levels were constant or slightly enhanced throughout the experiment, and diversity of phytoplankton remained unchanged, several species replacements did occur. Some fish populations, including young lake trout and white suckers, increased slightly in the intermediate stages of the experiment, perhaps in response to the abundance of chironomid prey. Early in the experiment, macrobenthic invertebrate productivity and chironomid emergence increased slightly but not above the natural range. Lake trout and fathead minnow recruitment failed, however, at pH levels near 5.6; within three years, the fathead minnow disappeared.

The first detectable effect of acidification occurred in the benthos. The population of the benthic crustacean *Mysis relicta* declined drastically at pH just below six. At pH 5.6, a population of the crayfish *Orconectes virilis* began to exhibit recruitment failure. Between pH 5.6 and 5.0, exoskeletons of adult crayfish failed to harden properly after molting, indicating recalcification and ion-balance problems. During the most severe stages of acidification, many eggs died and were subject to severe fungal infestation. The amphipod *Hyalella azteca* proved to be very sensitive to acidification; *Hyalella* populations were eliminated at pH below 5.8 (Grapentine, 1987). All fish recruitment ceased by 1983 when pH of the lake averaged 5.13, and *Orconectes, Mysis,* and *Hyalella* populations had become extinct. In spite of all these changes, adult lake trout (*Salvelinus namaycush*) populations remained at normal levels, but the physical condition of individuals declined, largely due to the disappearance of many invertebrate and fish prey species. A simple count of the adult fish population would miss the fact that wholesale changes were occurring in the benthos of the lake and that the fish were starving.

8.2.1.3. Little Rock Lake Studies

A whole-lake experiment at Little Rock Lake, Wisconsin, examined the effects of acidification on decomposition processes (Perry and Troelstrup, 1988). After 18 months of baseline study, the hourglass-shaped lake was separated at the point of constriction with a water-impermeable curtain. One basin was acidified in 0.5 pH-unit drops at two-year intervals from pH 6.1 to 4.6. Leaves of various species were placed in litter bags in the two basins and weighed at intervals to compare rates of decomposition. A range of species-specific responses to acidification was found: some leaf types decomposed more quickly, some more slowly, and some showed no difference in the acidified basin relative to the undisturbed basin. Acidification clearly affected decomposition rates, but the causes of the different species-specific responses remain unknown. The responses of the other components of the ecosystem currently are being analyzed.

8.2.2. Streams

Headwater streams can experience acute acid inputs during spring snow melt. This type of event is of short duration, but its effects on the community can be both profound and persistent. Short-term acidification experiments that simulate the effects of acid snow melt are whole-ecosystem manipulations that provide an ideal opportunity to study interrelationships between the benthic macroinvertebrates and the rest of the ecosystem.

8.2.2.1. Hubbard Brook Studies

Norris Brook, in the Hubbard Brook Experimental Forest, New Hampshire, was experimentally acidified in April 1977, and acid conditions were maintained through September 1977 (Hall et al., 1980; Likens, 1985). Ambient pH ranged from 6.5 to lows of 5.4 during rainstorms, prior to acidification. pH was experimentally lowered to about 4 by the addition of dilute sulfuric acid. The effects included a 75% decrease in the density of benthic macroinvertebrates, such as Diptera and Ephemeroptera, and a severe decline in the emergence of adult insects. The most obvious and immediate response of the stream ecosystem to sudden pH depression was an exodus of macrobenthic animals via downstream drift (Pratt and Hall, 1981). Drift samples taken shortly after application of the acid revealed increases in scraper, collector, and predator functional feeding groups, indicating a behavioral response to avoid the deteriorating conditions.

Although the overall taxonomic (generic-level) diversity of drift entering and leaving the acidified zone did not differ significantly, the diversity within functional or behavioral groupings did. For example, the Shannon-Weaver diversity of collector-gatherers in the drift leaving the treated reach was much greater than that entering it, indicating a net decrease in diversity among this functional group within the reach. Collector-gatherers are important because they collect fine particulate organic matter (FPOM) and incorporate it into biomass suitable for utilization by the rest of the community. The decreased standing stock and diversity of this group would, therefore, generate potentially profound changes, not only in the acidified region of the stream but also downstream, because the supply of FPOM increased.

The standing biomass of periphyton in the acidified reach of Norris Brook increased because of reduced grazing caused by the emigrating benthic macroinvertebrates (Hall et al., 1980). Overall nutrient flux of the acidified system was significantly diminished (Hall and Likens, 1981). This probably was caused by either increased uptake of nutrients from the water by the attached algae or downstream transport resulting from the exodus of acid-stressed invertebrates. Growth of the mayfly *Ephemerella funeralis* decreased dramatically, and recruitment of a new generation nearly ceased (Fiance, 1978). Many of the brook trout that were resident in the stream before acidification also migrated downstream. However, a subpopulation of brook trout was trapped in an acidified pool by low water levels for several months. These fish displayed no mortality or direct physiological deterioration (Hall et al., 1980). Again, major changes that resulted from acidification were not obvious from examination of fish, but were evident with benthic macroinvertebrates. This demonstrates the sensitivity of benthic macroinvertebrates and their value for biomonitoring purposes.

8.2.2.2. Other Stream pH Studies

Mackay and Kersey (1985) identified significant differences in the lotic insect communities of Ontario streams of varying pH. Communities in the most acid streams (annual range: pH 4.3 to 4.5) were greatly simplified relative to streams with pH 5.3 to 6.7. Ephemeroptera were missing from the two most acidic streams, and one of these streams also lacked Plecoptera. All groups were represented in the circumneutral streams, and greater evenness of species distributions was evident. Composition of the zoobenthos of acid streams clearly differed from that of circumneutral streams.

An experimental acid addition was conducted in one of these surveyed Ontario streams to determine if the effect of short-term (four-day) pulses of increased acidity varied seasonally (Hall, 1990). The experiment mimicked the pH reductions associated with rainstorms. Hall hypothesized that the acid pulse would have a greater effect on drifting invertebrates in the fall than the spring, because animals in the fall were both smaller-sized and more exposed to the current. The results supported his hypothesis. Thus, due to both behavior and life history, the impact of acid pulses on the benthic macroinvertebrates showed a clear seasonal difference.

An increase in soluble aluminum ions resulting from acid leaching of soils and aquatic sediments is a secondary effect of acid precipitation. A separate experiment at Norris Brook compared the effects of pH and aluminum on drift (Hall et al., 1987). Addition of aluminum chloride downstream from the acid application revealed that aluminum can cause a drift response at moderate pH levels that do not induce drift. Burton and Allan (1986) demonstrated the interaction effects of pH, aluminum, and organic matter on a number of lotic benthic macroinvertebrate species. Survival of all species was significantly reduced at pH 4.0 relative to survival at the control pH (7.0). When aluminum (500 μg) was added, it caused significantly higher mortality for four of the five species tested. When dissolved organic matter (42–47 mg C/l) was removed at the same time as the aluminum was added, the stonefly *Nemoura* sp. and the isopod crustacean *Asellus intermedius* were eliminated and survival of the caddis larva *Pychnopsyche guttifer* decreased to only 20%. The interaction between these variables was manifest in the responses of the benthic macroinvertebrates. Thus, secondary chemical effects of acid precipitation may indicate impending stress before the acute effects of proton concentration are noticeable.

8.2.2.3. *Coweeta Studies*

The impact of an insecticide on stream communities was tested by Cuffney et al. (1984) and Wallace et al. (1989). In 1980, a small headwater stream at Coweeta Hydrologic Laboratory, in western North Carolina, was treated with methoxychlor at a rate of 10 mg/l. This treatment caused catastrophic invertebrate drift (>12,000 individuals/m^3 of discharge). Biomass and densities of insects were reduced to <10% of those in control streams. Benthic macroinvertebrates in the treated stream changed from large-sized shredders to small-sized collector-gatherers and predators, compared with control streams and the pretreatment community. The new fauna was dominated by Oligochaeta, Chironomidae, and Turbellaria, and biomass of non-insect species increased dramatically in the treated stream. The authors suggested that these changes were related to levels of FPOM in the treated stream: "Methoxychlor treatment reduced the concentration, the amount exported, and the median particle size of transported particulate organic matter. . . ." (Cuffney et al., 1984, p. 153). They concluded that benthic insects regulate the production and transport of FPOM to downstream communities in these small woodland streams.

In 1985, this insecticide was applied to what previously had been the reference stream. Catastrophic drift resulted again. A total of >950,000 macroinvertebrates drifted, with a biomass of 70 g ash free dry mass from only 144 m^2 of stream bottom. Collector-gatherers (mainly Chironomidae) accounted for 63% of the drifting individuals, but biomass was dominated by shredders (48.9%). Benthic macroinvertebrate densities were reduced to 75% of pretreatment levels in December 1985 and 85% after a repeat treatment in March 1986. The community shifted from a species-rich fauna to one dominated by Chironomidae, copepods, Collembola, and Oligochaeta. The reduction of detritivore densities reduced leaf litter decomposition rates and seston concentrations (Wallace et al., 1982, 1986; Cuffney et al., 1984). Clearly, functioning of the ecosystem is dependent on the structure of the benthic macroinvertebrate community.

In each of these ecosystem experiments, the benthic macroinvertebrate community displayed early, and often dramatic, responses to disturbance. These responses can greatly influence the rest of the ecosystem by altering decomposition and, consequently, nutrient cycles, or by disrupting trophic relationships via extinction or emigration. The benthic macroinvertebrate community can, therefore, reveal key changes in the functioning of freshwater ecosystems.

8.3. Benthos as an Effective Indicator of Ecosystem Change

Two categories of reasons explain why the benthos should be very useful indicators of aquatic ecosystem change. The first deals with the nature of the benthos and benthic macroinvertebrate life histories. The second concerns the nature of aquatic ecosystems and the role of benthos within them.

8.3.1. Benthic Life-History Characteristics

Although some Chironomidae and Culicidae have short life spans (~20 days at temperatures >25°C), many benthic macroinvertebrates have only one generation per year. Some Megaloptera, Odonata, and Plecoptera live up to four or five years (Merritt and Cummins, 1984), and crayfish can live much longer. This is in contrast to the very short (days to weeks) life cycles of phytoplankton and zooplankton. Therefore, benthos will endure sustained exposure to environmental hazards that can lead to population and community changes. Wiederholm (1984) reviewed the uses of aquatic insects (mostly benthic) in pollution studies and recommended the use of benthic species for pollution assessment.

8.3.2. Organism-Sediment Interactions

The sediments are repositories of accumulated nutrients and toxins that are some of the principal causes of environmental deterioration in freshwater systems. Benthic macroinvertebrates live on and in the sediments. This tight coupling means that the zoobenthos likely will reflect changes in the chemical or pollution status of a lake or stream, especially because many benthic organisms are relatively sessile. Benthic macroinvertebrates bioaccumulate and biomagnify toxins such as heavy metals and pesticides (see Johnson et al., Chapter 4). Furthermore, benthic macroinvertebrates can affect cycling of a contaminant in freshwater ecosystems through bioturbation and sediment resuspension (Reynoldson, 1987).

8.3.3. Benthic Community Structure as an Indicator of Ecosystem Health

The structure of the benthic macroinvertebrate community reflects the state of the entire ecosystem. Many recent studies have demonstrated changes in the zoobenthos as a response to the pollution level of overlying waters. However, whole-system biological surveys are difficult and expensive to do in aquatic systems. As a practical alternative, several approaches to the rapid assessment of zoobenthos have been developed (see Resh and Jackson, Chapter 6).

An impressive case for the use of benthic macroinvertebrate taxa richness as an indicator of water quality is made by Lenat (1988). The North Carolina Division of Environmental Management devised a qualitative sampling scheme that is designed to maximize the collection of a wide array of benthic macroinvertebrate species in a reasonably short time. It is used presently for shallow streams (<1.5 m deep). The number of species collected is evaluated using separate criteria for Mountain, Piedmont, and Coastal Plain ecoregions (i.e., ecologically defined and reasonably uniform geographic zones). The method was tested on streams with different pollution levels in the French Broad River basin (Mountain Ecoregion). Taxa richness, as determined by this method, showed a strong relationship with water chemistry as measured by the Water Quality Index ($r^2 = 0.85$ for total taxa richness; $r^2 = 0.84$ for the combined taxa of Ephemeroptera, Plecoptera, and Trichoptera). Resh and Jackson (Chapter 6) also show that rapid assessment approaches using taxa richness are useful in detecting impacts in northern

California streams. These works demonstrate the value of zoobenthos in the rapid assessment of stream water quality.

The literature on community-level analyses and indices is voluminous; it is covered elsewhere in this book (Resh and Jackson, Chapter 6; Norris and Georges, Chapter 7), but some examples are considered here as well. Winner et al. (1980) used taxonomic composition to evaluate heavy metal pollution, and they showed that macroinvertebrate community structure changed predictably. They focused on the percentage of Chironomidae in the total community as a possible index. La Point et al. (1984) tested benthic community structure in 15 streams as a function of metal concentrations, and they concluded that benthic fauna can reflect changes in water quality. Like Winner et al. (1980), La Point et al. (1984) recommended using species richness as a preferred variable. Specht et al. (1984) analyzed changes in the benthic community downstream of fly ash settling ponds. As the ponds filled to >90%, discharge from them caused reductions in mayfly, caddisfly, and stonefly densities, species richness, and diversity in the receiving stream. Zanella (1982) demonstrated that Trichoptera species replacements occurred in the Sacramento River in the presence of heavy metal contamination. This study suggests that monitoring of individual species populations can be useful. Clements et al. (1988) placed sediment trays in the field, allowed them to be colonized by invertebrates, and then moved these artificial substrates to the laboratory to test for responses to copper exposure. After 96 hours of exposure, the relative abundance of Ephemeroptera decreased.

Species assemblages of benthic communities have been used to assess the trophic status of lakes (Brundin, 1958; Sæther, 1975, 1979; see also Johnson et al., Chapter 4). This method relies on the species composition of Chironomidae in the profundal and littoral zones, but its utility is severely limited by the taxonomic sophistication required and the time necessary for sorting and identifying chironomid samples. Most chironomid heads must be cleared, flattened, and mounted for identification to species. Although the use of chironomid assemblages may reflect community change accurately, the method probably will not be used widely.

Biomonitoring work also has used the concept of guilds (e.g., Severinghaus, 1981). The Index of Biotic Integrity (IBI), developed by Karr (1987) for aquatic systems, is based on fish guilds and their community structure.

8.4. The Benthos and Ecosystem Processes

Although most of the important processes of aquatic ecosystems occur throughout the system, some occur principally in the benthos. Primary production, herbivory, and nutrient flux in large lakes and rivers are principally pelagic processes (i.e., they take place in the open-water zone). However, dead algal cells, dead animals, and feces settle out of the water column to the bottom, which is the principal site of decomposition. Shallow lakes and streams have a high ratio of bottom surface area:water volume, so a high proportion of the water is in close contact with the bottom. Littoral zones of lakes are dominated by benthic processes, and shallow streams are regarded as principally benthic systems.

Decomposition is an ecosystem-level process, and it is critical to the functioning of every ecosystem. Through decomposition, dead organisms and feces are broken down into compounds that can be absorbed by plants and used in primary production. Decomposition is the vital link between life and death; without decomposition to regenerate nutrients and carbon dioxide from dead organic matter, the benthos eventually would become a sink for all organic matter in lakes and streams (and in the oceans). Breakdown of this organic matter happens primarily in the benthos.

Decomposition involves bacteria, fungi, and benthic invertebrates. Boling et al. (1975) analyzed and modeled the process of litter decomposition. Many species of benthic macroinvertebrates are specialized to feed on detritus (Cummins and Klug, 1979; Merritt and Cummins, 1984). Shredders (e.g., Tipulidae and many Plecoptera) feed on large particles like whole leaves and, in the process, tear the leaf into finer particles. This creates increased surface area for microbial colonization and growth. Fine-particle feeders (collectors: Ephemeroptera, Chironomidae) then consume these small particles that are enriched by the protein contained in the bacteria and fungi. The microbial loop is an important energy pathway (Pomeroy, 1974; Azam et al., 1983; Carpenter, 1988). In the benthos of lakes and streams (as well as the deep oceans), the microbial loop is the dominant energy pathway. Detritivore feeding on microbially conditioned organic matter is the principal energy pathway in streams (Fisher and Likens, 1973; Cummins, 1974; Cummins et al., 1973, 1989) and salt marshes.

Several stream studies have used decomposition to detect ecosystem changes in response to acid contamination. Hildrew et al. (1984) used the loss of tensile strength of cellulose cloth strips to measure decomposition rates in English streams. Cellulose decomposition was greatly inhibited at mean annual pH below 5.6 to 5.8, particularly in the autumn. Decomposition rates of autumn-shed leaves were lower in acidic than circumneutral Ontario streams (Mackay and Kersey, 1985). Changes in the macroinvertebrate community, such as sharp reductions in Ephemeroptera and Plecoptera in response to low pH, also were observed in these streams. These effects were seen by Hildrew et al. (1984) as well.

Allard and Moreau (1986) used litter bags in artificial channels in a lake outflow stream to test the effects of decreasing pH on decomposition of leaf litter. They lowered the pH from control levels (pH 6.2 to 7.0) to pH 4.0. The mass loss from leaves in the acidified channel was much less than the mass loss from the control channel. When alder (*Alnus rugosa*) leaves were transferred from the acidified to the control channel, decomposition rates increased, and abundance and diversity of benthic macroinvertebrates increased. Since no correlation existed between macroinvertebrate biomass and decomposition rates, Allard and Moreau (1986) suggested that microbial processes are more important than macroinvertebrate feeding to the overall decomposition process. They concluded that acidification of surface waters can seriously reduce decomposition rates.

A whole-ecosystem manipulation was undertaken at the U.S. Environmental Protection Agency's (USEPA) artificial stream facility at Monticello, Minnesota (Newman et al., 1987; Perry and Troelstrup, 1988; Newman and Perry, 1989), in order to measure directly the effects of chemical pollution on the decomposition process. Three stream channels were dosed with chlorine (10, 75, or 250 μg/l), two additional channels were treated with 3 μg/l of ammonia plus 10 or 250 μg/l Cl, and two channels were used as controls. Examination of mesh litter bags placed in treatment and reference channels showed that the rate of decomposition (as measured by weight loss) was reduced by the highest concentration of chlorine and both of the chlorine plus ammonia treatments. This effect coincided with reduced colonization and reduced functioning (respiration) of microbial populations in the litter, and reduced density and individual activity of benthic macroinvertebrate shredders. Although these effects appeared to be temporary, diminishing 11 days after treatment, the alteration in benthic animal populations and processes may hold severe consequences for the nutrient cycles and functions of the entire ecosystem.

The effects of pH 4 to 9 on leaf disc decomposition in a laboratory stream were studied without macroinvertebrates present (Thompson and Bärlocher, 1989). They compared these results to decomposition in two streams, having neutral to slightly acid conditions, with invertebrates present. Weight losses were correlated positively with both temperature and pH in the field. Weight loss was highest at neutral pH. However, in the laboratory stream, the response curve was hump-shaped. Peak weight loss occurred between pH 5.5 and 6.0. The relationship was: weight loss $= -52.3 + 36.1$ pH $+ -3.2$ pH^2 ($r^2 = 0.92$, $P < 0.001$). Weight loss due to fungal respiration was not uniformly affected by pH without macroinvertebrates present. This demonstrates that the role of macroinvertebrates in decomposition is pH dependent.

The role of salinity in lotic ecosystem functioning was tested by Reice and Herbst (1982) who compared decomposition in springs of different salinities in the Dead Sea basin in Israel. The three springs had constant temperatures of 26.5° C, and salinities of 2.54, 4.10, and 13.12 ppt. Leaves of the reed *Phragmites australis* were made into leaf bundles, incubated in the streams, and sampled after two, four, and six weeks. The fauna was similar in all three springs, so diversity did not explain the differences in decomposition rate. The decomposition rate in the stream of lowest salinity was very high (~10% mass loss per day). The decomposition rate of *Phragmites* decreased significantly ($P < 0.001$) as salinity increased. Salinity was a significant factor in controlling this ecosystem-level process.

Kelly et al. (1984) did not find a change in decomposition rate (as measured by methane and inorganic carbon release) in the Canadian ELA Lake 223 whole-lake acidification experiment, which lowered epilimnetic pH from 6.7 to 5.1. These authors suggested that microbial processes kept the pH in the sediments above 6.0. In laboratory studies using mixed, fresh epilimnetic sediment from Lake 114, Kelly et al. (1984) found that pH down to 4.0 did not affect decomposition rates of older organic carbon, but that pH below 5.25 depressed the decomposition rate of newly sedimented material. This points out that the nature of the organic matter being decomposed is important in controlling responses of decomposition to pH.

Analysis of benthic decomposition is a simple and accurate way to determine the state of health of an aquatic ecosystem, and monitoring of mass losses of leaf litter is a good measurement approach. Leafpack and leafbag decomposition studies are now common (for guides to methods, see Petersen and Cummins, 1974, and Reice, 1980, 1981); such studies have gained popularity because they are relatively easy to do, inexpensive, and they yield direct information on the process of decomposition. Ecosystems that are stressed usually will show much lower decomposition rates than similar, unstressed ecosystems.

8.5. Top-Down vs. Bottom-Up Control

The relative merits of the theories of top-down vs. bottom-up control of community structure in aquatic systems (Carpenter et al., 1985) have received much attention, but typically without regard to the zoobenthos. These two concepts place the control of community structure either with piscivorous fish (top-down) or nutrients (bottom-up). In top-down control of the pelagic community of a lake, ecosystem structure and function are seen as controlled by the cascading effects of top-level predation (Carpenter et al., 1985). For example, increasing the density of piscivores (e.g., bass) in a lake will decrease, in turn, the levels of vertebrate zooplanktivores (e.g., minnows), increase levels of zooplankton, and reduce phytoplankton density and productivity. It is suggested that manipulation of consumer populations may hold promise for ameliorating the effects of eutrophication. Conversely, bottom-up control views suggest that lake community structure is governed by nutrients, that is, the bottom of the food chain. Thus, lake enrichment stimulates algal growth, which, in turn, stimulates zooplankton, which then stimulates planktivorous fish, and finally, piscivores (Carpenter et al., 1985).

In reality, the zoobenthic community is a strong link in these trophic relationships. Research in the Cornell Ponds (Hall et al., 1970) supported both the top-down and bottom-up perspectives, because both fish predation and fertilization had significant effects on the zoobenthos. Carpenter et al. (1985) cited evidence that periphyton standing stock and productivity were influenced directly by fish grazing, which then would affect benthic macroinvertebrate herbivores. Vertebrate predators, such as fish, also prey on benthic macroinvertebrate predators and can greatly affect their densities. Reductions in macroinvertebrate predation can, in turn, relieve predatory pressure on other zoobenthic populations and completely alter community structure. Gilinsky (1984) showed that fish predation significantly influenced species richness and density of benthic macroinvertebrate populations in a pond, through its impact on benthic macroinvertebrate predators.

Top-down control of benthic community structure has been studied in lakes for nearly 40 years by people working on the effects of fish on the benthic community (see review by Healey 1984). Ball and Hayne (1952) used rotenone to remove bluegill sunfish, *Lepomis macrochirus*, from a four-hectare, southeastern Michigan lake. The volume of zoobenthos (especially amphipods and molluscs) increased significantly. Interpretation of these data is difficult because many benthic insect predators also were killed by the treatment. The importance of fish predation to the zoobenthos has been supported by a series of comparative studies on lakes with and without fish. Pope et al. (1973) compared *Chaoborus* species assemblages in 26 lakes in the Matamek region of northern Quebec, and they showed that the presence or absence of fish dramatically altered the species composition. These changes in *Chaoborus* spp. should produce cascading trophic effects on the rest of the benthos, but this has not been tested.

Dermott (1988) studied a series of four oligotrophic lakes, having gradients of depth, dissolved oxygen content, and fish abundance, in the Turkey Lakes Watershed, Ontario. The most important factor affecting benthic macroinvertebrate community structure was the presence or absence of fish. The highest-elevation lake in the catchment, Batchawana Lake, was fishless, and large benthic invertebrate predators (libellulid odonates and the amphipod *Crangonyx*) dominated the benthos. In other lakes of the catchment, these taxa were absent or rare when fish were present, apparently as a result of size selection by fish (see also Hall et al., 1970). In Little Turkey Lake, large benthic *Chironomus* spp. escaped fish predation by staying in the low-oxygen environment of the hypolimnion. The resultant *Chironomus* production gave Little Turkey Lake the highest mean biomass and productivity of the four lakes.

The Turkey Lakes study is made more complex when the role of pH is considered (R.J. Hall, Ontario Ministry of the Environment, Dorset, ON, personal communication). Lake Batchawana has low pH, and this may account for the absence of fish. *Chironomus* is an acid-tolerant species, which may explain its high production. Thus, laboratory and field experiments are needed to distinguish among the effects of pH, oxygen, and fish predation.

Bottom-up or input factors also may be important in some special instances. Effects of lake fertilization on benthic macroinvertebrate populations in a subarctic lake with naturally low nutrient supplies were examined by Aagaard (1982). Increased chironomid densities correlated with increased productivity in the lake following fertilization, suggesting that bottom-up factors may be important under the cold and oligotrophic conditions in arctic and subarctic lakes.

The influences of top-down and bottom-up control in streams are much less obvious. Top-down control of the zoobenthos by fishes has proven difficult to demonstrate in streams (Reice, 1983, unpublished data; Flecker and Allan, 1984; Reice and Edwards, 1986), but Gilliam et al. (1989) showed that high densities of fish in small isolated pools reduced the numbers of benthic macroinvertebrates. Reice (1991) showed that fish predation in a stream had no significant effect on zoobenthic species composition because of the replacement of macroinvertebrate species colonizing from the drift. However, total numbers of macroinvertebrates were reduced by fish predation. In the same study, detritus loading with coarse particulate organic matter (CPOM), a bottom-up control factor, also was shown to have relatively little effect on the structure of the benthic community. However, FPOM distribution in the sediments indeed may exert bottom-up control on benthic macroinvertebrate distribution (Rabeni and Minshall, 1977). Therefore, benthic macroinvertebrate community structure of streams may be best understood as a nonequilibrium system with individual taxa and the community responding to the disturbance regime and temporal variations in nutrient supply.

In short, the susceptibility of a particular aquatic community to top-down or bottom-up control cannot be generalized. The behavior of the community will depend on the nature of the system (lake or stream) and the particular environmental conditions to which it is subjected.

8.6. Conclusions

Assessment of ecosystem health can be achieved through a careful analysis of the benthic fauna and benthic processes. The state of an aquatic ecosystem cannot be fully understood without knowledge of the zoobenthos, which is a critical component of every aquatic ecosystem. It plays an essential role in the food chain, productivity, nutrient cycling, and decomposition. No aquatic ecosystem will function long without a healthy zoobenthic community. Knowledge of the state of the zoobenthic community is essential to the assessment of the health of the ecosystem.

For biomonitoring purposes, total taxa richness is the index of choice. However, resources permitting, species-specific population estimates also should be included. Litter decomposition studies give valuable information on the state of the ecosystem by measuring a critical process. A combination of litter decomposition and taxa richness measures will yield important insights into the state of the ecosystem.

The ecosystem perspective needs to be integrated into biomonitoring programs, but it is not a panacea. Community and ecosystem studies can demonstrate that a change took place. In order to identify cause and effect, a field experimental approach is optimal, but it is expensive and labor intensive. Other approaches, from laboratory bioassays and genetic analyses to population transplants in the field, all offer important clues to environmental impacts on freshwater ecosystems. What the ecosystem approach offers is the opportunity to understand not only species but also their interactions and roles in the whole system. The ecosystem perspective shifts our focus from small to large-scale phenomena, from population to community, and from structure to process. In this way, we can learn not only what species are present, but how they fit in and how they contribute to the functioning of the ecosystem.

References

Aagaard, K. 1982. Profundal chironomid populations during a fertilization experiment in Langvatn, Norway. *Holarctic Ecology* 5:325–31.

Allard, M. and G. Moreau. 1986. Leaf decomposition in an experimentally acidified stream channel. *Hydrobiologia* 139:109–17.

Azam, F., T. Fenchel, J.G. Field, J.S. Gray, L.A. Meyer-Reil, and F. Thingstad. 1983. The ecological role of water column-microbes in the sea. *Marine Ecology Progress Series* 10:257–63.

Ball, R.C. and D.W. Hayne. 1952. Effects of the removal of the fish population on the fish-food organisms of a lake. *Ecology* 33:41–8.

Boling, R.H., Jr., E.D. Goodman, J.A. Van Sickle, J.O. Zimmer, K.W. Cummins, R.C. Petersen, and S.R. Reice. 1975. Toward a preliminary model of detritus processing in a woodland stream. *Ecology* 56:141–51.

Brundin, L. 1958. The bottom faunistical lake type system and its application to the southern hemisphere. Moreover a theory of glacial erosion as a factor of productivity in lakes and oceans. *Internationale Vereinigung für Theoretische und Angewandte Limnologie Verhandlungen* 13:288–97.

Burton, T.M. and J.W. Allan. 1986. Influence of pH, aluminum, and organic matter on stream invertebrates. *Canadian Journal of Fisheries and Aquatic Sciences* 43:1285–9.

Cairns, J., Jr. 1981. Biological monitoring. Part VI—Future needs. *Water Research* 15:941–52.

Carpenter, S.R., ed. 1988. *Complex Interactions in Lake Communities*. Springer Verlag, New York.

Carpenter, S.R., J.F. Kitchell, and J.R. Hodgson. 1985. Cascading trophic interactions and lake productivity. *BioScience* 35:634–9.

Clements, W.H., D.S. Cherry, and J. Cairns, Jr. 1988. Structural alterations in aquatic insect communities exposed to copper in laboratory streams. *Environmental Toxicology and Chemistry* 7:715–22.

Courtemanch, D.L., S.P. Davies, and E.B. Laverty. 1989. Incorporation of biological information in water quality planning. *Environmental Management* 13:35–41.

Cuffney, T.F., J.B. Wallace, and J.R. Webster. 1984. Pesticide manipulation of a headwater stream: invertebrate responses and their significance for ecosystem processes. *Freshwater Invertebrate Biology* 3:153–71.

Cummins, K.W. 1974. Structure and function of stream ecosystems. *BioScience* 24:631–41.

Cummins, K.W. and M.J. Klug. 1979. Feeding ecology of stream invertebrates. *Annual Review of Ecology and Systematics* 10:147–72.

Cummins, K.W., R.C. Petersen, F.O. Howard, J.C. Wuycheck, and V.I. Holt. 1973. The utilization of leaf litter by stream detritivores. *Ecology* 54:336–45.

Cummins, K.W., M.A. Wilzbach, D.M. Gates, J.B. Perry, and W.B. Taliaferro. 1989. Shredders and riparian vegetation. *BioScience* 39:24–30.

Dermott, R.M. 1988. Zoobenthic distribution and biomass in the Turkey Lakes. *Canadian Journal of Fisheries and Aquatic Sciences* 45 (Supplement 1):107–14.

Fiance, S.B. 1978. Effects of pH on the biology and distribution of *Ephemerella funeralis* (Ephemeroptera). *Oikos* 31:332–9.

Fisher, S.G. and G.E. Likens. 1973. Energy flow in Bear Brook, New Hampshire: an integrative approach to stream ecosystem metabolism. *Ecological Monographs* 43:421–39.

Flecker, A.S. and J.D. Allan. 1984. The importance of predation, substrate and spatial refugia in determining lotic insect distributions. *Oecologia* (Berlin) 64:306–13.

Gilinsky, E. 1984. The role of fish predation and spatial heterogeneity in determining benthic community structure. *Ecology* 65:455–68.

Gilliam, J.F., D.F. Fraser, and A.M. Sabat. 1989. Strong effects of foraging minnows on a stream benthic invertebrate community. *Ecology* 70:445–52.

Grapentine, L.C. 1987. Consequences of environmental acidification to the freshwater amphipod *Hyalella azteca*. M.Sc. dissertation, Univ. of Manitoba, Winnipeg, MB.

Hall, D.J., W.E. Cooper, and E.E. Werner. 1970. An experimental approach to the production dynamics and structure of freshwater animal communities. *Limnology and Oceanography* 15:839–928.

Hall, R.J. 1990. Relative importance of seasonal, short-term pH disturbances during discharge variation on a stream ecosystem. *Canadian Journal of Fisheries and Aquatic Sciences* 47:2261–74.

Hall, R.J., C.T. Driscoll, and G.E. Likens. 1987. Importance of hydrogen ions and aluminium in regulating the structure and function of stream ecosystems: an experimental test. *Freshwater Biology* 18:17–43.

Hall, R.J. and G.E. Likens. 1981. Chemical flux in an acid-stressed stream. *Nature* (London) 292:329–31.

Hall, R.J., G.E. Likens, S.B. Fiance, and G.R. Hendrey. 1980. Experimental acidification of a stream in the Hubbard Brook Experimental Forest, New Hampshire. *Ecology* 61:976–89.

Healey, M. 1984. Fish predation on aquatic insects. In *The Ecology of Aquatic Insects,* eds. V.H. Resh and D.M. Rosenberg, pp. 255–88. Praeger Pubs., New York.

Hildrew, A.G., C.R. Townsend, J. Francis, and K. Finch. 1984. Cellulolytic decomposition in streams of contrasting pH and its relationship with invertebrate community structure. *Freshwater Biology* 14:323–8.

Karr, J.R. 1987. Biological monitoring and environmental assessment: a conceptual framework. *Environmental Management* 11:249–56.

Kelly, C.A., J.W.M. Rudd, A. Furutani, and D.W. Schindler. 1984. Effects of lake acidification on rates of organic matter decomposition in sediments. *Limnology and Oceanography* 29:687–94.

Kimball, K.D. and S.A. Levin. 1985. Limitations of laboratory bioassays: the need for ecosystem-level testing. *BioScience* 35:165–71.

La Point, T.W., S.M. Melancan, and M.K. Morris. 1984. Relationships among observed metal concentrations, criteria and benthic community structural responses in 15 streams. *Journal of the Water Pollution Control Federation* 56:1030–8.

La Point, T.W. and J.A. Perry. 1989. Use of experimental ecosystems in regulatory decision making. *Environmental Management* 13:539–44.

Lenat, D.R. 1988. Water quality assessment of streams using a qualitative collection method for benthic macroinvertebrates. *Journal of the North American Benthological Society* 7:222–33.

Likens, G.E. 1985. An experimental approach for the study of ecosystems. *Journal of Ecology* 73:381–96.

Mackay, R.J. and K.E. Kersey. 1985. A preliminary study of aquatic insect communities and leaf decomposition in acid streams near Dorset, Ontario. *Hydrobiologia* 122:3–11.

Merritt, R.W. and K.W. Cummins, eds. 1984. *An Introduction to the Aquatic Insects of North America,* 2nd ed. Kendall/Hunt Publishing, Dubuque, IA.

Newman, R.M. and J.A. Perry. 1989. The combined effects of chlorine and ammonia on litter breakdown in outdoor experimental streams. *Hydrobiologia* 184:69–78.

Newman, R.M., J.A. Perry, E. Tam, and R.L. Crawford. 1987. Effects of chronic chlorine exposure on litter processing in outdoor experimental streams. *Freshwater Biology* 18:415–28.

Perry, J.A. and N.H. Troelstrup, Jr. 1988. Whole ecosystem manipulation: a productive avenue for test system research? *Environmental Toxicology and Chemistry* 7:941–51.

Petersen, R.C. and K.W. Cummins. 1974. Leaf processing in a woodland stream. *Freshwater Biology* 4:343–68.

Pomeroy, L.R. 1974. The ocean's foodweb, a changing paradigm. *BioScience* 24:499–504.

Pope, G.F., J.C.H. Carter, and G. Power. 1973. The influence of fish on the distribution of *Chaoborus* spp. (Diptera) and density of larvae in the Matamek River system, Québec. *Transactions of the American Fisheries Society* 102:707–14.

Pratt, J.M. and R.J. Hall. 1981. Acute effects of stream acidification on the diversity of macroinvertebrate drift. In *Effects of Acidic Precipitation on Benthos,* ed. R. Singer, pp. 77–95. North American Benthological Society, Springfield, IL.

Rabeni, C.F. and G.W. Minshall. 1977. Factors affecting micro-distribution of stream benthic insects. *Oikos* 29:33–43.

Reice, S.R. 1980. The role of substratum in benthic macroinvertebrate microdistribution and litter decomposition in a woodland stream. *Ecology* 61:580–90.

Reice, S.R. 1981. Interspecific association in a woodland stream. *Canadian Journal of Fisheries and Aquatic Sciences* 38:1271–80.

Reice, S.R. 1983. Predation and substratum: factors in lotic community structure. In *Dynamics of Lotic Ecosystems,* eds. T.D. Fontaine, III, and S.M. Bartell, pp. 325–45. Ann Arbor Science Pubs., Ann Arbor, MI.

Reice, S.R. 1991. Effects of detritus loading and fish predation on leafpack breakdown and benthic macroinvertebrates in a woodland stream. *Journal of the North American Benthological Society* 10:42–56.

Reice, S.R. and R.L. Edwards. 1986. The effect of vertebrate predation on lotic macroinvertebrate communities in Québec, Canada. *Canadian Journal of Zoology* 64:1930–6.

Reice, S.R. and G. Herbst. 1982. The role of salinity in decomposition of leaves of *Phragmites australis* in desert streams. *Journal of Arid Environments* 5:361–8.

Reynoldson, T.B. 1987. Interactions between sediment contaminants and benthic organisms. *Hydrobiologia* 149:53–66.

Sæther, O.A. 1975. Nearctic chironomids as indicators of lake typology. *Internationale Vereinigung für Theoretische und Angewandte Limnologie Verhandlungen* 19:3127–33.

Sæther, O.A. 1979. Chironomid communities as water quality indicators. *Holarctic Ecology* 2:65–74.

Schindler, D.W., K.H. Mills, D.F. Malley, D.L. Findlay, J.A. Shearer, I.J. Davies, M.A. Turner, G.A. Linsey, and D.R. Cruikshank. 1985. Long-term ecosystem stress: the effects of years of experimental acidification on a small lake. *Science* 228:1395–401.

Severinghaus, W.D. 1981. Guild theory development as a mechanism for assessing environmental impact. *Environmental Management* 5:187–90.

Specht, W.L., D.S. Cherry, R.A. Lechleitner, and J. Cairns, Jr. 1984. Structural, functional, and recovery responses of stream invertebrates to fly ash effluent. *Canadian Journal of Fisheries and Aquatic Sciences* 41:884–96.

Steines, S.E. and R. Wharfe. 1987. A practical classification of unpolluted running waters in Kent and its application in water quality assessment. *Water Pollution Control* 1987:184–91.

Thompson, P.L. and F. Bärlocher. 1989. Effect of pH on leaf breakdown in streams and in the laboratory. *Journal of the North American Benthological Society* 8:203–10.

Wallace, J.B., G.J. Lugthart, T.F. Cuffney, and G.A. Schurr. 1989. The impact of repeated insecticidal treatments on drift and benthos of a headwater stream. *Hydrobiologia* 179:135–47.

Wallace, J.B., D.S. Vogel, and T.F. Cuffney. 1986. Recovery of a headwater stream from an insecticide-induced community disturbance. *Journal of the North American Benthological Society* 5:115–26.

Wallace, J.B., J.R. Webster, and T.F. Cuffney. 1982. Stream detritus dynamics: regulation by invertebrate consumers. *Oecologia* (Berlin) 53:197–200.

Wiederholm, T. 1984. Responses of aquatic insects to environmental pollution. In *The Ecology of Aquatic Insects,* eds. V.H. Resh and D.M. Rosenberg, pp. 508–57. Praeger Pubs., New York.

Winner, R.W., M.W. Boesel, and M.P. Farrell. 1980. Insect community structure as an index of heavy-metal pollution in lotic ecosystems. *Canadian Journal of Fisheries and Aquatic Sciences* 37:647–55.

Zanella, E. 1982. Shifts in caddisfly species composition in Sacramento River invertebrate communities in the presence of heavy metal contamination. *Bulletin of Environmental Contamination and Toxicology* 29:306–12.

9

Paleolimnological Biomonitoring Using Freshwater Benthic Macroinvertebrates

Ian R. Walker

9.1. Introduction

Aquatic systems have been subjected to a series of successive insults, including eutrophication, organotoxin and metal contamination, acidification, and perhaps, most recently, the beginnings of global climatic change. The spatial scale of perturbation has been elevated from local disruption of individual lakes and streams to regional and global effects. No aquatic system now can be assumed to exist in a truly pristine state.

The rate of environmental change has greatly outstripped our ability to collect baseline data. Without such data, it is impossible to assess accurately the timing or extent of disturbance, except through paleolimnology.

Within the past decade, paleolimnology has evolved from an esoteric discipline to a sophisticated, quantitative alternative to conventional biomonitoring (Smol, 1990). Recent advances in paleolimnological technique owe much to the challenges posed by the need to study lake acidification, especially to answer questions such as: (1) Did a lake or lakes acidify? and (2) How great was the change? Paleolimnological investigations have depicted accurately the timing and impact of acidification (e.g., Dixit et al., 1989a; Battarbee et al., 1990), and other environmental stresses such as eutrophication (Warwick, 1980) and climatic change (Hofmann, 1983). With paleolimnology, no preimpact investigations are necessary. Baseline data offered by a preimpact environmental assessment done two decades ago cannot compare with the thousand-year records of trends and natural variability preserved in sediments.

This chapter provides a brief overview of the use of paleolimnology in biomonitoring. Details of methods will not be discussed; instead, readers are referred to many recent review articles, foremost of which are the *Handbook of Holocene Palaeoecology and Palaeohydrology* (Berglund, 1986) and *Methods in Quaternary Ecology* (Warner, 1990).

The usefulness of paleolimnological biomonitoring is illustrated best with examples. A wood-pulp and paper mill was established in 1871 at Valkeakoski, a community in southern Finland adjacent to Lake Vanajavesi. Sulphate and sulphite pulp mills were added in 1880 and 1907, and a rayon fiber plant was established in 1943. Pulp and paper production increased gradually during the first half of the twentieth century, but tripled between 1950 and 1970 (Fig. 9.1A). The town's population grew from 2,800 in 1922 to 23,000 inhabitants in 1981. This industrial center was the major source of pollution for Lake Vanajavesi; however, recent pollution abatement measures prevented any further increase in wastewater loading.

In the spring of 1981, Finnish limnologists removed a sediment core from a 10 m-deep basin of Lake Vanajavesi (Kansanen and Jaakkola, 1985). Investigations of this core and three additional profiles, including geochemical analyses (Kansanen and Jaakkola, 1985) and analysis of the chitinous remains of chironomid larvae (Kansanen, 1985, 1986), provided a detailed record of how industrial developments had altered the lake (Fig. 9.1B).

Kansanen and Jaakkola (1985) established a detailed chronology for the sediments based upon varve (annual sediment layer) counts, correlation with historical events, and ^{137}Cs and ^{210}Pb dating. Late nineteenth to early twentieth century cultural impact was recorded by slight increases in Zn deposition. Increased pulp production was indicated by increasing sedimentary concentrations of dehydroabietic acid. Following construction of the rayon plant, Zn loading increased dramatically (Kansanen and Jaakkola, 1985).

Although geochemical analyses faithfully recorded disruptions of the natural chemical balance, especially following construction of the rayon plant, demonstration of environmental damage required biological evidence. Analyses of chironomid remains (Kansanen, 1985, 1986) indicated a preindustrial fauna typical of oligotrophic lakes. With expanded pulp production between 1880 and 1950, *Chironomus plumosus,* an indicator of eutrophy, increased greatly in abundance and soon dominated the profundal fauna. Anoxic conditions in the profundal zone of the basin became established in 1957, as indicated by well-preserved sediment laminae and by the declining abundance of profundal chironomid remains, especially those of *Ch. plumosus* (Fig. 9.1B).

The studies of Lake Vanajavesi portray the resolution potential of paleolimnological biomonitoring. Paleolimnological evidence demonstrated that early pulp production had seriously diminished water quality, but that the most deleterious biological effects resulted from continued growth of rayon and pulp and paper production after 1957. To achieve the same resolution, conventional biomonitoring would have required continuous sampling during the past century, along with an enormous budget to do so. The cost of paleolimnological assessment, by comparison, was small.

Similar resolution was achieved in recent studies on the timing of acidification and loss of fish populations in the Adirondack lakes of New York

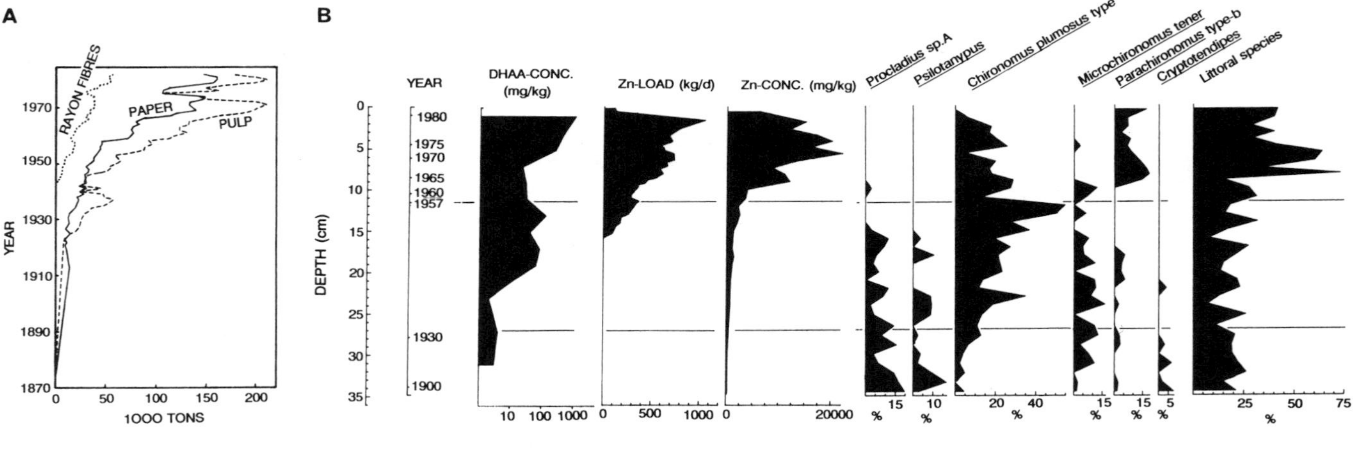

Figure 9.1. A: Historical summary of industrial production near Lake Vanajavesi, Finland. Modified from Kansanen and Jaakkola (1985). B: Summary of chronology for core KSI from Lake Vanajavesi. Adapted from Kansanen (1985) and Kansanen and Jaakkola (1985), reproduced by permission of the Finnish Zoological Publishing Board. Profiles indicate sediment dates, sediment dehydroabietic acid (DHAA) content, Zn loading (based on historical records), Zn concentration in sediment, and the relative abundance of common chironomid taxa.

State (Charles and Whitehead, 1989; Charles et al., 1990). These events were pinpointed on the basis of changing cladoceran (Paterson, 1985) and midge (Uutala 1986, 1990) populations, as well as by diatom analyses. Here, too, paleolimnological biomonitoring proved cost-effective relative to other means of assessment.

The usefulness of paleoecological techniques is not limited to the reconstruction of past environments. The effect of dredging on the Lake Mývatn fauna was predicted by examining ancient sediments in this Icelandic lake (Einarsson and Hafliðason, 1988; Einarsson et al. 1988). These sediments portray the biota that resided in the deep ancestral lake; the authors argued that conditions should be similar in Lake Mývatn following dredging.

9.2. The Sedimentary Record

The study of the heterogeneous constituents of sediment is the essence of paleolimnology. Sediments contain the following organic and inorganic materials derived both from a lake or a stream and its catchment: (1) clay, silt, sand, humus, and litter eroded from adjacent land and from the course of tributary streams; (2) autochthonous precipitates (i.e., precipitates produced within the lake or stream) composed of marl, iron oxides, sulfides and, in saline lakes, an array of evaporite minerals; (3) intact and partially decomposed remains of assorted organisms of aquatic and terrestrial origin; and (4) aerially dispersed substances, including soot, pollen, and several radioisotopes.

Each sedimentary constituent is a potentially valuable clue to past environmental conditions. The relative contributions of clay, silt, and sand indicate erosional intensity in the catchment and mechanical disturbance of the sediment surface by currents and waves at the site of deposition. The chemistry of allochthonous materials (i.e., materials originating outside of the lake or stream) reflects the geology of the catchment, the composition of soils, and the nature of pollutants (e.g., Mackereth, 1966; Mathewes and D'Auria, 1982; Engstrom and Wright, 1984; Engstrom and Hansen, 1985). Chemistry of autochthonous components yields insights into the past chemistry of water, climatic changes, lake productivity, and the composition of past algal communities (e.g., Swain, 1985; Edwards and Fritz, 1988; Last and Slezak, 1988; Sanger, 1988).

Although lakes are efficient traps for sediments, erosion prevails in the upper reaches of streams. Where sedimentation does take place in a stream, its shifting course will provide a constantly changing sedimentary environment in which sediment deposition may be interrupted by erosional events. The remains of aquatic organisms will be sorted in a manner similar to that for clay, silt, and sand prior to deposition. Given these difficulties, paleolimnological evidence of stream perturbations may be sought best in lakes, or other basins receiving stream flow, where remains of lotic biota may be deposited close to the entrance of inflowing water. In addition, sediments deposited in the many reservoirs that have been constructed along the course of streams should provide excellent records of stream biota. Because many lakes receive nutrient input from lotic systems, lacustrine biota can be used to monitor disturbance throughout the catchment.

Each species of fossil organism responds uniquely to the changing environmental milieu, reflecting past climate, nutrient loading, oxygen availability, pH, pollution, predation, or other conditions. Common fossils of aquatic organisms in lakes include: chrysophyte scales and cysts; diatom frustules; exoskeletal fragments of chaoborids, chironomids, cladocerans, molluscs, and ostracods; protozoan plates; sponge spicules; and rhizopod tests (Frey, 1964; Berglund, 1986; Gray, 1988). Scales, teeth, and bones of fish occasionally have been found in sediments (Frey, 1964), as have fragments of beetles, mites, and other terrestrial arthropods. In addition, pollen, spores, seeds, and vegetative macrofossils provide records of aquatic and terrestrial vegetation. Frey (1976, p. 2209) concluded that "All groups of aquatic organisms can leave at least some remains, sometimes under only highly specialized conditions. . . ."

9.3. Paleolimnological Methods

To unravel the history of a lake or other aquatic habitat, one or more cores or vertical sections of sediment are removed for analysis. Recent events are portrayed in surficial sediments, whereas older events are recorded in deeper layers. Thus, it is important to obtain cores or sections in which the layers of sediment have not been disturbed too greatly. Although sediment stratigraphy may be disturbed by natural processes (e.g., bioturbation), careless sampling procedures also must be avoided. Fossils, or other indicators, must be present in sufficient numbers to facilitate a detailed assessment of their distribution in the sediments. Finally, a depth-time or chronostratigraphic profile must be established for the sediment sequence. The vast majority of lake deposits satisfy these seemingly stringent requirements for paleolimnological biomonitoring.

Paleolimnologists favor cores from deepwater, lentic sites because the sediments at these sites will have been subjected to minimal disturbance. The rate of sediment deposition, and hence the time resolution, should be greater in deepwater than shallower environments (Hilton, 1985). Remains of many littoral organisms eventually are deposited in deep water, so a single deepwater core can provide data on planktonic and littoral organisms, as well as those of profundal habitats (Frey, 1988).

9.3.1. Sediment Collection

Recent changes, represented in the uppermost 0.5 m of lake sediments, will be of principal interest to most applied studies. However, it also may be necessary to analyze older deposits in order to determine natural variability of the fauna within the system, to assess the effects of climatic change, or to examine premodern human impact. In North America, the impact of native inhabitants rarely is evident, but in Europe, a long history of human intervention may be apparent (e.g., Hutchinson et al., 1970).

Designs of suitable equipment for coring lake deposits are readily available in the literature (reviewed by Aaby and Digerfeldt, 1986). Many of the corers originally developed as collectors of live benthos are suitable for paleolimnological sampling of surficial lake sediments (Fig. 9.2). Modified versions of the Kajak-Brinkhurst corer (Brinkhurst et al., 1969; Brinkhurst, 1974; Glew, 1989) have been adopted widely for sampling sediments deposited since European settlement in North America (Fig. 9.2D), but piston corers (Fig. 9.2A), at least theoretically, are superior (Aaby and Digerfeldt, 1986). Elimination of hydrostatic pressure on the inside of the corer and creation of a partial vacuum between the piston and the mud surface allow piston samplers to recover safely long columns of sediment. However, the Brown (1956) piston corer (Fig. 9.2A) must be lowered slowly or it will be preceded by a pressure wave that may disturb the uppermost sediment. Also, this corer is manipulated by rods and a cable, so it is awkward to use in deep water or from a small boat. Freeze corers (reviewed by O'Sullivan, 1983) also allow fine-resolution sampling of recent sediments. Sediment accumulation rates and ^{210}Pb dates estimated from a core may be biased if a small volume of sediment is lost through "core shortening" with pistonless (or other) corers, by disturbance from pressure waves, or by other causes (e.g., Blomqvist, 1985).

The Livingstone (1955) piston sampler is used most widely to examine deeper sediments for the study of long-term trends or natural variability within past communities. Aaby and Digerfeldt (1986) recommend Wright's (1967) modification of this equipment. As with Brown's design, these corers are manipulated by rods and cables. They must be deployed from a secure platform, such as a raft or lake ice, and they cannot be used in very deep water. The Mackereth (1958) corer provides means for collecting long cores from easily accessible deepwater sites, but fine stratigraphy in surficial sediments may be disrupted because this corer is difficult to maintain in a vertical position during retrieval. The percussion corers designed by Gilbert and Glew (1985) and Reasoner (1986), which are operated by a single cable, are highly portable and also are suitable for sampling deepwater sediments.

Unlike coring devices, some samplers, including the Russian sampler (Jowsey, 1966), are rotated to cut a vertical column from the sediment. The

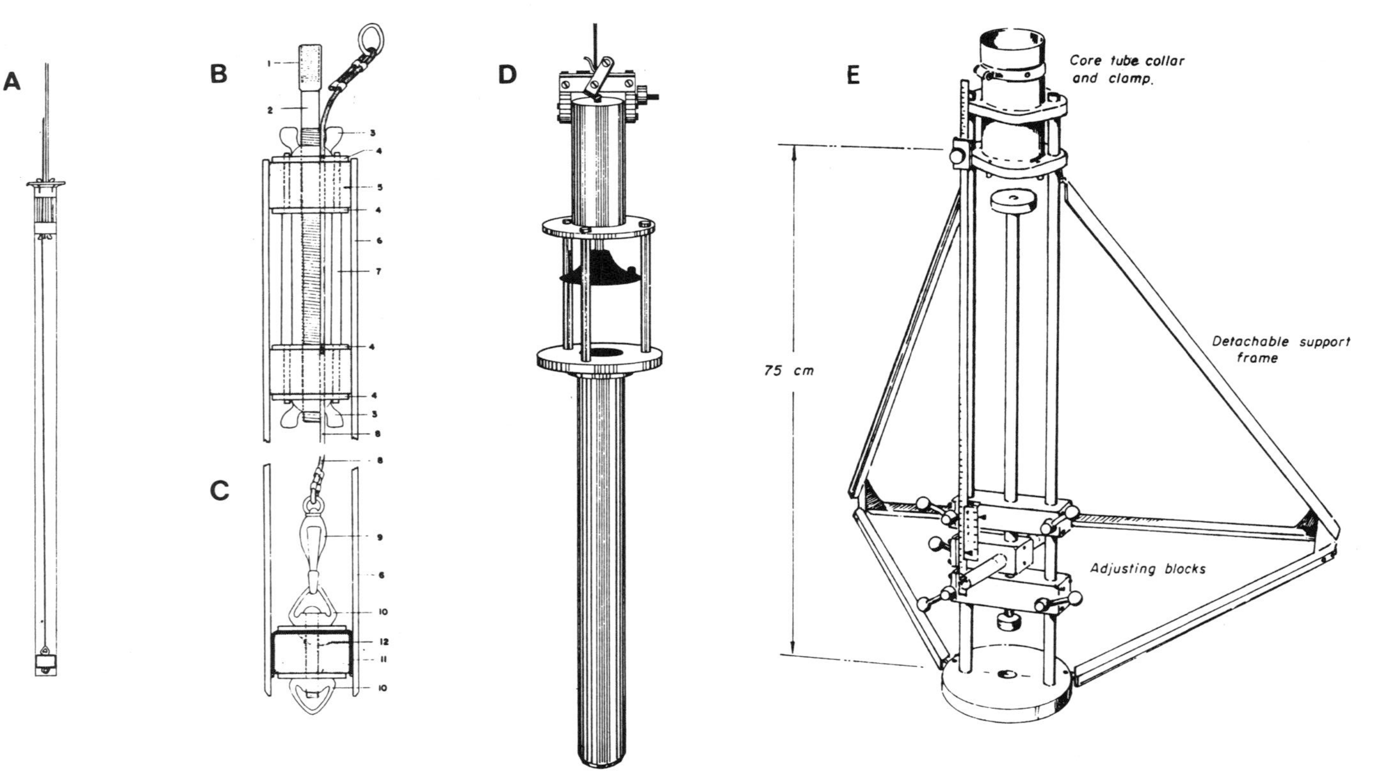

Figure 9.2. Examples of coring and extruding devices for sampling the uppermost meter of lake sediments. A–C: The Brown corer. A: Fully assembled; B: Detail of head piece; C: Detail of piston. From Brown (1956), reprinted by permission of The Ecological Society of America. D: The Kajak-Brinkhurst corer; version equipped with single core tube. From Brinkhurst (1974), reprinted by permission of R.O. Brinkhurst. E: The Glew extruder, without core tube in place. From Glew (1988), reprinted by permission of Kluwer Academic Publishers.

Russian sampler penetrates and samples lake sediments with less disturbance of stratigraphy than coring devices, so this sampler is favored for studies of varved sediments.

The samplers described will not provide suitable means for sampling all deposits. The coarse clastic materials deposited by many streams will be impervious to most coring devices. To penetrate sediments of the Fraser River Delta, Williams and Roberts (1987) relied upon a vibrating corer. In some instances, sediments can be sampled by hand from vertical sections of sediment, either naturally exposed along river banks or revealed by excavation (Aaby and Digerfeldt, 1986). It may be necessary to pump water from the excavation to facilitate sample collection. The exposure must be "cleaned" carefully to reveal fresh sediment prior to sample collection because sediment from other depths may contaminate the surface exposed during excavation. Samples may be collected as intact sediment columns, which are subsampled later in a laboratory, or small samples from each stratigraphic level can be collected while working at the exposure.

9.3.2. Sample Storage and Subsampling

The uppermost sediments of lakes usually are loose, uncompacted materials, which are more liquid than solid. Consequently, surficial lake cores must be handled with extreme care, so the detailed record of events sought by the paleoecologist is not disrupted. Surficial cores, unless frozen, must be transported vertically in the core tube until subsampling can be completed. If possible, cores should be extruded at the core site and must be extruded while still held in a vertical position. The vertical extruder developed by Glew (1988) (Fig. 9.2E) allows close-interval (e.g., 2.5 mm) sectioning of core sediments. Careful sectioning of surficial sediments is the key to reconstructing limnological changes that occurred in the past decade (e.g., Dixit et al., 1989a). Thin sections of lake or stream sediment are stored conveniently in plastic bags or vials.

Special techniques have been developed for the study of frozen sediment cores. The techniques used for the study of annually laminated sediments (reviewed by O'Sullivan, 1983) illustrate how maximum resolution of paleolimnological events may be achieved. The tape-peel technique devised by Simola (1977) and modified by Davidson (1988) will allow seasonal cycles in the deposition of siliceous plant and animal microfossils to be discerned within varves.

At depths $\gtrsim$1 m into lake mud, the sediment usually is sufficiently well-consolidated that vertical extrusion of wet sediments is not necessary. These cores can be extruded horizontally onto a clean surface, and a detailed description of them can be prepared. The outer surface of the core should be removed because some smearing inevitably occurs along the walls of the core tube. The core then can be sectioned immediately, or it can be kept intact and carefully wrapped. Intact cores and core subsamples normally are sealed carefully to prevent moisture loss, and they are stored at ~4° C in a dark room or in a refrigerator. Sediments that have dried can be extremely difficult to rehydrate for fossil analysis; however, some ostracod workers regularly freeze-dry samples before picking them.

9.3.3. Sediment Chronology

Sediment cores must be dated reliably in order to interpret accurately paleolimnological change. Sediment chronologies usually are developed from analyses of one or more radioisotopes, but correlation of events represented in cores with historical records provides an independent assessment. Olsson (1986) provided a valuable review of pertinent radiometric dating techniques; these should be performed only by experienced persons.

Quaternary geologists have adopted ^{14}C analyses as a favorite technique for dating organic materials from late-Pleistocene and more recent deposits. The new accelerator mass spectrometry method allows ^{14}C dates to be obtained from <1 mg of carbon (e.g., Brown et al., 1989). Unfortunately, this naturally occurring radioisotope is not suitable for dating sediments younger than 300 to 400 years (Stuiver, 1982), but two other radioactive isotopes, ^{210}Pb and ^{137}Cs, have provided means for dating more recent lake sediments. Other radioactive isotopes (e.g., ^{240}Pu, ^{32}Si, ^{228}Th, ^{232}Th) also may provide methods for sediment dating (e.g., Krishnaswamy et al., 1971; Olsson, 1986).

^{137}Cs deposited in recent sediments is a product of the nuclear age; bomb-produced ^{137}Cs fallout reached a maximum in 1963, although a lesser peak was evident in 1959 (Ritchie et al., 1973). The succession of maximum and minimum intervals of ^{137}Cs fallout has allowed a series of historical dates to be identified in lake sediments. Unfortunately, ^{137}Cs is mobile in sediments of acidic lakes and, in these locations, does not provide a reliable dating method (Davis et al., 1984; Heit and Miller, 1987). In addition, the Chernobyl disaster of 1987 has resulted in a tremendous flux of this radioisotope to sediments; this flux masks the 1963 peak in many European sediment cores. Another bomb-produced radioisotopic marker, ^{241}Am, is less prone to diffusion in sediments and will allow the 1963 peak to be identified (Appleby et al., 1990). The Chernobyl disaster, however, has provided a stratigraphic marker that will be useful for future paleolimnologists. Where well-defined ^{137}Cs or ^{241}Am peaks are identified, significant bioturbation of sediments cannot have occurred.

^{210}Pb is produced naturally through the atmospheric decay of ^{222}Rn, a member of the ^{238}U decay series. This "unsupported" ^{210}Pb is precipitated, bound to sediment particles, and deposited in lakes. It is abundant in sediments younger than ~100 years, but because of radioactive decay ($t_{1/2}$ = 22.26 yr), it does not persist in measurable amounts in older deposits. Assuming a constant supply rate of unsupported ^{210}Pb, it is possible to estimate, from a series of ^{210}Pb measurements, ages of sediments deposited in approximately the last 100 years (Appleby and Oldfield, 1978; Oldfield and Appleby, 1984). The constant rate of supply (CRS) model does not assume a monotonic decline of ^{210}Pb activity with depth, and appears to be superior to the constant initial concentration (CIC) model (Binford, 1990).

The presence or absence of bioturbation can influence significantly the ^{210}Pb profile and thus an inferred chronology of events. For example, Robbins et al. (1989) have assumed that sedimentation rates are constant in lakes and that the shape of the ^{210}Pb profile is influenced strongly by bioturbation. For Lake Erie, they estimated that sedimentation rates varied among sites from 0.01 to 0.86 g cm^{-2} yr^{-1} (median = 0.10 g cm^{-2} yr^{-1}); time resolution in the sediments varied from four to 34 years (median = 7 yr).

Because anthropogenic disturbance of lakes almost always is accompanied by increased sediment accumulation rates, most paleolimnologists are uncomfortable with the assumption of constant sedimentation rates. The astonishing detail revealed in many stratigraphic studies (e.g., Green, 1981; Dixit et al., 1989a) indicates good time resolution and suggests that the importance of bioturbation has been exaggerated; however, the importance of sediment mixing probably varies greatly among sites. Time resolution may be best in small, deep lakes and worst in large, shallow lakes such as Lake Erie.

For lakes studied as part of the Paleoecological Investigation of Recent Lake Acidification (PIRLA) project, Binford (1990) assumed that bioturbation had no impact on the ^{210}Pb profile; he attributed variations in ^{210}Pb to varying rates of sediment accumulation and ^{210}Pb decay. Alternative models are available for calculating ^{210}Pb dates where sedimentation rates have varied, and continuous bioturbation also has occurred, but an independent estimate of mixing depth is required (Binford, 1990). The ^{210}Pb dating method has been tested rigorously through analyses of varved lake deposits (e.g., Appleby et al., 1979; Brunskill and Ludlam, 1988); varved deposits may be more widespread than commonly thought, and they offer unparalleled chronological control. If the annual sediment layers are uniform in color, they are invisible to our eyes; however, other means of detection (e.g., x-ray analysis) may reveal fine structure.

Several chronostratigraphic markers may provide independent checks of age determinations. In southern Ontario, Canada, and the midwestern United States, an abrupt increase in *Ambrosia* (ragweed) pollen abundance appeared circa 1850, following extensive land clearance (e.g., Warwick, 1980). This important marker is not useful in many regions of North America and in Europe, but other palynological markers are evident. For example, *Alnus* (alder) pollen is abundant in postsettlement deposits in Atlantic Canada and coastal British Columbia (e.g., Mathewes and D'Auria, 1982; Walker and Paterson, 1983). The arrival of Russian thistle (*Salsola iberica*) and cultivated cereals at Devils Lake, North Dakota, has provided an independent assessment of ^{210}Pb profiles, and the potential temporal resolution of sediment profiles (Jacobson and Engstrom, 1989). Increased erosion following settlement sometimes is evident as changes in sediment lithology or geochemistry (e.g., Warwick, 1980; Burden et al., 1986). Charcoal and soot particles can serve as stratigraphic markers of fuel combustion and forest fires (e.g., Wik et al., 1986).

However, as Stoermer et al. (1985) noted, assumed relationships between historical events and core records introduces circularity to the dating process. It is dangerous to assume that lithological changes or biotic changes apparent in cores can be matched precisely with events described in historical literature. Inaccurate dates result in unreliable sedimentation and microfossil accumulation rates; inferences dependent upon these measures are especially prone to error when accurate chronologies are not assured. Use of several independent dating techniques allows the magnitude of dating errors to be evaluated (e.g., Kansanen and Jaakkola, 1985). Exact correspondence of dates among methods should not be expected, and ad hoc adjustments to improve the correspondence are strongly discouraged.

Dates inferred by correlation of core records with historical events provided one basis for the Bay of Quinte, Lake Ontario, chronology reported by Warwick (1980). The dates suggested by an erosion index and three radioisotopic methods (^{137}Cs and ^{14}C dating and unadjusted ^{210}Pb dating; Kipphut, 1978) differed, but it is not clear which method provided the most accurate chronology. Warwick (1980) described precautions that were used to reduce or avoid disturbance of the Bay of Quinte sediments; such disturbance can contribute to poorly resolved ^{137}Cs peaks and difficulties in developing ^{210}Pb chronologies. The CRS model of ^{210}Pb dating (Appleby and Oldfield, 1978) might have supplied more accurate dates, but it was not available at the time of the Bay of Quinte work. Warwick (1980) also noted that the "hardwater effect" adversely influenced Bay of Quinte ^{14}C dates. The hardwater effect occurs when carbon, dissolved from ancient carbonate rock, is incorporated into the tissues of aquatic organisms, causing these tissues to yield anomalously old ^{14}C dates.

Development of chronologies for stream deposits also involves difficulties. ^{137}Cs has been used to date reservoir sediments (Ritchie et al., 1973); ^{210}Pb dating also may be possible. However, accurate ^{210}Pb or ^{137}Cs dates probably cannot be obtained from most other stream deposits. Instead, Klink (1989) dated abandoned river channels by using a temporal series of River Rhine maps. He sampled sediments deposited at the time of abandonment, and thereby reconstructed recent changes in the Rhine fauna.

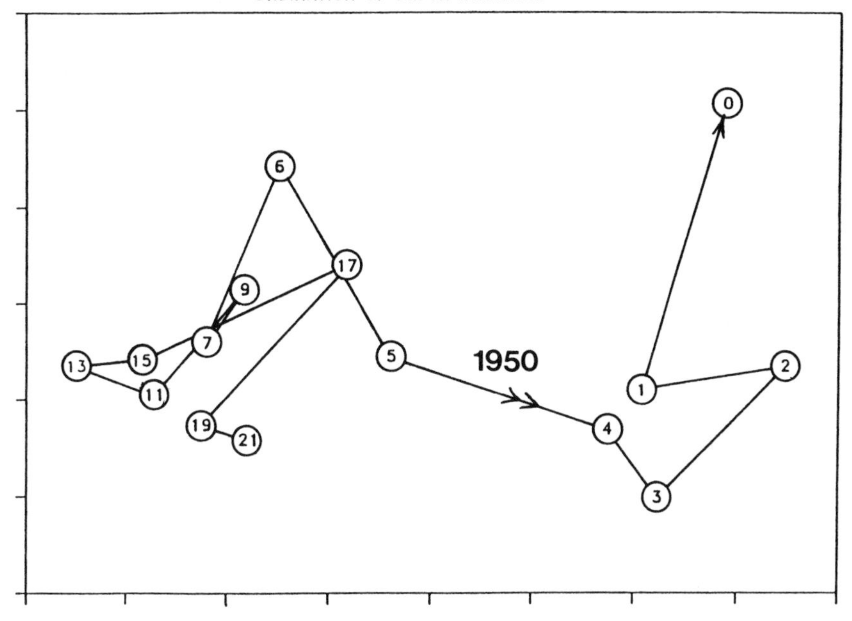

Figure 9.3. Ordination-based trajectory for the fossil chironomid fauna of Upper Wallface Pond, New York. Marked changes in the fauna occurred with lake acidification circa 1950. PIRLA = Paleoecological Investigation of Recent Lake Acidification project. Adapted from Uutala (1986), reproduced by permission of A.J. Uutala.

9.3.4. *Statistical Inference and Interpretation*

Until recently, fossil assemblages have been interpreted primarily from autecological knowledge of the most common taxa and subjective assessment by the analyst. However, palynologists and diatomists have led the search for more objective methods, and the application of ordination, multiple regression, and related techniques to surface sample data has elevated paleoecology to a quantitative science.

The natural or anthropogenically altered ontogeny of a system is easily recognizable when core data are portrayed in an ordination (e.g., Carney, 1982; Jacobson and Grimm, 1986; Uutala, 1986; Battarbee et al., 1988) (Fig. 9.3). Also, multiple regression analysis of surface sample data produces transfer functions that quantify the relationships between biota and

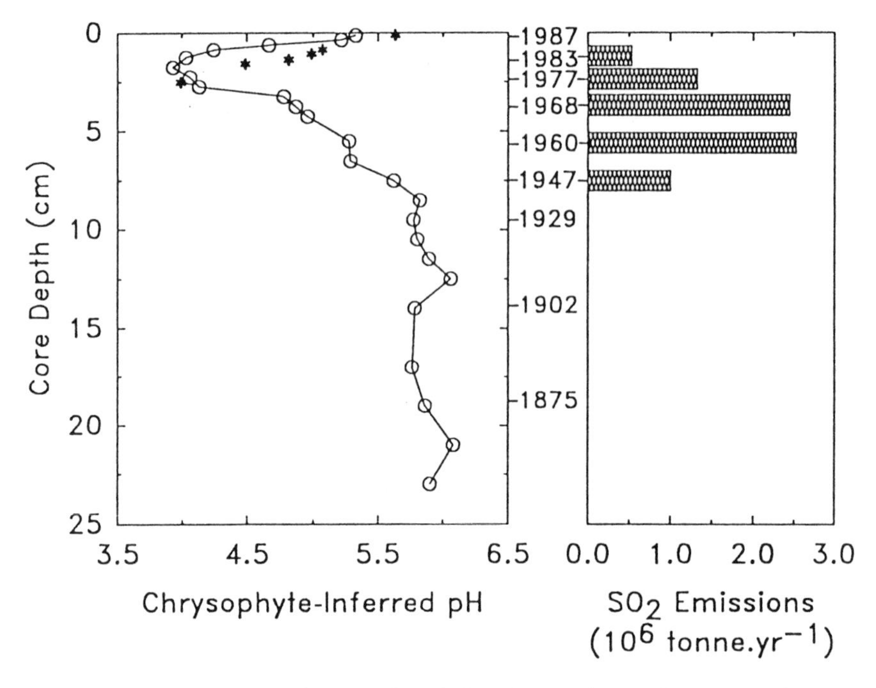

Figure 9.4. Comparison of chrysophyte-inferred pH estimates for Swan Lake, Ontario, actual lakewater pH measurements for Swan Lake (*), and historical records of SO_2 emissions at Sudbury, Ontario. Chrysophyte-inferred pH estimates are derived from a model based on multiple regression. From Dixit et al. (1989a), reproduced by permission of Fisheries and Oceans Canada and the Minister of Supply and Services Canada, 1990.

environmental variables. Thus, differences in the pollen flora of surficial lake sediments collected from prairie, tundra, and major forest regions, when analyzed using multivariate methods, permit the quantitative reconstruction of past climate (e.g., Vance, 1986). The pH tolerances of many diatom taxa also are apparent when compared among lakes. Analysis of these data through multiple regression, multivariate analysis followed by multiple regression, and related techniques has facilitated the quantitative reconstruction of recent pH histories for many lakes (Davis and Anderson, 1985; ter Braak and van Dam, 1989). Chrysophyte-inferred pH estimates for an Ontario lake, based on a multiple regression model, are shown in Fig. 9.4.

Multivariate techniques allow researchers to identify the timing and magnitude of biotic changes; the adoption of these or related interpretive techniques introduces objective means for environmental assessment. Although the paleobotanical world has been quick to adopt ordination methods and statistical inference procedures, zoologists generally have made less use of these methods (see Johnson et al., Chapter 4; Norris and Georges, Chapter 7; and Cooper and Barmuta, Chapter 11).

Canonical Correspondence Analysis (CCA), an ordination method recently developed by ter Braak (1986, 1987a), and the related procedures of Weighted Averaging (WA) regression and calibration (ter Braak, 1987b; ter Braak and Looman, 1987; Line and Birks, 1990), are state-of-the-art techniques for the analysis of paleoecological data. These procedures assume a unimodal model of species distributions across environmental gradients, whereas earlier ordination procedures (e.g., Principal Components Analysis, Redundancy Analysis) and regression and calibration techniques assumed a linear response by species to environmental variables. However, a linear response usually is seriously flawed when data are examined across broad environmental gradients (ter Braak and Prentice, 1988). WA models are being used extensively for diatom- (Stevenson et al., 1989; ter Braak and van Dam, 1989; Kingston and Birks, 1990) and chrysophyte-based (Dixit et al., 1989b) pH reconstructions. Independent reconstructions of several limnological variables (e.g., pH, alkalinity, Al, and dissolved organic carbon) are possible from the same paleolimnological data (e.g., Kingston and Birks, 1990). CCA and WA have been used by Walker et al. (1991a,b) to analyze the distributions of chironomids among lakes and to reconstruct past summer surface-water temperatures from fossil Chironomidae (see also Johnson et al., Chapter 4).

The power of these methods must not be overlooked. Distributional data for a variety of animal groups in surficial sediments are now available (e.g., Whiteside, 1970; Boubée, 1983; Cotten, 1985; Douglas and Smol, 1987; Crisman, 1989; Walker and Mathewes, 1989b), so a basis exists for the application of statistical inference procedures.

9.4. Paleolimnological Indicators

The array of indicators that may be used in the historical reconstruction of aquatic habitats is immense. However, despite the variety of available fossils, most paleolimnological studies have focused on diatoms, cladocerans, or chironomids (Frey, 1988). This chapter concentrates on the merits of using benthic macroinvertebrates, but the value of integrated studies as represented by the PIRLA program (Charles et al., 1986, 1990; Charles and Whitehead, 1986a, 1989; Whitehead et al., 1990), and the Surface Water Acidification Program (SWAP) (Battarbee et al., 1990), which used a variety of indicator groups, must be emphasized. Quite simply, benthic macroinvertebrates cannot be the best indicators of pelagic environments, and algal fossils offer no conclusions regarding aphotic profundal regions. Diatoms probably are the best indicators of past pH, chironomids may provide excellent evidence of eutrophication, and ostracods possibly provide the best evidence of salinity changes. However, until the indicator value of these organisms is assessed on the basis of the same surface sample data set, no means exist for objectively determining the relative indicator value of each group of organisms. The organisms listed above are useful for studies of lake sediments, whereas other macroinvertebrates, such as Simuliidae and Trichoptera, may be more useful when analyzing stream deposits (e.g., Williams and Morgan, 1977; Williams, 1981; Crosskey and Taylor, 1986). Although algal, fungal, bryophyte, and vascular plant fossils are not discussed in this chapter, many excellent reviews of these groups are available (Watts, 1978; Birks, 1980; Birks and Birks, 1980; Battarbee, 1984, 1986; Cronberg, 1986; Davis and Smol, 1986; Dickson, 1986; van Geel, 1986; Smol, 1987; Sherwood-Pike, 1988; Warner, 1989).

9.4.1. Chironomidae and Other Insecta

Of the many groups of freshwater benthic macroinvertebrates, Chironomidae currently are the most useful paleoenvironmental indicators in lake sediments. Their paleoecology has been reviewed by Walker (1987), Hofmann (1986a, 1988), Crisman (1988), and Frey (1988). Larval Chironomidae are rivaled only by Oligochaeta as the most abundant benthic macroinvertebrates of the littoral and profundal zones of lakes. Larvae also inhabit streams and moist soils. However, unlike the Oligochaeta, chironomids are well-represented by fossils in lake and stream sediments.

Head capsule concentrations in sediments of small temperate lakes usually exceed $100/cm^3$. Where a relatively large stream enters a lake, many of the chironomid fossils may be derived from stream habitats (e.g., Walker and Mathewes, 1987a). Recent taxonomic literature (e.g., Oliver and Roussel, 1983; Wiederholm, 1983) allows fossil head capsules to be identified to the generic level.

The difficulty of obtaining specific identifications poses a problem with many paleoecological indicator groups. Inevitably, some information is lost when specific identifications are not possible. Despite this difficulty, many generalizations are possible concerning the ecology of individual genera. For example, few Chironomini genera ever have been recorded in arctic/alpine environments. Such information can be used profitably in paleoenvironmental reconstructions.

Chironomids are excellent indicators of water quality. For example, *Chironomus anthracinus* and *Ch. plumosus* dominate the profundal zone of stratified eutrophic lakes in the Holarctic faunal region. *Tanytarsus* has been used as an indicator of oligotrophic lakes, although Thienemann's (1921) "*Tanytarsus*" lakes actually are based on a *Micropsectra*. Provided that a cold, hypolimnetic environment exists, *Heterotrissocladius* is a more reliable indicator of oligotrophy, and *Sergentia* and *Stictochironomus* are considered typical of the cool, profundal sediments of mesotrophic lakes. However, too much reliance should not be placed on the indicator value of single taxa; interpretations will be more reliable if based upon the response of whole communities.

A quantitative relationship, similar to the diatom-pH inference models developed by phycologists, should be established between chironomids and lake trophic state, which would greatly facilitate future paleolimnological reconstructions. Furthermore, chironomids should be studied together with other indicator groups for maximum effect. For example, the combined results of diatom and chironomid analyses (Warwick, 1980; Stoermer et al., 1985) provided a very detailed record of events in the Bay of Quinte.

The manner in which chironomid faunas are influenced by environmental factors is not entirely resolved. Low oxygen availability prevents many chironomids typical of oligotrophic lakes from inhabiting eutrophic lakes. Most oligotrophic taxa, unlike *Chironomus,* lack hemoglobin. However, in the more oligotrophic lakes, oxygen microstratification near the mud-water interface, food, or both may influence chironomid faunal composition (Brundin, 1951; Sæther, 1979).

Chironomid fossils have been used widely as indicators of recent eutrophication of lakes (e.g., Warwick, 1980; Brodin, 1982; Kansanen, 1985, 1986). Warwick's (1980) Bay of Quinte data provide a clear illustration of chironomid response to anthropogenic environmental change (Fig. 9.5). Initially (circa 1850), littoral chironomids declined as cold-stenothermous, profundal taxa with oligotrophic affinities increased. This response was attributed to rapidly accumulating mineral sediments, because of logging in the catchment, and to subsequent dilution of food particles by mineral matter (Warwick, 1980, 1989). Alternatively, an upward displacement of the thermocline, because of decreased water transparency, could account for the data (Walker and Mathewes, 1989a). Phytophilous chironomids also declined as turbidity eliminated macrophytes from deep water (Fig. 9.5). Following this initial response, the oligotrophic fauna was replaced abruptly by one typical of eutrophic environments (Warwick, 1980). Eutrophication of the Bay of Quinte could not be ascribed to a single cause; rapid population growth during the mid-nineteenth century, nutrient losses from the catchment following deforestation, and the advent of intensive agriculture may have been partially responsible, but evidence of advanced eutrophy is more recent. The installation of municipal sewage-disposal systems, application of commercial fertilizers to agricultural lands, and wastes contributed by industrial developments, including pulp and paper production, have had cumulative impacts on the lake biota.

Chironomids also are responsive to changes in lake acidity, salinity, and

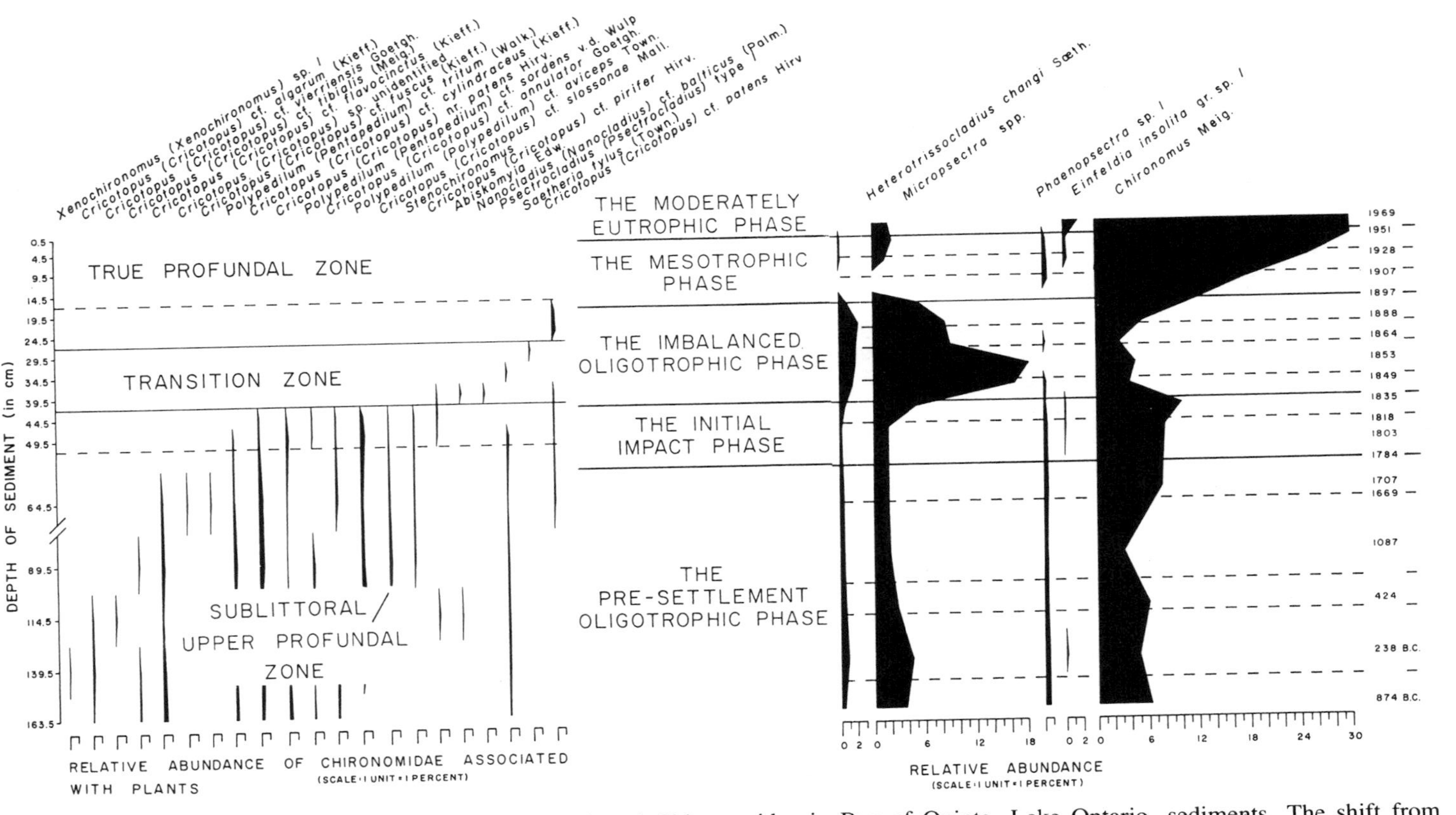

Figure 9.5. Changes in the relative abundance of selected Chironomidae in Bay of Quinte, Lake Ontario, sediments. The shift from sublittoral/upper profundal to true profundal conditions is apparent, as is recent eutrophication of the Bay. Adapted from Warwick (1980), reproduced by permission of Fisheries and Oceans Canada and the Minister of Supply and Services Canada, 1990.

temperature, as indicated by recent ecological and paleoecological studies (e.g., Mossberg and Nyberg, 1979; Timms et al., 1986; Walker and Mathewes, 1989b; Brodin, 1990; Hammer et al., 1990; Walker et al., 1991b). Acidification of temperate lakes often is accompanied by the increased abundance of some *Chironomus, Sergentia, Tanytarsus, Cladotanytarsus, Psectrocladius,* and/or *Zalutschia* species (e.g., Brodin, 1986; Uutala, 1986; Bilyj and Davies, 1989). In Florida, *Glyptotendipes, Labrundinia, Polypedilum,* and *Stictochironomus* are better indicators of low pH (Crisman, 1989). In addition, abnormal chironomid mouthparts and antennae can serve as indicators of toxic stress (see Johnson et al., Chapter 4), and Warwick (1980) and Klink (1989) have correlated increases in morphological abnormalities of chironomid remains with recent pollution loading.

The potential of chironomids and other aquatic organisms as climatic indicators has provided a focus for recent paleoecological debate (Hann and Warner, 1987; Walker and Mathewes, 1987a,b; Warner and Hann, 1987; Walker and Mathewes, 1989a,b,c; Warwick, 1989; Walker, 1990; Smol et al., 1991; Walker et al., 1991a,b). Some chironomid distributions appear, in part, to be determined climatically. For example, many littoral Chironomidae common in low-elevation temperate lakes are excluded from lakes in colder climates (Walker and Mathewes, 1989b; Walker, 1990). These taxa disappear with increasing latitude and altitude in subarctic and subalpine forest regions. Across these regions, summer surface water temperature declines from ~20° C to ~10° C; temperatures within this range may be critical for the development of temperate taxa. Air temperature, humidity, and precipitation also will influence the survival and reproductive success of adult insects. WA regression/calibration models of this climatic-faunistic trend should permit quantitative reconstructions of past climate (Walker et al., 1991a,b). Chironomids and other aquatic biota have short life spans and disperse efficiently, so they may provide more sensitive indications of climatic change than other lines of evidence, including pollen analyses (Williams, 1981; Smol et al., 1991).

In addition to chironomids, remains of several other Diptera often are recorded in lake sediments. Larvae of the planktonic phantom midge *Chaoborus americanus* are excellent indicators of fishless lakes (Johnson and McNeil, 1988; Johnson et al., 1990), so *C. americanus* has been used to indicate declining fish populations in acidified North American lakes (Uutala, 1986, 1990). Fossils of larval biting midges (Ceratopogonidae) are common, but seldom are identified. Larval black fly (Simuliidae) fossils have been found in sediments of lakes that receive stream inflows (Walker and Mathewes, 1987a; Currie and Walker, 1992).

Remains of insects other than Diptera are common in lake and stream sediments and have had great value to Quaternary paleoecologists and paleoclimatologists (e.g., Morgan and Morgan, 1980; Williams, 1988, 1989).

Relative to Chironomidae, the concentrations of Coleoptera, Hemiptera, and Trichoptera in sediments are low (e.g., five to 35 individuals/dm^3; Lemdahl and Liedberg-Jönsson, 1988). These low concentrations have not prevented paleoentomologists from providing detailed reconstructions of glacial, interglacial, and postglacial fluvial and lacustrine environments (e.g., Ashworth, 1977; Williams and Morgan, 1977; Williams et al., 1981; Miller and Morgan, 1982; Morgan et al., 1985; Wilkinson, 1984, 1987). Although most inferences from these indicators pertain to changes in climate, terrestrial and aquatic vegetation, stream size, current velocity, or depositional environment, Wilkinson (1987) noted that the medieval trichopteran fauna of the River Severn at Stourport, England, could not exist in the polluted habitat of the present river.

Klink (1989) has demonstrated the power of paleoecology by reconstructing the former fauna and pollution history of the River Rhine. He has recovered and identified remains of Ephemeroptera, Plecoptera, Heteroptera, Trichoptera, and Diptera from Rhine sediments, and he has revealed a dramatic decline in Simuliidae remains relative to those of Chironomidae since the eighteenth century (Fig. 9.6A). The decline may be related to changes in the quality of suspended solids on which black fly larvae feed and fluctuation in current velocities induced by waves from passing ships (Klink, 1989). The occurrence of deformed *Chironomus* remains has risen dramatically since 1950 (Fig. 9.6B) and possibly is related to pollution by heavy metals or chlorinated hydrocarbons.

9.4.2. Cladocera

Cladoceran remains are more abundant than those of Chironomidae in lake sediments. The usefulness of Cladocera as paleoenvironmental indicators has been reviewed by Frey (1986, 1988), Hofmann (1987), Whiteside and Swindoll (1988), and Hann (1989). Planktonic taxa principally include the genera *Bosmina* and *Daphnia,* although *Chydorus* gr. *sphaericus* is pelagic in many polluted lakes. *Chydorus* gr. *sphaericus* is one species group within the Chydoridae, a diverse tychoplanktonic family. As Whiteside and Swindoll (1988, p. 407) noted, ". . . the Chydoridae are mostly weak swimmers, and they exist either on vegetation or benthic surfaces in the littoral region." Most cladoceran fossils are remains of Bosminidae, Chydoridae, and Daphniidae. Fossil ephippia, mandibles, and claws usually are the only evidence of past daphnid populations. Headshields, mandibles, postabdominal claws, and shells of *Bosmina* are well-represented as fossils. Chydorids are identifiable from shells and head shields.

Cladocera are responsive to a variety of environmental influences. Cladoceran diversity increases with decreasing alkalinity, conductivity, or pH. Cotten (1985, as cited by Frey 1988) distinguished among several lake types

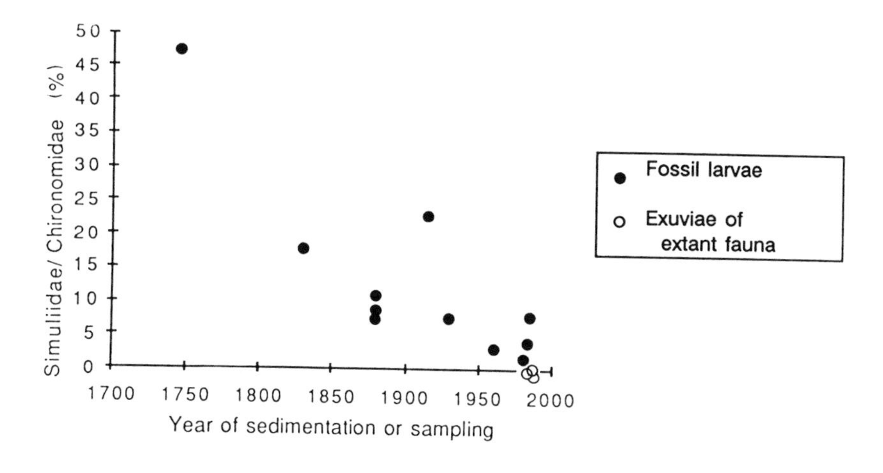

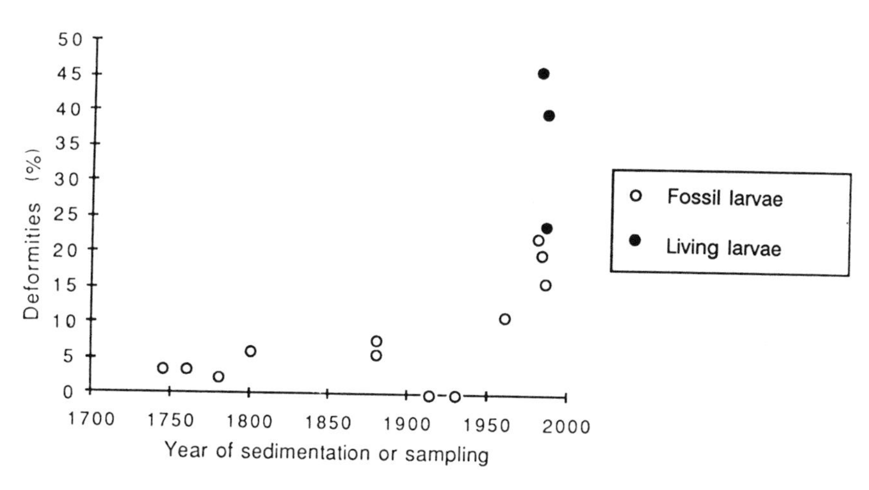

Figure 9.6. Reconstruction of pollution history of the River Rhine. A: Recent changes in the abundance of Simuliidae (relative to Chironomidae). B: Incidence of mouthpart deformities in remains of larval *Chironomus*. From Klink (1989), reprinted by permission of John Wiley & Sons, Ltd.

on the basis of cladoceran fossils. These lake types also were distinct with regard to pH, nutrients, and water color (Frey, 1988; Huttunen et al., 1988). *Bosmina* species are sensitive to changes in lake trophic state (i.e., productivity) as well as trophic structure (i.e., predator-prey interactions). The abundance of *Bosmina,* relative to chydorid fossils, is one indication of the extent of littoral versus pelagic habitat. Chydoridae decrease in relative abundance as water depth increases; however, littoral habitat and, therefore, the abundance of chydorids also can be reduced by decreasing water transparency, either due to suspended mineral matter or increased algal biomass. Chydorids also are responsive to changes in substrate type and macrophytes within the littoral zone.

Cladocera have not been used extensively in applied paleoecological studies; however, they can be used as indicators of anthropogenic eutrophication (e.g., Parise and Riva, 1982; Boucherle and Züllig, 1983). Boucherle and Züllig (1983) provided a detailed reconstruction of the anthropogenic eutrophication of three Swiss lakes and the recovery of these lakes once more effective sewage-disposal procedures were implemented.

Steinberg et al. (1984) and Krause-Dellin and Steinberg (1986) used chydorid Cladocera as quantitative indicators of acidification, although Hofmann (1986b) expressed reservations concerning the method. Paleolimnological investigations using chydorids, chrysophytes, and diatoms indicated the recent acidification of a Bavarian lake, Kleiner Arbersee (Steinberg et al., 1988) (Fig. 9.7). In North America, the recent increase of chydorids with acidification of Adirondack lakes may be an indication of increased water transparency (a consequence of decreased dissolved organic carbon following acidification) and/or declining fish populations (Paterson, 1985; Charles et al., 1990). Attempts are being made to reconstruct changes in water color, in addition to pH, using cladoceran fossils (Huttunen et al., 1988).

Studies of cladoceran distributions in surficial lake sediments have been done by De Costa (1964), Whiteside (1970), Synerholm (1979), Crisman (1980), and Cotten (1985). These studies and the quantitative evaluation of chydorid relationships to environmental factors (e.g., Huttunen et al., 1988) will strengthen future paleoecological interpretations.

9.4.3. Ostracoda

Ostracoda are represented in sediments by fossils of their calcareous shells. Although the taxonomy of living ostracods is based upon many characters that are not preserved readily (e.g., antennae, thoracic legs), the shells are identifiable and in alkaline deposits often are sufficiently abundant for detailed paleoecological analysis. The paleoecology of freshwater Ostracoda has been reviewed by Löffler (1986), Forester (1987), De Deckker and Forester (1988), and Delorme (1989).

Unlike Cladocera, Ostracoda inhabit profundal as well as littoral environments. Their fossils, and those of the Chironomidae, offer the best means for determining past conditions beneath the photic zone of lakes. Ostracod taxa are especially responsive to salinity fluctuations, temperature variations, and to the anionic composition of saline water (Forester, 1987; De

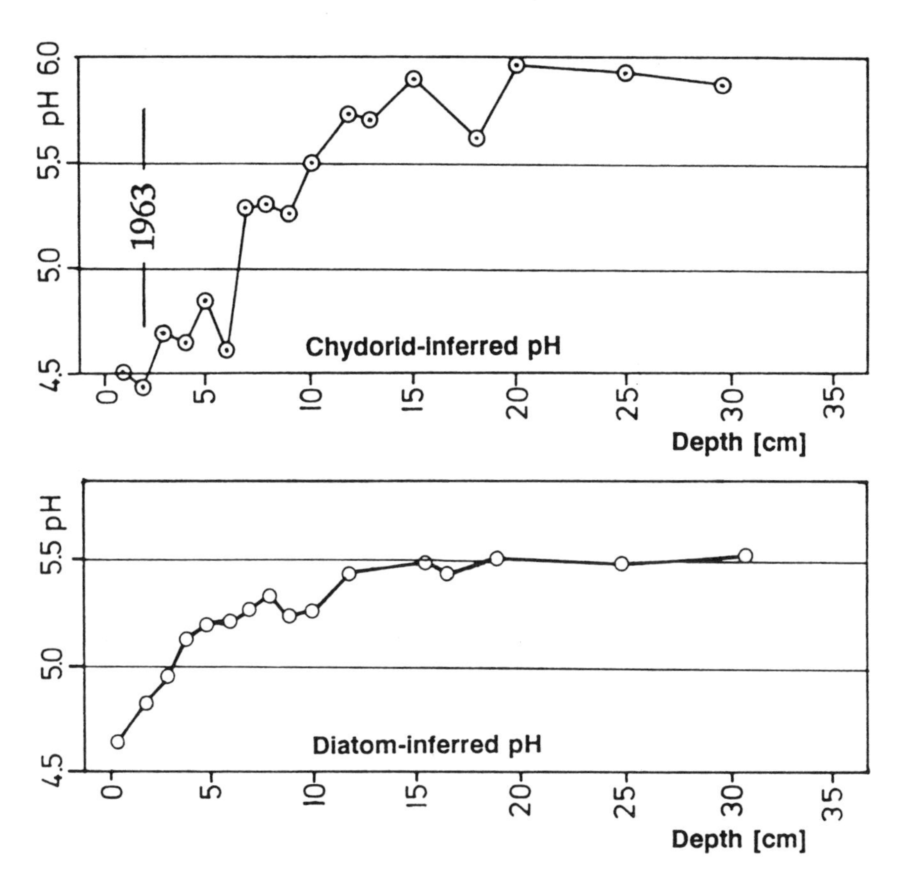

Figure 9.7. Comparison of diatom-inferred and chydorid-inferred pH histories for Kleiner Arbersee, Germany. From Steinberg et al. (1988), reprinted by permission of Kluwer Academic Publishers.

Deckker and Forester, 1988; Delorme, 1989). Thus, their fossils have been used extensively in paleoclimatological studies (e.g., Delorme and Zoltai, 1984). Stark (1976) also noted the importance of dissolved oxygen to species distributions.

Ostracoda have been used infrequently in applied paleoecological studies, but they offer considerable future promise. Stark (1976), for example, has noted recent changes related to impoundment in the fauna of Elk Lake, Minnesota. The impact of salt mining on organisms of an Austrian lake is recorded by the recent extinction of ostracods and other zoobenthos in areas now covered by alkaline sludge (Löffler, 1983) (Fig. 9.8). The fossil ostracod record from Lake Erie has been interpreted as indicating periodic anoxia even prior to settlement by European immigrants (Delorme, 1982).

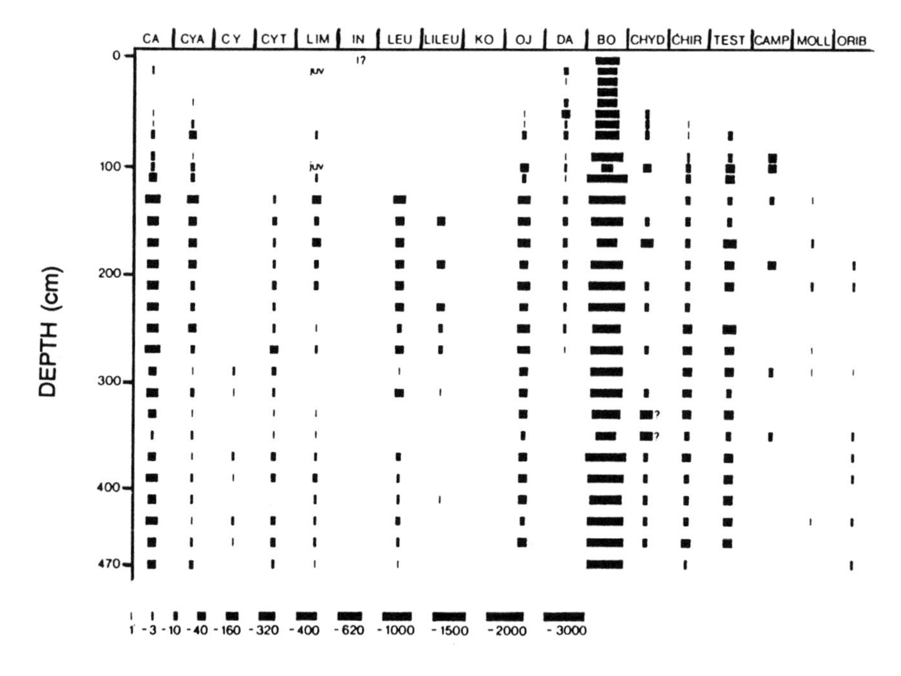

Figure 9.8. Decline of benthic taxa in response to alkaline sludge deposition in Traunsee, Austria. Ostracod taxa include *Candona* (CA), *Cypria ophthalmica* (CYA), *Cyclocypris* (CY), *Cytherissa lacustris* (CYT), *Limnocythere sanctipatricii* (LIM), *L. inopinata* (IN), *Leucocythere mirabilis* (LEU), *Limnocythere* and *Leucocythere* juveniles (LILEU), *Kovalevskiella* (KO), and ostracod juveniles (OJ). Also illustrated are Cladocera [*Daphnia* ephippia (DA), *Bosmina* (BO), and chydorids (CHYD)], chironomids (CHIR), Testacea (TEST), the diatom *Campylodiscus* (CAMP), molluscs (MOLL), and mites [Oribatei (ORIB)]. *Daphnia* and *Bosmina* are planktonic, so they have not been affected to the same extent as the other fauna. Adapted from Löffler (1986), reprinted by permission of Kluwer Academic Publishers.

Ostracod remains may provide evidence of a number of environmental perturbations: thermal pollution; past water chemistry (De Deckker and Forester, 1988), including toxic contaminants; postsettlement agriculture and irrigation; and evaporation from impounded lakes that yields more saline waters. Few Ostracoda inhabit strongly acidic lakes, and their remains are not preserved readily in sediments having a pH lower than 8.3 (Delorme, 1989). Relatively large volumes of sediment (>50 ml) are required for fossil analyses.

Delorme (1989) has accumulated over 6,000 samples of Ostracoda from Canadian inland waters; these samples provide detail on the environmental requirements of ostracod taxa. This database has allowed Delorme and Zoltai (1984) to assess past climatic and limnological conditions.

9.4.4. Mollusca

Quaternary paleoecologists long have noted the presence of mollusc remains in sediments and have included these in their interpretations of past environments (Crisman, 1978; Birks and Birks, 1980; Ložek, 1986; Miller and Bajc, 1989). Mollusca are common in littoral habitats, and their shells are preserved well in calcareous sediments. These fossils can provide information regarding water chemistry, lake trophic state, oxygen concentration, and water level (e.g., Crisman, 1978). Mollusca do not inhabit strongly acidic waters, and their remains rarely are preserved in sediments of weakly acidic lakes.

The remains of molluscs are not readily transported offshore, so sediment cores for molluscan analysis are obtained best near the lower limit of the littoral zone. Large volumes of sediment usually are required to obtain sufficient fossil material.

The value of molluscs to applied paleoecology has been demonstrated by Birks et al. (1976). In this study, pulmonate snails, which breathe atmospheric O_2, completely replaced clams and prosobranch snails as a result of eutrophication in St. Clair Lake, Minnesota. According to Crisman (1978), similar results have been obtained by Covich (1976) and Stark (1976). Thus, Mollusca potentially are valuable paleoindicators of human impact on aquatic systems.

9.4.5. Other Zoobenthic Paleoindicators

Although the aforementioned fossils are the most widely used zoobenthic paleoindicators, others such as bryozoan statoblasts, sponge spicules, protozoan plates, rhizopod tests, and turbellarian egg cocoons also are useful.

9.4.5.1. Bryozoa

The abundance of bryozoan statoblasts may be related to phytoplankton biomass and the extent of littoral habitat (Crisman et al., 1986). Unfortunately, the low concentrations of these remains limit their value for applied studies.

9.4.5.2. Oligochaeta

Very little effort has been devoted to oligochaete paleoecology, although cocoons sometimes are preserved in lake sediments. The abrupt extinction of *Spirosperma ferox* from an Italian lake circa 1930 to 1940 coincided with the advent of high Cu loading (Bonacina et al., 1986).

9.4.5.3. *Porifera*

The gemmules and siliceous spicules of sponges are preserved well in most lake sediments, and recent studies have demonstrated their value to paleolimnology (Harrison 1988a,b). These fossils allow inferences to be made of light conditions, turbidity, temperature, pH, specific conductance, and concentrations of dissolved Si. Spicules are available in high concentrations in most lake sediments. Larger volumes of sediment are needed for analysis of gemmules, which are required for species determinations. Unfortunately, sponge diversity is low in sediments.

9.4.5.4. *Protozoa*

Many Protozoa of the subphylum Sarcodina construct tests by agglutination of foreign particles or by autogenic secretion of proteinaceous, siliceous, or calcareous components. Agglutinated tests of the Class Rhizopodea are preserved intact in the sediments of most lakes. Apart from siliceous plates, remains of autogenic tests rarely are preserved (Tolonen, 1986; Douglas and Smol, 1987, 1988; Medioli and Scott, 1988; Warner, 1988). Little is known about the ecology of lacustrine taxa, but Schönborn (1967) has identified several taxa as indicators of oligotrophic, eutrophic, or dystrophic lakes. Eutrophication in the western basin of Lake Erie was traced through the abundance of fossil *Difflugia tricuspis* (Medioli and Scott, 1988).

9.4.5.5. *Turbellaria*

Many types of egg cocoons from neorhabdocoele Turbellaria can be recognized in sediments (Harmsworth, 1968), but few of these fossils can be assigned to described species (Warner, 1989). However, it may be possible to use these remains as paleoecological indicators, even without the association of cocoons with mature Turbellaria. Multivariate statistical analyses, discussed earlier in this chapter, when applied to analysis of surface sample data, will permit correlations between the cocoons and various environmental influences to be recognized. These cocoon-environment relationships then could be used in applied paleoecological studies.

9.5. Conclusions

Paleolimnological studies offer benthic ecologists an opportunity to reconstruct the past biota of lakes and streams and biotic responses to recent anthropogenic environmental change; a historical perspective often is necessary as proof of presumed impacts. Paleolimnological biomonitoring also offers means by which very recent limnological change can be discerned, including events of the past 10 years. Although much recent work has focused on lake acidification, paleolimnological studies can test hypotheses related to eutrophication, metal contamination, thermal pollution, climatic change, or other anthropogenic disturbances of aquatic systems. Moreover, considerable attention now is being devoted to the possibility of rehabilitating lakes and streams to reestablish "natural conditions." In the absence of detailed historical information, paleoecology should play a leading role in assessing what "natural conditions" previously existed (e.g., Buskens, 1989).

Global climatic warming appears to be the new frontier for paleolimnology (Smol et al., 1991). Lakes, streams, and their biota respond to climatic change, as is evident through fluctuations in water level, current velocity, salinity and productivity, and changes in floristic and faunistic composition of their communities. Limnological changes that occurred at the end of the last glaciation (e.g., Walker and Mathewes, 1987a) are the best available analog for changes that can be expected with present global warming. The termination of the Younger Dryas climatic interval, for example, was accompanied by a 7° C warming in Greenland over just 50 years (Dansgaard et al., 1989). Climatic models have been constructed to depict present and future climate (e.g., Webb et al., 1987), but it is only on the basis of paleoclimatic data that the accuracy of the models' predictions can be assessed. If the climatic models cannot accurately simulate past climate, they will provide erroneous projections for future trends.

A common misconception persists that biological analyses are not amenable to quality assurance/quality control (QA/QC) considerations. However, a high standard of QA/QC *is* achievable. In many respects, paleolimnological studies dealing with lake acidity have taken the lead in establishing and implementing QA/QC standards for biomonitoring (J.P. Smol, Department of Biology, Queen's University, Kingston, ON, personal communication). The application of recently developed statistical inference techniques allows objective interpretation, which together with carefully standardized research protocols (e.g., Charles and Whitehead, 1986b), are the keys to achieving high QA/QC standards.

Laboratories should standardize the taxonomic treatment used in their identification of fossils and monitor this consistency by adopting inter- and intralaboratory checks on identifications (Kingston, 1986). Taxonomic publications (e.g. Camburn et al., 1984–86) are important initiatives for the standardization of taxonomic work and for meeting QA/QC guidelines. Just as standard chemical solutions and standard weights and measures permit calibration of laboratory equipment, these protocols calibrate biological analyses. The quality of all biological assessments is only as good as the taxonomy used (see also Resh and McElravy, Chapter 5).

Paleolimnological methods have been used more extensively as biomonitoring tools for lakes than for streams. Although stream studies are less advanced, an unexploited biomonitoring resource exists in deposits of formerly lotic habitats such as reservoirs and beaver dams. This potentially rich fossil stream fauna deserves more attention.

Careful preparation is necessary in advance of paleolimnological studies because the science is evolving rapidly and methods presently in use soon become obsolete. Advice of experts should be solicited for assistance with sampling, dating of sediments, choice of paleoindicators, identification of fossils, and statistical analysis because the field has become too broad for any individual to master.

Acknowledgments

I thank S.S. Dixit, J.C. Kingston, J.P. Smol, A.J. Uutala, N.E. Williams, and an anonymous reviewer for their criticisms of earlier drafts of this manuscript. I am also grateful to M.S.V. Douglas and J.R. Glew for their respective comments regarding protozoan paleoecology and sampling techniques. I was assisted by a Natural Sciences and Engineering Research Council of Canada Postdoctoral Fellowship during preparation of this review.

References

Aaby, B. and G. Digerfeldt. 1986. Sampling techniques for lakes and bogs. In *Handbook of Holocene Palaeoecology and Palaeohydrology,* ed. B.E. Berglund, pp. 181–94. John Wiley, Chichester, England.

Appleby, P.G. and F. Oldfield. 1978. The calculation of lead-210 dates assuming a constant rate of supply of unsupported ^{210}Pb to the sediment. *Catena* 5:1–8.

Appleby, P.G., F. Oldfield, R. Thompson, P. Huttunen, and K. Tolonen. 1979. ^{210}Pb dating of annually laminated lake sediments from Finland. *Nature* (London) 280:53–5.

Appleby, P.G., N. Richardson, P.J. Nolan, and F. Oldfield. 1990. Radiometric dating of the United Kingdom SWAP sites. *Philosophical Transactions of the Royal Society of London* (Series B) 327:233–8.

Ashworth, A.C. 1977. A late Wisconsinan Coleopterous assemblage from southern Ontario and its environmental significance. *Canadian Journal of Earth Sciences* 14:1625–34.

Battarbee, R.W. 1984. Diatom analysis and the acidification of lakes. *Philosophical Transactions of the Royal Society of London* (Series B) 305:451–77.

Battarbee, R.W. 1986. Diatom analysis. In *Handbook of Holocene Palaeoecology and Palaeohydrology,* ed. B.E. Berglund, pp. 527–70. John Wiley, Chichester, England.

Battarbee, R.W., R.J. Flower, A.C. Stevenson, V.J. Jones, R. Harriman, and P.G. Appleby. 1988. Diatom and chemical evidence for reversibility of acidification of Scottish lochs. *Nature* (London) 332:530–2.

Battarbee, R.W., J. Mason, I. Renberg, and J.F. Talling, eds. 1990. Palaeolimnology and lake acidification. *Philosophical Transactions of the Royal Society of London* (Series B) 327:227–445.

Berglund, B.E., ed. 1986. *Handbook of Holocene Palaeoecology and Palaeohydrology*. John Wiley, Chichester, England.

Bilyj, B. and I.J. Davies. 1989. Descriptions and ecological notes on seven new species of *Cladotanytarsus* (Chironomidae: Diptera) collected from an experimentally acidified lake. *Canadian Journal of Zoology* 67:948–62.

Binford, M.W. 1990. Calculation and uncertainty analysis of ^{210}Pb dates for PIRLA project lake sediment cores. *Journal of Paleolimnology* 3:253–67.

Birks, H.H. 1980. Plant macrofossils in Quaternary lake sediments. *Ergebnisse der Limnologie* 15:1–60.

Birks, H.H., M.C. Whiteside, D.M. Stark, and R.C. Bright. 1976. Recent paleolimnology of three lakes in northwestern Minnesota. *Quaternary Research* 6:249–72.

Birks, H.J.B. and H.H. Birks. 1980. *Quaternary Palaeoecology*. Edward Arnold, London.

Blomqvist, S. 1985. Reliability of core sampling of soft bottom sediment—an *in situ* study. *Sedimentology* 32:605–12.

Bonacina, C., G. Bonomi, and C. Monti. 1986. Oligochaete cocoon remains as evidence of past lake pollution. *Hydrobiologia* 143:395–400.

Boubée, J.A.T. 1983. Past and present benthic fauna of Lake Maratoto with special reference to the Chironomidae. Ph.D. dissertation, Univ. of Waikato, Hamilton, New Zealand.

Boucherle, M.M. and H. Züllig. 1983. Cladoceran remains as evidence of change in trophic state in three Swiss lakes. *Hydrobiologia* 103:141–6.

Brinkhurst, R.O. 1974. *The Benthos of Lakes*. MacMillan Press, London.

Brinkhurst, R.O., K.E. Chua, and E. Batoosingh. 1969. Modifications in sampling procedures as applied to studies on the bacteria and tubificid oligochaetes inhabiting aquatic sediments. *Journal of the Fisheries Research Board of Canada* 26:2581–93.

Brodin, Y. 1982. Palaeoecological studies of the recent development of the Lake Växjösjön. IV. Interpretation of the eutrophication process through the analysis of subfossil chironomids. *Archiv für Hydrobiologie* 93:313–26.

Brodin, Y.W. 1986. The postglacial history of Lake Flarken, southern Sweden, interpreted from subfossil insect remains. *Internationale Revue der gesamten Hydrobiologie* 71:371–432.

Brodin, Y.W. 1990. Midge fauna development in acidified lakes in northern Europe. *Philosophical Transactions of the Royal Society of London* (Series B) 327:295–8.

Brown, S.R. 1956. A piston sampler for surface sediments of lake deposits. *Ecology* 37:611–3.

Brown, T.A., D.E. Nelson, R.W. Mathewes, J.S. Vogel, and J.R. Southon. 1989.

Radiocarbon dating of pollen by accelerator mass spectrometry. *Quaternary Research* 32:205–12.
Brundin, L. 1951. The relation of O_2-microstratification at the mud surface to the ecology of the profundal bottom fauna. *Institute of Freshwater Research Drottningholm Report* 32:32–42.
Brunskill, G.J. and S.D. Ludlam. 1988. The variation of annual ^{210}Pb flux to varved sediments of Fayetteville Green Lake, New York from 1885 to 1965. *Internationale Vereinigung für Theoretische und Angewandte Limnologie Verhandlungen* 23:848–54.
Burden, E.T., G. Norris, and J.H. McAndrews. 1986. Geochemical indicators in lake sediment of upland erosion caused by Indian and European farming, Awenda Provincial Park, Ontario. *Canadian Journal of Earth Sciences* 23:55–65.
Buskens, R.F.M. 1989. Monitoring of chironomid larvae and exuviae in the Beuven, a soft water pool in the Netherlands, and comparisons with palaeolimnological data. In Advances in chironomidology. Proceedings of the Tenth International Symposium on Chironomidae, Debrecen, Hungary, July 25–28, 1988. Part 2. Faunistics, population dynamics, ecology, production and community structure, ed. G. Dévai. *Acta Biologica Debrecina Oecologica Hungarica* 3:41–50.
Camburn, K.E., J.C. Kingston, and D.F. Charles, eds. 1984–86. *PIRLA Diatom Iconograph*. PIRLA Unpublished Report Series, Report 3. Indiana University, Bloomington, IN.
Carney, H.J. 1982. Algal dynamics and trophic interactions in the recent history of Frains Lake, Michigan. *Ecology* 63:1814–26.
Charles, D.F., M.W. Binford, E.T. Furlong, R.A. Hites, M.J. Mitchell, S.A. Norton, F. Oldfield, M.J. Paterson, J.P. Smol, A.J. Uutala, J.R. White, D.R. Whitehead, and R.J. Wise. 1990. Paleoecological investigation of recent lake acidification in the Adirondack Mountains, NY. *Journal of Paleolimnology* 3:195–241.
Charles, D.F. and D.R. Whitehead. 1986a. The PIRLA project: Paleoecological Investigation of Recent Lake Acidification. *Hydrobiologia* 143:13–20.
Charles, D.F. and D.R. Whitehead, eds. 1986b. *Paleoecological Investigation of Recent Lake Acidification (PIRLA): Methods and Project Description*. EPRI EA-4906. Electric Power Research Institute, Palo Alto, CA.
Charles, D.F. and D.R. Whitehead, eds. 1989. *Paleoecological Investigation of Recent Lake Acidification (PIRLA): 1983–1985*. EPRI EN-6526. Electric Power Research Institute, Palo Alto, CA.
Charles, D.F., D.R. Whitehead, D.S. Anderson, R. Bienert, K.E. Camburn, R.B. Cook, T.L. Crisman, R.B. Davis, J. Ford, B.D. Fry, R.A. Hites, J.S. Kahl, J.C. Kingston, R.G. Kreis, Jr., M.J. Mitchell, S.A. Norton, L.A. Roll, J.P. Smol, P.R. Sweets, A.J. Uutala, J.R. White, M.C. Whiting, and R.J. Wise. 1986. The PIRLA project (Paleoecological Investigation of Recent Lake Acidification): preliminary results for the Adirondacks, New England, N. Great Lakes states, and N. Florida. *Water, Air, and Soil Pollution* 30:355–65.
Cotten, C.A. 1985. Cladoceran assemblages related to lake conditions in eastern Finland. Ph.D. dissertation, Indiana Univ., Bloomington, IN.
Covich, A. 1976. Recent changes in molluscan species diversity of a large tropical lake (Lago de Peten, Guatemala). *Limnology and Oceanography* 21:51–9.

Crisman, T.L. 1978. Reconstruction of past lacustrine environments based on the remains of aquatic invertebrates. In *Biology and Quaternary Environments,* ed. D. Walker, pp. 69–101. Australian Academy of Science, Canberra, Australia.
Crisman, T.L. 1980. Chydorid cladoceran assemblages from subtropical Florida. In *Evolution and Ecology of Zooplankton Communities,* ed. W.C. Kerfoot, pp. 657–68. Special Symposium Vol. 3, American Society of Limnology and Oceanography. Univ. Press of New England, Hanover, NH.
Crisman, T.L. 1988. The use of subfossil benthic invertebrates in aquatic resource management. In *Aquatic Toxicology and Hazard Assessment: 10th Volume,* eds. W.J. Adams, G.A. Chapman, and W.G. Landis, pp. 71–88. American Society for Testing and Materials Special Technical Publication 971. American Society for Testing and Materials, Philadelphia, PA.
Crisman, T.L. 1989. A preliminary assessment of chironomid distributions in acidic Florida lakes. In *Paleoecological Investigation of Recent Lake Acidification (PIRLA): 1983–1985,* eds. D.F. Charles and D.R. Whitehead, pp. 14.1–14.10. EPRI EN-6526. Electric Power Research Institute, Palo Alto, CA.
Crisman, T.L., U.A.M. Crisman, and M.W. Binford. 1986. Interpretation of bryozoan microfossils in lacustrine sediment cores. *Hydrobiologia* 143:113–8.
Cronberg, G. 1986. Blue-green algae, green algae and chrysophyceae in sediments. In *Handbook of Holocene Palaeoecology and Palaeohydrology,* ed. B.E. Berglund, pp. 507–26. John Wiley, Chichester, England.
Crosskey, R.W. and B.J. Taylor. 1986. Fossil blackflies from Pleistocene interglacial deposits in Norfolk, England (Diptera: Simuliidae). *Systematic Entomology* 11:401–12.
Currie, D.C. and I.R. Walker. 1992. Recognition and palaeohydrologic significance of fossil black fly larvae, with a key to the Nearctic genera (Diptera: Simuliidae). *Journal of Paleolimnology* 7:37–54.
Dansgaard, W., J.W.C. White, and S.J. Johnsen. 1989. The abrupt termination of the Younger Dryas climate event. *Nature* (London) 339:532–4.
Davidson, G.A. 1988. A modified tape-peel technique for preparing permanent qualitative microfossil slides. *Journal of Paleolimnology* 1:229–34.
Davis, R.B. and D.S. Anderson. 1985. Methods of pH calibration of sedimentary diatom remains for reconstructing history of pH in lakes. *Hydrobiologia* 120:69–87.
Davis, R.B., C.T. Hess, S.A. Norton, D.W. Hanson, K.D. Hoagland, and D.S. Anderson. 1984. ^{137}Cs and ^{210}Pb dating of sediments from soft-water lakes in New England (U.S.A.) and Scandinavia, a failure of ^{137}Cs dating. *Chemical Geology* 44:151–85.
Davis, R.B. and J.P. Smol. 1986. The use of sedimentary remains of siliceous algae for inferring past chemistry of lake water—problems, potential and research needs. In *Diatoms and Lake Acidity,* eds. J.P. Smol, R.W. Battarbee, R.B. Davis, and J. Meriläinen, pp. 291–300. Junk Pubs., Dordrecht, The Netherlands.
De Costa, J.J. 1964. Latitudinal distribution of chydorid Cladocera in the Mississippi Valley, based on their remains in surficial lake sediments. *Investigations of Indiana Lakes and Streams* 6:65–101.
De Deckker, P. and R.M. Forester. 1988. The use of ostracods to reconstruct con-

tinental palaeoenvironmental records. In *Ostracoda in the Earth Sciences,* eds. P. De Deckker, J.-P. Colin, and J.-P. Peypouquet, pp. 175–99. Elsevier, Amsterdam, The Netherlands.

Delorme, L.D. 1982. Lake Erie oxygen: the prehistoric record. *Canadian Journal of Fisheries and Aquatic Sciences* 39:1021–9.

Delorme, L.D. 1989. Methods in Quaternary ecology #7. Freshwater ostracodes. *Geoscience Canada* 16:85–90.

Delorme, L.D. and S.C. Zoltai. 1984. Distribution of an arctic ostracod fauna in space and time. *Quaternary Research* 21:65–73.

Dickson, J.H. 1986. Bryophyte analysis. In *Handbook of Holocene Palaeoecology and Palaeohydrology,* ed. B.E. Berglund, pp. 627–43. John Wiley, Chichester, England.

Dixit, S.S., A.S. Dixit, and J.P. Smol. 1989a. Lake acidification recovery can be monitored using chrysophycean microfossils. *Canadian Journal of Fisheries and Aquatic Sciences* 46:1309–12.

Dixit, S.S., A.S. Dixit, and J.P. Smol. 1989b. Relationship between chrysophyte assemblages and environmental variables in seventy-two Sudbury lakes as examined by canonical correspondence analysis (CCA). *Canadian Journal of Fisheries and Aquatic Sciences* 46:1667–76.

Douglas, M.S.V. and J.P. Smol. 1987. Siliceous protozoan plates in lake sediments. *Hydrobiologia* 154:13–23.

Douglas, M.S.V. and J.P. Smol. 1988. Siliceous protozoan and chrysophycean microfossils from the recent sediments of *Sphagnum* dominated Lake Colden, NY., U.S.A. *Internationale Vereinigung für Theoretische und Angewandte Limnologie Verhandlungen* 23:855–9.

Edwards, T.W.D. and P. Fritz. 1988. Stable-isotope paleoclimate records for southern Ontario, Canada: comparison of results from marl and wood. *Canadian Journal of Earth Sciences* 25:1397–1406.

Einarsson, Á. and H. Hafliðason. 1988. Predictive paleolimnology: effects of sediment dredging in Lake Mývatn, Iceland. *Internationale Vereinigung für Theoretische und Angewandte Limnologie Verhandlungen* 23:860–9.

Einarsson, Á., H. Hafliðason, and H. Óskarsson. 1988. *Mývatn: saga lífríkis og gjóskutímatal í Syðriflóa.* Náttúruverndarráð 17, Reykjavik, Iceland.

Engstrom, D.R. and B.C.S. Hansen. 1985. Postglacial vegetational change and soil development in southeastern Labrador as inferred from pollen and chemical stratigraphy. *Canadian Journal of Botany* 63:543–61.

Engstrom, D.R. and H.E. Wright, Jr. 1984. Chemical stratigraphy of lake sediments as a record of environmental change. In *Lake Sediments and Environmental History,* eds. E.Y. Haworth and J.W.G. Lund, pp. 11–67. Leicester Univ. Press, Leicester, England.

Forester, R.M. 1987. Late Quaternary paleoclimate records from lacustrine ostracodes. In *North America and Adjacent Oceans During the Last Deglaciation, Vol. K–3*, eds. W.F. Ruddiman and H.E. Wright, Jr., pp. 261–76. *The Geology of North America.*Geological Society of America, Boulder, CO.

Frey, D.G. 1964. Remains of animals in Quaternary lake and bog sediments and their interpretation. *Ergebnisse der Limnologie* 2:1–114.

Frey, D.G. 1976. Interpretation of Quaternary paleoecology from Cladocera and midges, and prognosis regarding usability of other organisms. *Canadian Journal of Zoology* 54:2208–26.

Frey, D.G. 1986. Cladocera analysis. In *Handbook of Holocene Palaeoecology and Palaeohydrology*, ed. B.E. Berglund, pp. 667–92. John Wiley, Chichester, England.

Frey, D.G. 1988. Littoral and offshore communities of diatoms, cladocerans and dipterous larvae, and their interpretation in paleolimnology. *Journal of Paleolimnology* 1:179–91.

Gilbert, R. and J. Glew. 1985. A portable percussion coring device for lacustrine and marine sediments. *Journal of Sedimentary Petrology* 55:607–8.

Glew, J.R. 1988. A portable extruding device for close interval sectioning of unconsolidated core samples. *Journal of Paleolimnology* 1:235–9.

Glew, J.R. 1989. A new trigger mechanism for sediment samplers. *Journal of Paleolimnology* 2:241–3.

Gray, J., ed. 1988. *Paleolimnology: Aspects of Freshwater Paleoecology and Biogeography*. Elsevier, Amsterdam, The Netherlands.

Green, D.G. 1981. Time series and postglacial forest ecology. *Quaternary Research* 15:265–77.

Hammer, U.T., J.S. Sheard, and J. Kranabetter. 1990. Distribution and abundance of littoral benthic fauna in Canadian prairie saline lakes. *Hydrobiologia* 197:173–92.

Hann, B.J. 1989. Methods in Quaternary ecology #6. Cladocera. *Geoscience Canada* 16:17–26.

Hann, B.J. and B.G. Warner. 1987. Late Quaternary Cladocera from coastal British Columbia, Canada: a record of climatic or limnologic change? *Archiv für Hydrobiologie* 110:161–77.

Harmsworth, R.V. 1968. The developmental history of Blelham Tarn (England) as shown by animal microfossils, with special reference to the Cladocera. *Ecological Monographs* 38:223–41.

Harrison, F.W. 1988a. Methods in Quaternary ecology #4. Freshwater sponges. *Geoscience Canada* 15:193–8.

Harrison, F.W. 1988b. Utilization of freshwater sponges in paleolimnological studies. *Palaeogeography, Palaeoclimatology, Palaeoecology* 62:387–97.

Heit, M. and K.M. Miller. 1987. Cesium-137 sediment depth profiles and inventories in Adirondack lake sediments. *Biogeochemistry* 3:243–65.

Hilton, J. 1985. A conceptual framework for predicting the occurrence of sediment focusing and sediment redistribution in small lakes. *Limnology and Oceanography* 30:1131–43.

Hofmann, W. 1983. Stratigraphy of subfossil Chironomidae and Ceratopogonidae (Insecta: Diptera) in late glacial littoral sediments from Lobsigensee (Swiss Plateau). Studies in the late Quaternary of Lobsigensee 4. *Revue de Paléobiologie* 2:205–9.

Hofmann, W. 1986a. Chironomid analysis. In *Handbook of Holocene Palaeoecology and Palaeohydrology*, ed. B.E. Berglund, pp. 715–27. John Wiley, Chichester, England.

Hofmann, W. 1986b. Developmental history of the Großer Plöner See and the Schöhsee (north Germany): cladoceran analysis, with special reference to eutrophication. *Archiv für Hydrobiologie Supplement* 74:259–87.

Hofmann, W. 1987. Cladocera in space and time: analysis of lake sediments. *Hydrobiologia* 145:315–21.

Hofmann, W. 1988. The significance of chironomid analysis (Insecta: Diptera) for paleolimnological research. *Palaeogeography, Palaeoclimatology, Palaeoecology* 62:501–9.

Hutchinson, G.E., E. Bonatti, U.M. Cowgill, C.E. Goulden, E.A. Leventhal, M.E. Mallett, F. Margaritora, R. Patrick, A. Racek, S.A. Roback, E. Stella, J.B. Ward-Perkins, and T.R. Wellman. 1970. Ianula: an account of the history and development of the Lago di Monterosi, Latium, Italy. *Transactions of the American Philosophical Society* 60(4):1–178.

Huttunen, P., J. Meriläinen, C. Cotten, and J. Rönkkö. 1988. Attempts to reconstruct lakewater pH and colour from sedimentary diatoms and Cladocera. *Internationale Vereinigung für Theoretische und Angewandte Limnologie Verhandlungen* 23:870–3.

Jacobson, G.L., Jr. and E.C. Grimm. 1986. A numerical analysis of Holocene forest and prairie vegetation in central Minnesota. *Ecology* 67:958–66.

Jacobson, H.A. and D.R. Engstrom. 1989. Resolving the chronology of recent lake sediments: an example from Devils Lake, North Dakota. *Journal of Paleolimnology* 2:81–97.

Johnson, M.G., J.R.M. Kelso, O.C. McNeil, and W.B. Morton. 1990. Fossil midge associations and the historical status of fish in acidified lakes. *Journal of Paleolimnology* 3:113–27.

Johnson, M.G. and O.C. McNeil. 1988. Fossil midge associations in relation to trophic and acidic state of the Turkey lakes. *Canadian Journal of Fisheries and Aquatic Sciences* 45(Supplement 1):136–44.

Jowsey, P.C. 1966. An improved peat sampler. *New Phytologist* 65:245–8.

Kansanen, P.H. 1985. Assessment of pollution history from recent sediments in Lake Vanajavesi, southern Finland. II. Changes in the Chironomidae, Chaoboridae and Ceratopogonidae (Diptera) fauna. *Annales Zoologici Fennici* 22:57–90.

Kansanen, P.H. 1986. Information value of chironomid remains in the uppermost sediment layers of a complex lake basin. *Hydrobiologia* 143:159–65.

Kansanen, P.H. and T. Jaakkola. 1985. Assessment of pollution history from recent sediments in Lake Vanajavesi, southern Finland. I. Selection of representative profiles, their dating and chemostratigraphy. *Annales Zoologici Fennici* 22:13–55.

Kingston, J.C. 1986. Diatom analysis—basic protocol. In *Paleoecological Investigation of Recent Lake Acidification (PIRLA): Methods and Project Description,* eds. D.F. Charles and D.R. Whitehead, pp. 6.1–6.11. EPRI EA-4906. Electric Power Research Institute, Palo Alto, CA.

Kingston, J.C. and H.J.B. Birks. 1990. Dissolved organic carbon reconstructions from diatom assemblages in PIRLA project lakes, North America. *Philosophical Transactions of the Royal Society of London* (Series B) 327:279–88.

Kipphut, G.W. 1978. An investigation of sedimentary processes in lakes. Ph.D. dissertation, Columbia Univ., New York.

Klink, A. 1989. The lower Rhine: palaeoecological analysis. In *Historical Change of Large Alluvial Rivers: Western Europe,* ed. G.E. Petts, pp. 183–201. John Wiley, Chichester, England.

Krause-Dellin, D. and C. Steinberg. 1986. Cladoceran remains as indicators of lake acidification. *Hydrobiologia* 143:129–34.

Krishnaswamy, S., D. Lal, J.M. Martin, and M. Meybeck. 1971. Geochronology of lake sediments. *Earth and Planetary Science Letters* 11:407–14.

Last, W.M. and L.A. Slezak. 1988. The salt lakes of western Canada: a paleolimnological overview. *Hydrobiologia* 158:301–16.

Lemdahl, G. and B. Liedberg-Jönsson. 1988. A late Weichselian palaeoenvironmental reconstruction, based on biostratigraphic studies at a site in SW Skåne, S Sweden. In *Palaeoclimatic and Palaeoecological Studies Based on Subfossil Insects from Late Weichselian Sediments in Southern Sweden,* G. Lemdahl, Appendix III. LUNDQUA Thesis 22, Department of Quaternary Geology, Lund Univ., Lund, Sweden.

Line, J.M. and H.J.B. Birks. 1990. WACALIB version 2.1—a computer program to reconstruct environmental variables from fossil assemblages by weighted averaging. *Journal of Paleolimnology* 3:170–3.

Livingstone, D.A. 1955. A lightweight piston sampler for lake deposits. *Ecology* 36:137–9.

Löffler, H. 1983. Changes of the benthic fauna of the profundal zone of Traunsee (Austria) due to salt mining activities. *Hydrobiologia* 103:135–9.

Löffler, H. 1986. Ostracod analysis. In *Handbook of Holocene Palaeoecology and Palaeohydrology,* ed. B.E. Berglund, pp. 693–702. John Wiley, Chichester, England.

Ložek, V. 1986. Mollusca analysis. In *Handbook of Holocene Palaeoecology and Palaeohydrology,* ed. B.E. Berglund, pp. 729–40. John Wiley, Chichester, England.

Mackereth, F.J.H. 1958. A portable core sampler for lake deposits. *Limnology and Oceanography* 3:181–91.

Mackereth, F.J.H. 1966. Some chemical observations on post-glacial lake sediments. *Philosophical Transactions of the Royal Society of London* (Series B) 250:165–213.

Mathewes, R.W. and J.M. D'Auria. 1982. Historic changes in an urban watershed determined by pollen and geochemical analyses of lake sediment. *Canadian Journal of Earth Sciences* 19:2114–25.

Medioli, F.S. and D.B. Scott. 1988. Lacustrine thecamoebians (mainly arcellaceans) as potential tools for palaeolimnological interpretations. *Palaeogeography, Palaeoclimatology, Palaeoecology* 62:361–86.

Miller, B.B. and A.F. Bajc. 1989. Methods in Quaternary ecology #8. Non-marine molluscs. *Geoscience Canada* 16:165–75.

Miller, R.F. and A.V. Morgan. 1982. A postglacial coleopterous assemblage from Lockport Gulf, New York. *Quaternary Research* 17:258–74.

Morgan, A., A.V. Morgan, and S.A. Elias. 1985. Holocene insects and paleoecology of the Au Sable River, Michigan. *Ecology* 66:1817–28.

Morgan, A.V. and A. Morgan. 1980. Beetle bits—the science of paleoentomology. *Geoscience Canada* 7:22–9.

Mossberg, P. and P. Nyberg. 1979. Bottom fauna of small acid forest lakes. *Institute of Freshwater Research Drottningholm Report* 58:77–87.

Oldfield, F. and P.G. Appleby. 1984. Empirical testing of ^{210}Pb-dating models for lake sediments. In *Lake Sediments and Environmental History,* eds. E.Y. Haworth and J.W.G. Lund, pp. 93–124. Leicester Univ. Press, Leicester, England.

Oliver, D.R. and M.E. Roussel. 1983. *The Insects and Arachnids of Canada. Part 11. The Genera of Larval Midges of Canada; Diptera: Chironomidae*. Agriculture Canada Publication 1746, Ottawa, ON.

Olsson, I.U. 1986. Radiometric dating. In *Handbook of Holocene Palaeoecology and Palaeohydrology,* ed. B.E. Berglund, pp. 273–312. John Wiley, Chichester, England.

O'Sullivan, P.E. 1983. Annually-laminated lake sediments and the study of Quaternary environmental changes—a review. *Quaternary Science Reviews* 1:245–313.

Parise, G. and A. Riva. 1982. Cladocera remains in recent sediments as indices of cultural eutrophication of Lake Como. *Schweizerische Zeitschrift für Hydrologie* 44:277–87.

Paterson, M.J. 1985. Paleolimnological reconstruction of cladoceran response to presumed acidification of four lakes in the Adirondack Mountains (New York). M.A. dissertation, Indiana Univ., Bloomington, IN.

Reasoner, M.A. 1986. An inexpensive, lightweight percussion core sampling system. *Géographie physique et Quaternaire* 40:217–9.

Ritchie, J.C., J.R. McHenry, and A.C. Gill. 1973. Dating recent reservoir sediments. *Limnology and Oceanography* 18:254–63.

Robbins, J.A., T. Keilty, D.S. White, and D.N. Edgington. 1989. Relationships among tubificid abundances, sediment composition, and accumulation rates in Lake Erie. *Canadian Journal of Fisheries and Aquatic Sciences* 46:223–31.

Sæther, O.A. 1979. Chironomid communities as water quality indicators. *Holarctic Ecology* 2:65–74.

Sanger, J.E. 1988. Fossil pigments in paleoecology and paleolimnology. *Palaeogeography, Palaeoclimatology, Palaeoecology* 62:343–59.

Schönborn, W. 1967. Taxozönotik der beschalten Süsswasser—Rhizopoden: eine raumstrukturanalytische Untersuchung über Lebensraumerweiterung und Evolution bei der Mikrofauna. *Limnologica* 5:159–207.

Sherwood-Pike, M.A. 1988. Freshwater fungi: fossil record and paleoecological potential. *Palaeogeography, Palaeoclimatology, Palaeoecology* 62:271–85.

Simola, H. 1977. Diatom succession in the formation of annually laminated sediment in Lovojärvi, a small eutrophicated lake. *Annales Botanici Fennici* 14:143–8.

Smol, J.P. 1987. Methods in Quaternary ecology #1. Freshwater algae. *Geoscience Canada* 14:208–17.

Smol, J.P. 1990. Paleolimnology: recent advances and future challenges. In Scientific perspectives in theoretical and applied limnology, eds. R. de Bernardi, G. Giussani, and L. Barbanti. *Memorie dell'Istituto Italiano di Idrobiologia* 47:253–76.

Smol, J.P., I.R. Walker, and P.R. Leavitt. 1991. Paleolimnology and hindcasting climatic trends. *Internationale Vereinigung für Theoretische und Angewandte Limnologie Verhandlungen* 24:1240–6.

Stark, D.M. 1976. Paleolimnology of Elk Lake, Itasca State Park, northwestern Minnesota. *Archiv für Hydrobiologie Supplement* 50:208–74.

Steinberg, C., K. Arzet, and D. Krause-Dellin. 1984. Gewässerversauerung in der Bundesrepublik Deutschland im Lichte paläolimnologischer Studien. *Naturwissenschaften* 71:631–4.

Steinberg, C., H. Hartmann, K. Arzet, and D. Krause-Dellin. 1988. Paleoindication of acidification in Kleiner Arbersee (Federal Republic of Germany, Bavarian Forest) by chydorids, chrysophytes, and diatoms. *Journal of Paleolimnology* 1:149–57.

Stevenson, A.C., H.J.B. Birks, R.J. Flower, and R.W. Battarbee. 1989. Diatom-based pH reconstruction of lake acidification using canonical correspondence analysis. *Ambio* 18:228–33.

Stoermer, E.F., J.A. Wolin, C.L. Schelske, and D.J. Conley. 1985. Postsettlement diatom succession in the Bay of Quinte, Lake Ontario. *Canadian Journal of Fisheries and Aquatic Sciences* 42:754–67.

Stuiver, M. 1982. A high-precision calibration of the AD radiocarbon time scale. *Radiocarbon* 24:1–26.

Swain, E.B. 1985. Measurement and interpretation of sedimentary pigments. *Freshwater Biology* 15:53–75.

Synerholm, C.C. 1979. The chydorid cladocera from surface lake sediments in Minnesota and North Dakota. *Archiv für Hydrobiologie* 86:137–51.

ter Braak, C.J.F. 1986. Canonical correspondence analysis: a new eigenvector technique for multivariate direct gradient analysis. *Ecology* 67:1167–79.

ter Braak, C.J.F. 1987a. The analysis of vegetation-environment relationships by canonical correspondence analysis. *Vegetatio* 69:69–77.

ter Braak, C.J.F. 1987b. Calibration. In *Data Analysis in Community and Landscape Ecology,* eds. R.H.G. Jongman, C.J.F. ter Braak, and O.F.R. van Tongeren, pp. 78–90. Centre for Agricultural Publishing and Documentation (Pudoc), Wageningen, The Netherlands.

ter Braak, C.J.F. and C.W.N. Looman. 1987. Regression. In *Data Analysis in Community and Landscape Ecology,* eds. R.H.G. Jongman, C.J.F. ter Braak, and O.F.R. van Tongeren, pp. 29–77. Centre for Agricultural Publishing and Documentation (Pudoc), Wageningen, The Netherlands.

ter Braak, C.J.F. and I.C. Prentice. 1988. A theory of gradient analysis. *Advances in Ecological Research* 18:271–317.

ter Braak, C.J.F. and H. van Dam. 1989. Inferring pH from diatoms: a comparison of old and new calibration methods. *Hydrobiologia* 178:209–23.

Thienemann, A. 1921. Seetypen. *Naturwissenschaften* 9:343–6.

Timms, B.V., U.T. Hammer, and J.W. Sheard. 1986. A study of benthic communities in some saline lakes in Saskatchewan and Alberta, Canada. *Internationale Revue der gesamten Hydrobiologie* 71:759–77.

Tolonen, K. 1986. Rhizopod analysis. In *Handbook of Holocene Palaeoecology and Palaeohydrology,* ed. B.E. Berglund, pp. 645–66. John Wiley, Chichester, England.

Uutala, A.J. 1986. Paleolimnological assessment of the effects of lake acidification on Chironomidae (Diptera) assemblages in the Adirondack region of New York. Ph.D. dissertation, State Univ. of New York College of Environmental Science and Forestry, Syracuse, NY.

Uutala, A.J. 1990. *Chaoborus* (Diptera: Chaoboridae) mandibles–paleolimnological indicators of the historical status of fish populations in acid-sensitive lakes. *Journal of Paleolimnology* 4:139–51.

Vance, R.E. 1986. Aspects of the postglacial climate of Alberta: calibration of the pollen record. *Géographie physique et Quaternaire* 40:153–60.

van Geel, B. 1986. Application of fungal and algal remains and other microfossils in palynological analyses. In *Handbook of Holocene Palaeoecology and Palaeohydrology,* ed. B.E. Berglund, pp. 497–505. John Wiley, Chichester, England.

Walker, I.R. 1987. Chironomidae (Diptera) in paleoecology. *Quaternary Science Reviews* 6:29–40.

Walker, I.R. 1990. Modern assemblages of arctic and alpine Chironomidae as analogues for late-glacial communities. *Hydrobiologia* 214:223–7.

Walker, I.R. and R.W. Mathewes. 1987a. Chironomidae (Diptera) and postglacial climate at Marion Lake, British Columbia, Canada. *Quaternary Research* 27:89–102.

Walker, I.R. and R.W. Mathewes. 1987b. Chironomids, lake trophic status, and climate. *Quaternary Research* 28:431–7.

Walker, I.R. and R.W. Mathewes. 1989a. Much ado about dead Diptera. *Journal of Paleolimnology* 2:19–22.

Walker, I.R. and R.W. Mathewes. 1989b. Chironomidae (Diptera) remains in surficial lake sediments from the Canadian Cordillera: analysis of the fauna across an altitudinal gradient. *Journal of Paleolimnology* 2:61–80.

Walker, I.R. and R.W. Mathewes. 1989c. Early postglacial chironomid succession in southwestern British Columbia, Canada, and its paleoenvironmental significance. *Journal of Paleolimnology* 2:1–14.

Walker, I.R., R.J. Mott, and J.P. Smol. 1991a. Allerød-Younger Dryas lake temperatures from midge fossils in Atlantic Canada. *Science* 253: 1010-2.

Walker, I.R. and C.G. Paterson. 1983. Post-glacial chironomid succession in two small, humic lakes in the New Brunswick-Nova Scotia (Canada) border area. *Freshwater Invertebrate Biology* 2:61–73.

Walker, I.R., J.P. Smol, D.R. Engstrom, and H.J.B. Birks. 1991b. An assessment of Chironomidae as quantitative indicators of past climatic change. *Canadian Journal of Fisheries and Aquatic Sciences* 48:975–87.

Warner, B.G. 1988. Methods in Quaternary ecology #5. Testate Amoebae (Protozoa). *Geoscience Canada* 15:251–60.

Warner, B.G. 1989. Methods in Quaternary ecology #10. Other fossils. *Geoscience Canada* 16:231–42.

Warner, B.G., ed. 1990. *Methods in Quaternary Ecology*. Geoscience Canada Reprint Series 5. Geological Association of Canada, St. John's, NF.

Warner, B.G. and B.J. Hann. 1987. Aquatic invertebrates as paleoclimatic indicators? *Quaternary Research* 28:427–30.

Warwick, W.F. 1980. Palaeolimnology of the Bay of Quinte, Lake Ontario: 2800 years of cultural influence. *Canadian Bulletin of Fisheries and Aquatic Sciences* 206:1–117.

Warwick, W.F. 1989. Chironomids, lake development and climate: a commentary. *Journal of Paleolimnology* 2:15–7.

Watts, W.A. 1978. Plant macrofossils and Quaternary paleoecology. In *Biology and Quaternary Environments*, eds. D. Walker and J.C. Guppy, pp. 53–67. Australian Academy of Science, Canberra, Australia.

Webb, T., F.A. Street-Perrott, and J.E. Kutzbach. 1987. Late-Quaternary paleoclimatic data and climate models. *Episodes* 10:4–6.

Whitehead, D.R., D.F. Charles, and R.A. Goldstein. 1990. The PIRLA project (Paleoecological Investigation of Recent Lake Acidification): an introduction to the synthesis of the project. *Journal of Paleolimnology* 3:187–94.

Whiteside, M.C. 1970. Danish chydorid Cladocera: modern ecology and core studies. *Ecological Monographs* 40:79–118.

Whiteside, M.C. and M.R. Swindoll. 1988. Guidelines and limitations to cladoceran paleoecological interpretations. *Palaeogeography, Palaeoclimatology, Palaeoecology* 62:405–12.

Wiederholm, T., ed. 1983. Chironomidae of the Holarctic region. Keys and diagnoses. Part 1—Larvae. *Entomologica scandinavica Supplement* 19:1–457.

Wik, M., I. Renberg, and J. Darley. 1986. Sedimentary records of carbonaceous particles from fossil fuel combustion. *Hydrobiologia* 143:387–94.

Wilkinson, B. 1984. Interpretation of past environments from sub-fossil caddis larvae. In *Proceedings of the Fourth International Symposium on Trichoptera, Clemson, SC, July 11–16, 1983*, ed. J.C. Morse, pp. 447–52. Junk Pubs., The Hague, The Netherlands.

Wilkinson, B.J. 1987. Trichoptera sub-fossils from temperate running water sediments. In *Proceedings of the Fifth International Symposium on Trichoptera, Lyon, France, July 21–26, 1986*, eds. M. Bournaud and H. Tachet, pp. 61–5. Junk Pubs., Dordrecht, The Netherlands.

Williams, H.F.L. and M.C. Roberts. 1987. Mid-Holocene paleosurfaces in vertical accretion deposits, Fraser River delta, British Columbia. In *Current Research, Part A*, pp. 757–61. Paper 87–1A. Geological Survey of Canada, Ottawa, ON.

Williams, N.E. 1981. Aquatic organisms and palaeoecology: recent and future trends. In *Perspectives in Running Water Ecology*, eds. M.A. Lock and D.D. Williams, pp. 289–303. Plenum Press, New York.

Williams, N.E. 1988. The use of caddisflies (Trichoptera) in palaeoecology. *Palaeogeography, Palaeoclimatology, Palaeoecology* 62:493–500.

Williams, N.E. 1989. Factors affecting the interpretation of caddisfly assemblages from Quaternary sediments. *Journal of Paleolimnology* 1:241–8.

Williams, N.E. and A.V. Morgan. 1977. Fossil caddisflies (Insecta: Trichoptera) from the Don Formation, Toronto, Ontario, and their use in paleoecology. *Canadian Journal of Zoology* 55:519–27.

Williams, N.E., J.A. Westgate, D.D. Williams, A. Morgan, and A.V. Morgan. 1981. Invertebrate fossils (Insecta: Trichoptera, Diptera, Coleoptera) from the Pleistocene Scarborough Formation at Toronto, Ontario, and their paleoenvironmental significance. *Quaternary Research* 16:146–66.

Wright, H.E., Jr. 1967. A square-rod piston sampler for lake sediments. *Journal of Sedimentary Petrology* 37:975–6.

10

Toxicity Studies Using Freshwater Benthic Macroinvertebrates

Arthur L. Buikema, Jr. and J. Reese Voshell, Jr.

10.1. Introduction

As a science, toxicology historically is based on physiology. It arose as a discipline in the 1800s in response to the development of synthetic organic chemicals (Zapp, 1980). The first "toxicity" experiments were published in 1816 when 15 species of freshwater molluscs reportedly survived longer in 2% than in 4% saline solutions (B.G. Anderson, 1980). Studies on the survival of freshwater invertebrates exposed to metals and organic compounds appeared in the mid-1890s (B.G. Anderson, 1980).

Many factors affect an organism's response to stress. In 1871, it was reported that survival of various arthropods in seawater was inversely proportional to temperature (B.G. Anderson, 1980). Twenty years later it was demonstrated that acclimation to environmental conditions also affected organism survival. Acclimation to test conditions, especially temperature, is one of the most important factors considered in contemporary toxicity testing. In the early 1930s, E.C.L. Naumann published the first extensive studies on factors that affected an organism's sensitivity to chemicals; he examined the effects of a wide variety of stressors and natural variables on the acute mortality of *Daphnia* (B.G. Anderson, 1980). Naumann suggested that water temperature be maintained at 20° C during testing, and he also noted that egg production was affected by toxicants.

In the 1940s and 1950s, scientists noted differences in the biota of streams that did and did not receive wastewaters. Although it was easy to document damage, the development of a predictive measure before damage occurred became a goal. Much of the early toxicological research examined effects on fish because they were large, visible, and attracted public and political interest. During this time, scientists demonstrated the utility of using acute toxicity tests for predicting potential damage from industrial discharge (Hart et al., 1945; Doudoroff et al., 1951; Cairns, 1957; Henderson, 1957; Tarzwell, 1958). J.B. Sprague entrenched the future use of toxicity tests in his three-part series on toxicity testing and application of testing results (Sprague, 1969, 1970, 1971).

It soon became apparent that, to protect fish, it would be necessary also to protect other members of the community. For example, Patrick (1949) and Patrick et al. (1968) demonstrated that algae and macroinvertebrates often were more sensitive than fish to pollutants. Interest then developed in tests on lower-food-chain aquatic organisms because these organisms represented a greater proportion of total biomass than fish and because often they were eaten by fish.

Although the use and types of toxicity tests have expanded since the early 1950s, little has been added to the conceptual framework of acute toxicity tests in the last 50 years. Most of the expansion in the types of organisms used in toxicity tests and the standardization of methods appears to have coincided with the passage of environmental legislation (e.g., in the United States this expansion coincided with passage of the Clean Water Act; Federal Insecticide, Fungicide and Rodenticide Act; Toxic Substances Control Act; and Resource and Conservation Recovery Act).

Toxicity tests have been used to answer many types of questions (e.g., Sprague, 1973; Baker and Crapp, 1974; Brungs and Mount, 1978); typical ones include:

1. At which concentration is the material toxic to the test organism?
2. What are the effects of a test material if an organism is exposed for a portion or all of its life cycle?
3. Which waste, product, or sediment is most toxic?
4. Which organism is most or least sensitive?
5. Under which conditions is a waste or product most or least toxic?
6. Is toxicity altered once a material enters the environment?
7. Does a waste or product exceed a regulatory standard?
8. How much of a receiving system is affected by toxicants?
9. What are the short-term effects of an episodic spill?
10. Can we explain the ecological impacts of episodic releases, spills, and other accidents?
11. Can we use this information in risk assessment to predict ecological effects?

In the 1970s and 1980s, toxicity testing has emphasized the best application of existing methods to achieve management goals (Buikema et al., 1982). Information from toxicity tests has been used to manage pollution in order to: (1) regulate discharge; (2) compare toxicants, animal sensitivities, or factors affecting toxicity; and (3) predict environmental effects. The last goal has been the most difficult to achieve.

The purpose of this chapter is to evaluate the state-of-the-art of toxicity tests using freshwater macroinvertebrates. "State-of-the-art" applies to the methods and types of information currently used by scientists and regulatory agencies. Although interpretations by the two groups are not identical, neither are they mutually exclusive (Jenkins et al., 1989); both groups have influenced the development of toxicity test methods and their use. Although this review will focus on trends in toxicity testing in the United States, these trends are believed to be common in other technologically developed countries.

10.2. Types of Toxicity Tests

Toxicity tests form a continuum of complexity, from tests conducted on single species to those conducted on whole ecosystems (Boyle, 1983). Toxicity tests at either end of the continuum are easy to categorize, the former always being conducted indoors in small vessels and the latter being conducted outdoors in natural bodies of water. However, differences exist in the levels of complexity of toxicity tests between the two ends of the continuum.

Three basic types of toxicity tests use macroinvertebrates: acute single species tests, chronic single species tests, and multispecies tests (Table 10.1). Single species tests are the most frequently used. Much has been done to confirm and expand their utility in biomonitoring, but they remain simplistic and environmentally unrealistic (Boyle, 1983); it is difficult to predict the effects of a chemical on an ecosystem from the effects on its parts (e.g., Kimball and Levin, 1985). In contrast, multispecies tests use artificial assemblages of organisms or components of natural communities and allow studies of community dynamics after relatively long-term or episodic exposures.

All toxicity tests require basic ecological information about populations. For acute tests on macroinvertebrates, this information is important for culturing and maintaining test animals and designing standardized test methods. For chronic and multispecies toxicity tests, basic ecological information usually is not available. Moreover, even if such data were available, its interpretation and application often is problematical.

10.2.1. Acute Toxicity Tests

Acute toxicity tests are short-term tests that last 48 or 96 hours (Table 10.1). Their length usually is limited by starvation, because the early life-stage animals used are not fed during the test. Tests usually are conducted

Table 10.1. Comparison of different types of toxicity tests using macroinvertebrates.

Types of Tests	Duration	Organism Source	Endpoints
Single species: acute	48 to 96 h	Cultured, purchased, collected	Mortality, growth (shell deposition), behavior
Single species: chronic			
Life cycle	One generation	Cultured	Mortality, growth, reproduction
Sub-life cycle	Egg through early life stage	Cultured/reared	Mortality, growth, behavior, metamorphosis
Life history	Egg to death or cessation of reproduction	Cultured	Mortality, growth, reproduction, behavior, metamorphosis
Multispecies			
Artificial assemblage	Days to weeks to months	Cultured	Mortality, growth, competition, predator-prey interactions, biomass, density, diversity, taxa richness, community metabolism
Natural assemblage	Days to weeks to months	Collected	Taxa richness, community metabolism, secondary production, behavior, mortality, growth, competition, predator-prey interactions, biomass, density, diversity, functional feeding groups

with single species (e.g., APHA, 1985) but simultaneous exposures also have been done. For example, Ewell et al. (1986) exposed seven taxa (isopods, daphnids, amphipods, snails, flatworms, oligochaetes, and fathead minnows) to 27 chemicals. Test species were chosen on the basis of ecological importance, diversity of life form, and ease of culture. The acute toxicity values that resulted were comparable to those obtained in individual tests, and reproducibility of results was good. This approach offers reductions in the amount of chemical needed and the amount discarded at the end of the experiment and increases the number of data points collected in one experiment. However, a problem exists in ensuring that sufficient numbers of all organisms are available at the same time for the test.

10.2.1.1. Modifiers of Toxicity

Many variables can modify the sensitivity of an organism to a chemical, and consequently the organism's LC50 or EC50 (Lethal or Effective Concentration that kills or affects 50% of a population within a defined period of time) and its fiducial limits, or endpoints such as growth and reproduction. These variables, which include life-history aspects, are summarized in Buikema and Benfield (1979), Maciorowski and Clarke (1980), Buikema et al. (1982), Parrish (1985) and in methods publications such as APHA (1985), Horning and Weber (1985), Peltier and Weber (1985), and ASTM (1988).

Buikema and Benfield (1979) acknowledged that it would be impossible to incorporate all aspects of an organism's biology into a test design, but they expressed concern over the simplistic nature of many toxicity tests. Variables such as temperature, light, salinity, pH, water hardness, body size, age, stage of molt, nutritional state, diet, season of the year, and animal activity may alter significantly the outcome of a test (Buikema and Benfield, 1979). Without control of these and other test conditions, the results more likely would be an artifact of test conditions rather than an indication of toxicant effect. Environmental regulation would be difficult if it was based on toxicity test results that did not control modifiers of toxicity.

10.2.1.2. Availability of Test Organisms

Perusal of various professional journals (e.g., Water Environment Research, formerly Journal of the Water Pollution Control Federation, annual reviews), methods publications by professional societies (e.g., APHA, 1985) and regulatory agencies (e.g., Horning and Weber, 1985; Peltier and Weber, 1985), and activities of consensus organizations (e.g., ASTM, 1988) indicates that approximately 200 freshwater test organisms have been used or are recommended for acute toxicity testing. Approximately 50 of these organisms are macroinvertebrates, <30% of which can be cultured easily in the laboratory. Most of the macroinvertebrates that can be cultured belong to five taxa: Amphipoda, Malacostraca, Ephemeroptera, Diptera, and Oligochaeta (Buikema and Benfield, 1979; Buikema et al., 1982).

Most of the animals recommended as test organisms are collected and then reared or held. Rearing is used in studies where gravid females are brought into the laboratory and the offspring are used in the test. The term "holding" applies when various life stages are brought into the laboratory, acclimated to test conditions, and then used in a test. In practice, these tests use specific, sensitive, early life stages; however, many of them do not incorporate life-history information in the test design because such information is not available. Moreover, additional species could be used if basic ecological information was available (Buikema and Benfield, 1979).

Site-specific toxicity information may be important to understand the effect of an effluent on a receiving system; to obtain this information, toxicity tests often are conducted over one or more years, and the test design incorporates seasonal changes in receiving-stream temperature and water quality. The use of indigenous macroinvertebrates is advantageous for obtaining this information because the organisms already are acclimated to ambient conditions (Buikema et al., 1982). However, several practical problems arise when using indigenous organisms. First, an organism may not be amenable to laboratory test conditions (e.g., riverine organisms cannot survive in static or low-flow conditions). Second, because of a complex life cycle, an appropriate age or size of the organism may not be available on demand. This creates a problem if the objective is to examine toxicity over time. Third, if the organisms are relatively rare, the cost of collection may be prohibitive. Fourth, collection of indigenous organisms from an already stressed population may add additional stress and hinder extrapolation of laboratory-derived data to the field. Fifth, many jurisdictions may require a collecting permit, which is often difficult to obtain, especially if a water body contains a protected fishery.

10.2.1.3. Selection of a Test Organism

Selection of a test organism depends on the question that is being asked and how the test data will be used. The assessment of toxicity, the control of effluent discharges, or the registration of a new chemical require the collection of consistent and comparable data. In these cases, a "routine" test organism such as *Daphnia* can be used. However, it may be necessary to modify test methods to suit site-specific information needs; for example, to regulate a discharge may require data for indigenous organisms tested under a variety of environmental conditions. In all situations, the data are used to protect aquatic resources.

To best protect these resources, the U.S. Environmental Protection Agency (USEPA, 1979) and others (e.g., Rosenberger et al., 1978) have recommended the use of an "ideal" test organism, which has the following characteristics:

1. It should represent an ecologically or economically important group (in terms of taxonomy, trophic level, realized niche, or use).
2. It should occupy a trophic position leading to humans or other "important" species.
3. It should have adequate background data (e.g., physiology, genetics, taxonomy, ecology).
4. It should be available widely, amenable to laboratory testing, easily maintained, genetically stable, and not prone to disease, infection, or physical damage.
5. Its response characteristics should be comparable to those of indigenous species.
6. It should be sensitive to chemicals and exhibit a consistent response to the same test chemical.
7. It should have an endpoint that is easily identified and measured.

Few potential test organisms meet all of these criteria. Furthermore, some of the criteria may be inconsistent with each other (Buikema et al., 1982). For example, few test methods exist for economically important species that are food organisms leading to humans, most organisms are not easily maintained or cultured, and to assume genetic uniformity is presumptuous. In reality, sufficient ecological information still does not exist for most currently used test species.

Test organisms that have response characteristics comparable to indigenous species may be difficult to find. However, several authors have attempted to correlate acute toxicity values between species. For example, Mayer and Ellersieck (1986) analyzed data for several species, mostly daphnids and fish, and found that correlations were highest between species within a family. A similar pattern would be expected for benthic macroinvertebrates. LeBlanc (1984) and Suter and Vaughn (1985) concluded that the more distant two organisms are taxonomically, the greater the difference in sensitivity to a chemical; the research of Slooff et al. (1986) supported this conclusion.

10.2.1.4. Test Standardization

The goal of obtaining consistent and defensible data requires standardized test procedures. This requirement was realized in the early 1950s (e.g., Doudoroff et al., 1951) and has influenced the development of methods for acute toxicity testing since the mid-1970s (e.g., APHA, 1985; Horning and Weber, 1985; Peltier and Weber, 1985; ASTM, 1988). Uniformity of procedures has made tests easier to conduct by different personnel and laboratories, replication has increased the accuracy of data, and both have facilitated the comparison of data (Davis, 1977).

However, in spite of attempts at standardization, considerable variability exists among test results because of differences in sources of test organisms and dilution water, general laboratory conditions, technician expertise, etc. "Round-robin" studies, in which simultaneous toxicity tests are conducted by independent laboratories using the same toxicant and animals of the same genetic strain and source, suggest that it is important to inform a participating laboratory that it is involved in such a test (e.g., Buikema, 1983; Grothe and Kimerle, 1985). However, when toxicity tests are conducted by different laboratories, even using standardized test methods and the best technical expertise, the problem of interpreting variable test results always will exist.

Standardized test procedures are not always necessary; the purpose of the toxicity test will determine the best experimental design (Buikema et al., 1982). A comparison of data requires an inflexible test procedure, whereas standardization would be overly restrictive in predicting the effects of a chemical on a population, community, or ecosystem. Tests designed to answer questions about specific receiving systems may have to deviate substantially from a standardized procedure. For example, most toxicants are tested in the laboratory using constant, known concentrations of chemical. However, in nature, organisms usually are exposed to short-term, single or periodic exposures that vary in concentration; the responses of organisms will vary with different exposure regimes.

Inflexible test methods stifle innovative and creative work, especially if regulatory agencies do not use information that does not coincide with their priorities (Davis, 1977). Regulatory agencies also tend to be conservative; it can take up to 20 years between the development of a test method and its use (Jenkins et al., 1989). Much of this delay results from a desire to use test methods that have been confirmed and validated, but a fear also exists that new methods will not provide data that can be used in risk assessment. If inappropriate data are submitted to an agency and a chemical is not registered as a result, the subsequent threat of a lawsuit by the registrant also will be an issue because it can affect agency budgets and personnel.

Long-lived organisms have been ignored in the development of toxicity tests in order to obtain data quickly, reliably, and economically (Buikema et al., 1982). However, naturally long generation times should not, *a priori*, preclude the use of such organisms in the laboratory. For example, a basic understanding of the influences of temperature and photoperiod on the life history of the grass shrimp, *Palaemonetes pugio*, allowed its generation time to be reduced from 12 to three months in the laboratory (Little, 1968).

An obvious need exists to develop standardized test methods for macroinvertebrates, but this development has been slow for two reasons. First, this type of research is considered to be either too basic or too applied; the consequence has been a lack of funding. Second, toxicity data often are inaccessible because they are difficult to publish in the primary literature, so they appear in the grey literature or they are deemed to be proprietary.

10.2.1.5. Number of Species Tested

Various federally sponsored studies and publications indicate a lack of unanimity on how many species should be tested to protect waters for the propagation of aquatic life. The problem is not knowing how many—and which—species adequately represent the whole range of biological sensitivities (Mount, 1982). For most chemicals, Mount (1982) noted that the five most commonly tested species, which included a gammarid amphipod, represented a large part of the range of sensitivity. Mayer and Ellersieck (1986) suggested the use of three test organisms, one of which was a macroinvertebrate. Birge and Black (1982) recommended that from three to five species could be used to establish a biological response range. Toxicity values for algae were within two orders of magnitude of the most sensitive species 95% of the time for three randomly selected species and 99% of the time for five randomly selected species (Blanck et al., 1984). The minimum number of acute tests required for protection of aquatic life appears to be five species, but the number will vary depending upon regulatory mandates. Inclusion of macroinvertebrates is important in identifying the range of biological sensitivities.

To prohibit point and non-point discharges of toxic substances, standardized test procedures were developed for use by regulatory agencies and permittees for self-monitoring (e.g., Horning and Weber, 1985; Peltier and Weber, 1985). The emphasis has been on tests that are quick, economical, and reproducible and that can be compared among and within effluents. Although many test organisms could be used, those of known stock and acclimation history are strongly encouraged (Peltier and Weber, 1985). This emphasis has limited the test species primarily to daphnids and fathead minnows. Therefore, the objective of regulatory agencies to obtain comparative data precludes the use of many suggested organisms, so macroinvertebrates may be ignored. However, once a sensitive organism has been identified and a substantial database has been accumulated on it, a new test method or test organism is adopted by regulatory agencies (Jenkins et al., 1989).

The registration of new compounds also requires toxicity data. Because thousands of chemicals are in use and new ones are being produced each year, cost-effective tests, the results of which can be used in ranking chemicals for safety, are being emphasized. The registration process uses a tiered test approach. It starts with simple tests and becomes progressively more complex and expensive if a test endpoint identifies an environmental concern (e.g., acute toxicity below 1 mg/l). Tier I acute toxicity tests may include fish and *Daphnia*; Tier II tests may use a freshwater gammarid amphipod. Testing proceeds to a higher tier involving more complex tests if additional concerns are identified. Because tests at lower tiers may not give sufficient information to protect aquatic systems (Mayer and Ellersieck, 1986), aquatic field studies are part of Tier IV, the highest level; macroinvertebrates also are important components of higher tier tests.

The most extensive use of macroinvertebrate toxicity data in the U.S.A. occurs under the section of the Clean Water Act concerning development of water quality criteria (e.g., Larson and Hyland, 1987), which is a non-regulatory, scientific assessment of concentrations of chemicals that should not unacceptably affect freshwater organisms and their uses. The data requirements for deriving numerical, national water-quality criteria (Stephan, et al., 1985) include defensible data for a minimum of eight species representing different taxa, life forms, and trophic levels. At a minimum, the species include: (1) a salmonid; (2) a warm-water fish; (3) one other member of the Chordata; (4) a planktonic crustacean; (5) a benthic crustacean; (6) an aquatic insect; (7) a nonarthropod or chordate; and (8) a family in any order of insect or any phylum not already represented. For example, development of a short-term exposure criterion for zinc (Larson and Hyland, 1987) used 173 sets of data of which 27% were for invertebrates. A total of 43 taxa were represented, 47% of which were invertebrates. Of the 10 most sensitive genera, five were invertebrates: three macroinvertebrates and two daphnids. Therefore, data for several life forms, including macroinvertebrates, were used to derive a concentration of zinc that is assumed to protect 95% of all species.

It has been suggested that requiring additional single species tests may not provide much new information. Mayer and Ellersieck (1986) reviewed data obtained from 4,901 acute toxicity tests conducted on 66 species of organisms exposed to 410 organic chemicals; most of the chemicals were pesticides. They concluded that in 88% of the cases, the lowest toxicity value was obtained by testing only three organisms: *Daphnia* spp., *Gammarus fasciatus,* and the rainbow trout (*Salmo gairdneri*). Addition of other species only increased the probability of obtaining a lower toxicity value by 2.5% per organism. Their review indicates that the cost of adding another test or species would not substantially improve the database, but such a conclusion would be fallacious, given that regulatory agencies are required to maintain and restore the integrity of the environment and promote the protection and propagation of aquatic life. If protection is the primary concern, then additional tests should be required, regardless of the cost, in order to reduce the uncertainty of protecting the environment. Slooff (1983) also pointed out that pollution tolerance of macroinvertebrates is chemical-specific; not all animals respond the same way to toxicants. Even benthic survey results are not predictive of a particular type of pollution.

10.2.1.6. Test Endpoints

The acute toxicity test endpoints used should be easy to identify and measure. Usually the endpoint is death or immobilization (Buikema et al., 1982), although shell deposition or glochidial transformation has been used in mollusc tests (APHA, 1985). Other endpoints also have been suggested for short-term tests and include: changes in general organismal activity (Buikema and Benfield, 1979); changes in drift (Wiley and Kohler, 1984); impairment of net construction (Petersen and Petersen, 1985); and morphological deformities (Cushman, 1984) (see also Johnson et al., Chapter 4). The use of pupal cast skins has been suggested as an endpoint in long-term tests (Wilson and McGill, 1977). With the possible exceptions of death, immobilization, and alterations in cast pupal skins, the usefulness of most of the endpoints listed in predicting ecological effects remains unproven.

10.2.1.7. Value of Macroinvertebrates in Acute Toxicity Testing

Considerable data are available that indicate the value of macroinvertebrates in toxicity testing (e.g., Mayer and Ellersieck, 1986; Larson and Hyland, 1987). Testing of additional species of macroinvertebrates may provide important information, depending on the question being asked and the chemical being tested; macroinvertebrates also are important if sediments are used in toxicity tests.

Insecticide toxicity data (Mayer and Ellersieck, 1986) showed that insects as a group, and stoneflies in particular, were the most sensitive of several species tested; crustaceans were next. When relative sensitivities were compared among 40 chemicals, *Daphnia* or rainbow trout were most sensitive only 32% of the time; for nine of the chemicals, the following macroinvertebrates were the most sensitive organisms tested: *Gammarus* (five species), stoneflies (two species), a crayfish, and a chironomid (Mayer and Ellersieck, 1986).

Nine families of macroinvertebrates and six families of vertebrates were represented in tests with DDT, malathion, and endrin (Mayer and Ellersieck, 1986). From most to least sensitive, these families were: Perlidae, Pteronarcidae, Gammaridae, Daphniidae, Centrarchidae, Palaemonidae, Salmonidae, Astacidae, Cypridae, Asellidae, Ictaluridae, Cyprinidae, Rhagionidae, Bufonidae, and Hylidae. Macroinvertebrates represented the most sensitive extreme of a biological response range.

Insects were generally the most sensitive organisms 50% of the time when acute toxicity values obtained for a great range of insecticides were ranked

(Mayer and Ellersieck, 1986). Three families of stoneflies (Perlidae, Perlodidae, and Pteronarcidae) and one family of mayflies (Baetidae) were especially sensitive. The extreme sensitivity of stoneflies and baetids to pesticides is consistent with the intolerance of these groups to pollution under field conditions (Hilsenhoff, 1987). Toxicity values for other insect families (Ephemeridae, Coenagrionidae, Gomphidae, Hydropsychidae, Limnephilidae, Culicidae, Rhagionidae, and Tipulidae) usually were similar to those obtained for the more commonly tested vertebrates and invertebrates. The family Chironomidae was most sensitive to selected chemicals, but overall its responses were similar to the more commonly used organisms.

The frequency distributions of acute toxicity values for different macroinvertebrate taxa exposed to a wide variety of chemicals often were bimodal (Mayer and Ellersieck 1986). Peaks occurred at <100 μg/l for insecticides and at $>1,000$ μg/l for herbicides, fungicides, and industrial and other chemicals. The stonefly *Pteronarcys* was the most sensitive organism to insecticides, whereas *Gammarus pseudolimnaeus* and *Chironomus plumosus* were the most sensitive organisms to the other chemicals. However, *G. fasciatus* showed equal sensitivity over a range of concentrations from 1 to 10,000 μg/l. Again, their results indicated that macroinvertebrates are the most sensitive organisms that can be used in toxicity testing to define a range of biological sensitivities.

Most laboratory toxicity tests are conducted only with chemicals dissolved in water, although the use of sediments may provide a greater degree of environmental realism and increase the ability to predict environmental effects. Sediments often affect the toxicity of chemicals (Chapman, 1986). For example, sediments can: (1) provide a refuge for organisms (e.g., Oseid and Smith, 1974; Wallace et al., 1975; Buikema and Benfield, 1979); (2) mitigate toxicity by adsorbing chemicals (e.g., Cairns et al., 1984; Solomon, 1986); or (3) be a source of toxicity (e.g., Mauck and Olson, 1977; Malueg et al., 1984; Giesy et al., 1988; Schuytema et al., 1988). For these reasons, sediments should be considered in toxicity tests and macroinvertebrates are ideal organisms for sediment toxicity tests; however, few organisms are available for use because test methods still need to be developed.

10.2.1.8. Future of Acute Toxicity Testing

Acute toxicity tests will continue to be emphasized for ranking chemicals and managing pollution. Research should not be directed so much at refining existing test methods, but rather at using more sensitive species or developing new methods. Many test variables only have a nominal impact on toxicity values; for example, the range in acute toxicity values obtained under a multitude of test conditions and for various chemicals varied less than fivefold more than 80% of the time (Mayer and Ellersieck, 1986). Research should be directed at identifying macroinvertebrates that can provide relevant information for predicting ecological effects and for protecting aquatic resources. Slooff et al. (1986) suggested that additional acute tests be developed, especially those having physiological and ecological functions, and that they be used instead of the more expensive and complex higher-tiered tests.

Our ability to protect the aquatic environment requires finding sensitive species that can be easily handled and cultured in the laboratory, and it requires continued development of acute toxicity tests using macroinvertebrates. Winner et al. (1980), Sheehan (1980), Slooff (1983), Mayer and Ellersieck (1986), and Hilsenhoff (1987) provide guidelines for the selection of organisms to be used in acute toxicity tests. Highest priorities should be directed at the insects, especially stoneflies, mayflies, and caddisflies. Expansion of testing with the burrowing mayfly *Hexagenia* should be paramount (Fremling and Mauck, 1980). Selected dipterans also should be examined. Although many chironomids are considered to be tolerant of pollution (Winner et al., 1980; Hilsenhoff, 1987), they make ideal test organisms because they are important to the functioning of natural communities, they are easy to culture (R.L. Anderson, 1980; Giesy et al., 1988), and they provide easily measured, sensitive responses (Giesy et al., 1988).

Initially, macroinvertebrates should be used in acute studies even if indigenous species have to be collected and acclimated to test conditions. Many of these animals subsequently can be cultured in the laboratory when sufficient life-history information is available; chronic studies then should be possible. Given current levels of effort, chronic tests with these organisms will be possible within 10 to 20 years.

10.2.2. Chronic Toxicity Tests

The second basic type of test that uses macroinvertebrates, the chronic toxicity test, involves three types of sublethal responses: (1) life cycle; (2) sub-life cycle; and (3) life history (Buikema and Benfield, 1979) (Table 10.1). In the life-cycle test, an organism completes a generation (e.g., passes from egg to egg). The sub-life cycle test is shorter and usually emphasizes sensitive early life stages. In the life-history test, an organism progresses from egg or early developmental stage until death or cessation of reproduction. Duration of chronic toxicity studies ranges from 2% to 100% of an organism's life span. Most such tests are conducted on easily cultured animals such as amphipods, chironomids, and crayfish.

10.2.2.1. Test Endpoints

Chronic toxicity test endpoints typically involve long-term mortality and impairment of growth and reproduction. Other possible endpoints include:

changes in behavior (Buikema and Benfield, 1979); impairment of net-building activities (Petersen and Petersen, 1985); morphological deformities (Cushman, 1984); and impaired metamorphosis (Wilson and McGill, 1977) (see also Johnson et al., Chapter 4). With the possible exceptions of death and impaired growth and reproduction, the usefulness of most of the endpoints in predicting ecological effects remains unproven.

10.2.2.2. Use of Macroinvertebrates in Chronic Toxicity Testing

The development of chronic toxicity tests began after enactment of legislation to achieve pollution management goals. Requirements for chronic toxicity data are similar worldwide because few standardized test methods are available. Given the primary goal, to protect surface and ground waters from point-source pollution, short-term (4–7 d) chronic studies on algae, the crustacean *Ceriodaphnia,* and fathead minnows (Horning and Weber, 1985) are used in the United States. However, protection of water quality for the propagation of aquatic life requires the establishment of a long-term exposure criterion; yet the data are limited. For example, for zinc, a commonly tested metal, data were available for 13 chronic tests conducted on nine species (Larson and Hyland, 1987). One test used a caddisfly, four tests used *Daphnia magna,* and the remaining tests used fish. It is obvious that other commonly tested species of macroinvertebrates should have been included (e.g., no chronic data were available for a gammarid amphipod).

Problems also exist in interpreting chronic toxicity data. Many chronic tests have been conducted with animals that are held or reared, rather than cultured. Consequently, other than survival, it is not known how information on early life stages can be related to long-term effects on growth and reproduction, or how nutrition or stage of molt cycle affect an organism's sensitivity to pollutants. Furthermore, the ecological significance of impaired reproduction is not well-understood.

10.2.2.3. Future of Chronic Toxicity Tests

Chronic toxicity test data are used to derive a long-term exposure criterion for the protection of aquatic resources (e.g., Larson and Hyland, 1987). More chronic toxicity data are needed to protect aquatic resources, but this will not occur until methods have been developed for other macroinvertebrates. More species could be listed for use in chronic toxicity testing if basic biological and ecological information was available for them. Species that are amenable to culturing (e.g., gastropods, oligochaetes, and triclads) are especially valuable (Buikema and Benfield, 1979). For example, oligochaetes recently have been used for testing the sensitivity of detritivores to sediments (e.g., Schuytema et al., 1988) and to pure chemicals (Ewell et al., 1986).

10.2.3. Multispecies Toxicity Tests

The third type of toxicity test that uses macroinvertebrates is the multispecies test (Table 10.1). It is simply any test containing two or more species that interact (e.g., predator-prey interaction) (Cairns, 1985); these tests are more environmentally realistic than the other two basic types of tests. Multispecies tests are often natural subsets of a more complex ecosystem (e.g., a riffle community from a river). Two categories of multispecies tests exist: microcosms and mesocosms; microcosms will be emphasized in this section.

10.2.3.1. Definitions

The term *microcosm* has been used broadly to describe model or experimental ecosystems, from flask cultures to artificial streams and even whole lakes (Giddings, 1980; Giesy and Odum, 1980; Levin et al., 1984; Lundgren, 1985). However, the term should be restricted to small-scale systems in which colonization of the biotic assemblage either is completely controlled (i.e., *gnotobiotic*; e.g., Woltering, 1985; Taub et al., 1986) or is restricted to a subset of the complete natural assemblage found in the system that is being modeled (e.g., Maki, 1980; Giddings, 1981; Clements et al., 1989). *Mesocosms* are large, outdoor experimental systems that resemble natural ecosystems more closely than microcosms (see Section 10.3, "Mesocosms," below for details).

Microcosms can be situated either indoors or outdoors (typically indoors) and have a volume <10 m^3. The freshwater literature includes examples of benthic macroinvertebrate communities being modeled successfully in microcosms ranging from 0.1 to 0.6 m^3 (Maki, 1980; Hansen and Garton, 1982a,b; Lynch et al., 1982, 1985; Stephenson and Kane, 1984; Selby et al., 1985). Microcosms typically are not self-sustaining, so the duration of experiments usually ranges from several weeks to several months. Our concept of microcosm is synonymous with the "laboratory-scale microcosm" of Giesy and Odum (1980, p. 5), the "laboratory microcosm" of Boyle (1983, p. 407), and the "laboratory system" of Lundgren (1985, p. 159).

10.2.3.2. Single vs. Multispecies Tests

The usefulness of toxicity data for single species has been debated by many authors (Cairns, 1983, 1986; Odum, 1984; Kimball and Levin, 1985). It is unlikely that community effects can be predicted adequately from toxicity tests using single species. From a systems viewpoint, effects at the ecosystem level cannot be predicted from studies conducted on its component parts (Kimball and Levin, 1985; Maciorowski, 1988; however, see the arguments of Hansen and Garton, 1982a; Adams et al., 1983; and Slooff et al., 1986). Additionally, the impact of a chemical in the environment may

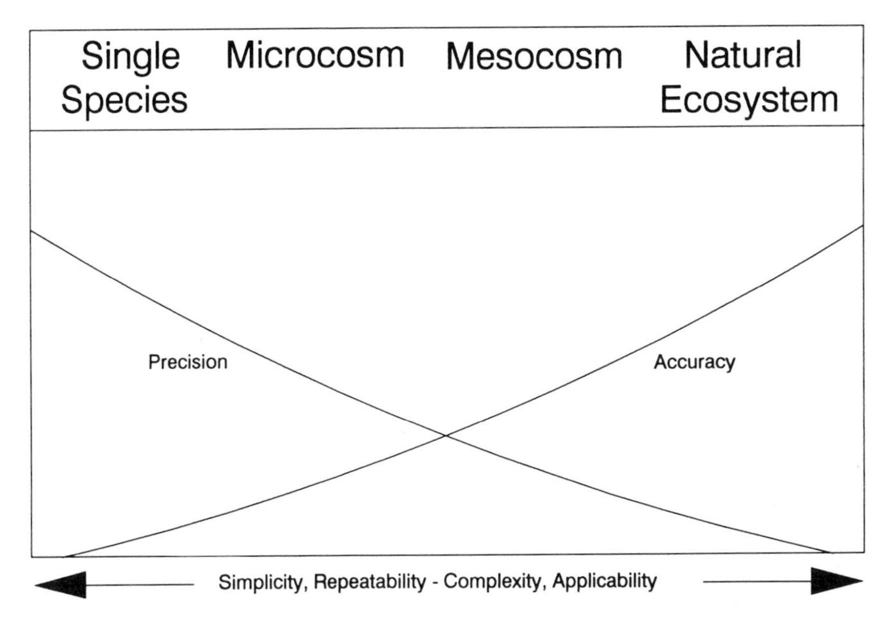

Figure 10.1. Characteristics of different levels of toxicity testing. Adapted from Boyle (1983), reproduced by permission of the American Society for Testing and Materials.

occur at lower concentrations than predicted from standardized single species tests (e.g., Dewey, 1986). To be predictive, toxicity tests must be more realistic environmentally.

10.2.3.3. Environmental Realism

The greatest realism possible is obtained by conducting experiments in natural, whole ecosystems (Fig. 10.1; see also Cooper and Barmuta, Chapter 11). However, use of whole ecosystems presents formidable problems for the following reasons: (1) replication is difficult; (2) experiments are subject to the vagaries of natural conditions and, hence, are difficult to control fully; (3) public access to manipulated bodies of water may be impeded; (4) such bodies of water may be contaminated by the substances that are to be tested; and (5) the complexity of natural ecosystems usually means higher costs for these studies. The use of multispecies tests mitigates some of these problems while being more environmentally realistic.

10.2.3.4. Test Endpoints

Multispecies tests allow the study of community dynamics under long-term exposure and recovery after episodic events. Endpoints for multispecies assemblages are summarized by Kreuger et al. (1988) and include: changes in taxa richness, species diversity, and density; shifts in dominant species; changes in predator-prey interactions (Clements et al., 1989); and shifts in functional feeding groups (e.g., Clements et al., 1988a; see also Resh and Jackson, Chapter 6). These tests can be useful for evaluating the impacts of chemicals on interspecies relationships (including trophic and predator-prey interactions) and system-level responses. This potential for greater environmental realism has encouraged much current research in ecotoxicology (Giesy, 1980; Hammons, 1981; Cairns, 1985).

Although multispecies tests increase environmental realism, the results should be used with caution; it still may be difficult to predict ecosystem-level effects (Slooff et al., 1986). Not all laboratory and field studies show good correlations. For example, Livingston (1988) compared the responses of leaf-pack macroinvertebrate communities to toxic wastes both in the laboratory and field. Local stream features, type of toxicant, the community index used, and duration of exposure to toxicants were important variables that influenced reliability of laboratory to field comparisons. The proportion of pollution-intolerant species present also affected the response of laboratory populations. Species richness was a robust indicator of stress effects in both test situations. A principle drawback of this study was the expense, which, however, was much lower than would be expected for a replicated mesocosm study. Additionally, multispecies tests may use organisms from different communities and these communities can respond differently to the same toxicant (Harass and Taub, 1985). Criteria for verification and extrapolation of results from multispecies tests have been proposed by Livingston and Meeter (1985), but more research is needed to validate these criteria.

10.2.3.5. Precision and Accuracy in Testing

It must be remembered that both microcosms and mesocosms are only *models* of ecosystems. Because of the different levels of complexity contained in these models, each has advantages and disadvantages for toxicity testing (Fig. 10.1). Microcosm tests are simpler and cheaper than mesocosm tests and are subject to greater control by the investigator, so they exhibit greater precision (i.e., the degree of mutual agreement among individual measurements). Precision certainly is important to those responsible for regulating toxic substances because if within-treatment variability is too great, the effects of a toxic substance will appear to be insignificant. However, having a high degree of precision does not mean that a test realistically predicts what the effect of a stressor will be in a natural ecosystem. Mesocosms offer the advantage of greater *accuracy* than microcosms (i.e., degree of agreement with the true value).

Therefore, multispecies tests provide a compromise between precision and accuracy for toxicity testing (Voshell, 1989). Reasonably realistic model ecosystems that can be controlled, manipulated, and replicated can simulate the responses of whole, natural ecosystems to contaminants (deNoyelles and Kettle, 1985; La Point and Perry, 1989).

10.2.3.6. Microcosm Tests in the Regulatory Process

Within the tier-test strategy (e.g., Urban and Cook, 1986), considerable interest has developed in using microcosm tests as intermediates between chronic single species tests and mesocosm tests because the latter are so expensive. However, microcosm studies may not provide information needed by regulatory agencies to refute the presumption of harm, nor will registrants obtain information needed to predict ecosystem-level effects. Unless microcosms can be shown to have a high degree of similarity with the ecosystems of interest, their predictive value is "dubious" (Heath, 1980, p. 346).

Microcosm tests are not currently an integral part of any risk assessment because of uncertainties in standardization, interpretation of results, and the position of such tests in tier testing (Loewengart and Maki, 1985; Mount, 1985). Microcosms may be useful for certain regulatory decisions, such as the registration of a new pesticide; the results could be used to determine if a more expensive mesocosm or field study is required. Microcosm tests need to be standardized if the results are used to rank chemicals, but standardization may limit the usefulness of microcosms because their design will change depending on the chemical in question and its mode of action (i.e., herbicides will require a different design than insecticides).

Regardless of the problems, Cairns (1983) concluded that because the objective is to protect communities and ecosystems, toxicity tests beyond single species must be considered. Currently, microcosm tests may be useful as screening tools at intermediate tiers of testing to precede use of the more expensive mesocosm or field studies.

10.2.3.7. Microcosm Tests as Higher-Tiered Toxicity Tests

The following discussion concentrates on microcosm test designs that use macroinvertebrates. Emphasis is on two major types of tests: (1) natural (or food-chain) assemblages of organisms; and (2) artificial (or gnotobiotic) assemblages.

10.2.3.7.1. Natural Assemblages

Natural assemblages are preferable to artificial ones because species interactions in the former are not assumed or fabricated. Interactions among species in natural assemblages theoretically are established among trophic levels (*sensu strictu* Merritt and Cummins, 1984). Examples of exposures of natural assemblages to toxicant stress include the studies of Burks and Wilhm (1977) and Clements et al. (1988a). In both experiments, artificial substrates were colonized by macroinvertebrates in a natural habitat, transferred to an artificial stream, and exposed to a toxicant for a defined period of time. Changes in total numbers, number of taxa, and community composition indicated community level responses to stress (e.g., Clements et al., 1988a). These studies were conducted in small-scale streams that probably were not as amenable to colonization as larger streams.

Maki (1980) studied the effect of the lampricide 3-trifluoromethyl-4-nitrophenol (TFM) on the structure and function of large (8 m × 0.6 m) indoor streams that received surface water, substrate, and biota from a nearby natural source. After an eight-week stabilization period, some of the streams were exposed to TFM; control and treated streams were observed for up to six months for deleterious effects and subsequent recovery. Numbers and drift rates of benthic organisms initially were reduced and then recovered. Measurements of diversity, taxa richness, and secondary production indicated that deleterious long-term effects did not occur. Results from these large laboratory microcosms accurately reflected effects observed in the field.

Other studies have examined population interactions. For example, Clements et al. (1989) noted that low concentrations of copper affected predator-prey interactions of the stonefly *Paragnetina media* and the caddisflies *Chimarra* sp. and *Hydropsyche morosa*; predation by the stonefly on *Hydropsyche* increased at 6 μg/l of copper. The authors speculated that this response may have resulted from the increased time that the caddisflies spent outside their retreats to repair nets. Predation on tube-building chironomids also increased as time outside the tube increased (Hershey, 1987). Such an impact could not have been predicted from single-species toxicity tests.

However, these test results should be interpreted carefully to avoid including artifacts of experimental design. For example, first, although Clements et al. (1989) provided evidence that the reduction in numbers of prey organisms probably resulted from increased predation, the prey organisms could not escape from the test system. In nature, increased drift rate (from predator avoidance) and mortality also would be factors in explaining changes in communities. Second, in a comparison of effects of heavy metals in field and experimental stream studies, Clements et al. (1988b) found that the total numbers of organisms and taxa decreased with exposure time in the experimental studies; that is, no invasion or reproduction occurred. This finding implied that the test system may have exerted its own stress.

10.2.3.7.2. Artificial (Gnotobiotic) Assemblages

Gnotobiotic microcosms typically are small-scale lentic systems. The Taub microcosm (Taub, 1985; Harass and Taub, 1985; Stay et al., 1988) is replicable and includes producers, decomposers, and omnivores; taxa such as daphnids, rotifers, amphipods, ostracods, and protozoans are represented. Changes in amphipod and ostracod abundance can be used to predict field responses, but scale may be a problem. For example, the capacity of ponds to recover from a 0.5 mg/l dose of atrazine could not be predicted from these kinds of microcosms (Stay et al., 1988).

However, Woltering's (1985) use of a gnotobiotic assemblage was more successful. He studied the effects of dieldrin on artificial assemblages of attached algae, organic sediments, snails, amphipods, and guppies over an 18-month period. The study design incorporated predator-prey and competitor populations and exploitation of the fish populations to imitate predation; test endpoints emphasized population-level variables such as biomass. In a standard, single-species toxicity test, guppy survival, growth, and fecundity were affected by a sublethal dose of dieldrin. When guppy populations were exposed to this sublethal level in a multispecies test, different results were obtained. With no exploitation, guppy populations recovered to a pre-exposure population structure even though dieldrin was still being introduced into the systems. At a 40% exploitation rate, the guppy population could not adapt to dieldrin and eventually became extinct while amphipod density increased. At 10% and 20% exploitation rates, intermediate shifts in fish and amphipod biomass were observed. Toxicant effects on these variables could not have been predicted from single-species tests.

10.3. Mesocosms

The current use and growing importance of mesocosms in toxicity testing requires that they be considered in detail here. The purpose of this section is to provide an up-to-date summary of their design, experimental use, and needs for future research.

10.3.1. Definitions

The term *mesocosm* was proposed by Banse (1982) for experimental systems larger than benchtop containers but smaller than, and isolated from, any subunit of the natural environment (Grice and Reeve, 1982). Mesocosms typically have a volume >10 m^3 (but see Cooper and Barmuta, Chapter 11). E.P. Odum defined these ". . . middle-sized worlds . . ." (1983, p. 68) as ". . . bounded and partially enclosed outdoor experimental setups . . . falling between laboratory microcosms and the large, complex, real world macrocosms" (Odum, 1984, p. 558). Mesocosms contain representative portions of complete natural assemblages found in systems that are being modeled, except perhaps for fish. Mesocosms should be capable energetically of supporting reproducing populations of fishes (Boyle, 1983), but this may be difficult to achieve in lotic situations (Giesy and Allred, 1985). Mesocosms can be either physical enclosures of a portion of a natural ecosystem or human-made structures such as ponds or stream channels. Experiments in them generally last for several months, but may extend for one or more years. Our concept of mesocosms is synonymous with the "large-scale microcosms" of Bowling et al. (1980, p. 224) and Giesy and Odum (1980, p. 8); the "experimental ecosystems" of Grice and Reeve (1982, p. 1); the "controlled ecosystems" of Parsons (1982, p. 411); the "semicontrolled artificial ecosystems" of Boyle (1983, p. 407); and the "large-scale model ecosystems" of Lundgren (1985, p. 157).

10.3.2. Physical Design

10.3.2.1. Types and Examples

Mesocosms can be categorized by their physical characteristics. Lotic mesocosms can be open (water passes through channels only once with no recirculation), partially closed (some water is released and some is recirculated), or closed (water is continuously recirculated with no release).

Lentic mesocosms are of two types: enclosures and experimental ponds. Enclosures using flexible synthetic material separate individual sections of a natural lake or large pond. They can be placed in the open-water zone by dropping walls from the surface to the bottom (limnocorrals), or they can include and extend out from the shoreline (littoral enclosures). Experimental ponds are permanent structures that are built either by excavation or erection of walls.

Examples of mesocosms that have been used to determine the effects of pollutants on benthic macroinvertebrates are described in Table 10.2. Almost all published studies using *lotic* mesocosms have included benthic macroinvertebrates. However, many studies using *lentic* mesocosms, particularly limnocorrals, have emphasized plankton and fish; these studies have not been included in Table 10.2. A similar lack of attention to macrobenthic organisms also was noted for marine studies using enclosures that extended from the surface to the bottom (Banse, 1982).

10.3.2.2. Environmental Realism and Safety

Each type of mesocosm has advantages and disadvantages that must be evaluated according to the objectives of a study and practical restrictions set by fiscal resources and environmental safety. Table 10.2 indicates that most lotic mesocosms have been designed as open systems, a desirable feature in terms of environmental realism, but an undesirable feature in terms of the release of toxic substances to the natural environment. In fact, an appropriate pollution discharge permit could be required for many such studies.

Table 10.2. Summary of physical characteristics of mesocosms in which benthic macroinvertebrates have been studied. Dimensions: L = long, W = wide, D = deep. (—) indicates that information was not provided by the authors.

Type/ Description	Location	Number of Units Available	Dimensions (m unless otherwise noted)	Volume (m^3)	Source of Water	Substratum	References
I: Lotic							
A. Open							
Cinderblock channels with PVC film liner	Channels Facility, Savannah River Ecology Laboratory, U.S. Environmental Protection Agency (USEPA), Aiken, SC	6	98.2 L, including 1 run 92 L × 0.6 W × 0.2 D and 2 pools 3.1 L × 1.5 W × 0.9 D	19**	Well	Washed quartz sand	Bowling et al. (1980), Giesy and Allred (1985)
Concrete channels	Browns Ferry Biothermal Research Station, Tennessee Valley Authority-USEPA, Athens, AL	12	112 L × 4.3 W, including 6 riffles 0.3 D and 6 pools 1.2 D	530	Reservoir	Riffles: 5–15 cm crushed limestone; pools: reservoir sediments	Armitage (1980), Rodgers (1980, 1982, 1983)
Man-made channels	Monticello Ecological Research Station, USEPA, Monticello, MN	8	520 L, including 8 riffles 30.5 L × 2.4 W × 0.1–0.2 D and 9 pools 30.5 L × 3.6 W × 0.8–0.9 D	568	River	Riffles: 2–5 cm gravel; pools: mud (fine sandy loam to coarse sand)	Nordlie and Arthur (1981), Arthur et al. (1982), Newman et al. (1987)
Man-made channels	Same as above	8	122 L, including 2 riffles and 1 pool as above	135**	River	Same as above	Arthur et al. (1983), Zischke et al. (1983, 1985), Hedtke and Arthur (1985)
Man-made channels	Same as above	8	245 L, including 4 riffles and 4 pools as above	270**	River	Same as above	Eaton et al. (1985)

(continued)

Table 10.2. (Continued)

Type/ Description	Location	Number of Units Available	Dimensions (m unless otherwise noted)	Volume (m^3)	Source of Water	Substratum	References
Divisions of natural channel	Sierra Nevada Aquatic Research Laboratory, University of California-Santa Barbara, Convict Creek, CA	4	340–500 L, including different numbers and sizes of riffles and pools	—	Stream	Coarse sand to pebbles and cobble	Leland et al. (1986, 1989)
Man-made channels	Troy Experimental Channels, Grande Ronde River, Troy, OR	2	62.3 L × 6.0 W, including riffles and runs	—	River	River gravel	Corrarino and Brusven (1983)
B. Closed							
4 man-made channels modified into 2 recycling stream systems	Monticello Ecological Research Station, USEPA, Monticello, MN	2	1000 L, including 16 riffles and 18 pools as above	1500	River or well	Same as above	Cooper and Stout (1982, 1985), Stout and Cooper (1983)
C. Partially closed							
Streams with Hypalon liner	National Fisheries Contaminant Research Center, U.S. Fish and Wildlife Service (USFWS), Columbia, MO	3	50 L, including 3 riffles 10 L × 1 W × 0.1 D and 2 pools 10 L × 2 W × 0.3 D	15**	Well	15 cm river gravel	Fairchild et al. (1987)
II: Lentic							
A. Littoral enclosures							
Scrimweve walls on 3 sides	Environmental Resarch Laboratory, USEPA, Duluth, MN	12	10 L × 5 W × 1.1 D (maximum)	25**	Ponds	Sediment, macrophytes	Brazner et al. (1988), Siefert et al. (1989)
B. Experimental ponds							
4 concrete walls	Imperial Chemical Industry Plant Protection Division, Berkshire, UK	8*	5 L × 5 W × 1 D	25**	—	Sediment, macrophytes	Hill (1985)

2 concrete walls, 2 earth sides	Shell Research Ltd., Kent, UK	12	10 L × 5 W × 1 D	40	—	Sediment, macrophytes	Crossland (1984), Crossland and Bennett (1984), Crossland and Hillaby (1985), Crossland and Wolff (1985)
Earthen dikes on hardpan clay base	Cornell University, Ithaca, NY	50	0.07 ha surface × 1.3 D (maximum)	650	Excavated upland swamp through trickle filter to holding canal	Sediment, macrophytes	Hall et al. (1970)
Unlined excavations	University of California-Riverside, Riverside, CA	12	17 L × 8 W × 0.2–0.3 D	30–40**	Irrigation canal	Sediment, macrophytes	Hurlbert et al. (1970, 1972)
Earthen	National Fisheries Contaminant Research Center, USFWS, Columbia, MO	6*	0.04 ha surface × 2.0 D (maximum)	450**	—	Sediment, macrophytes	Macek et al. (1972)
Earthen	Same as above	12*	0.08 ha surface × 1.5 D (maximum)	744	Well	Sediment, macrophytes	Boyle (1980), Mauck et al. (1976)
Earthen	Same as above	14*	0.08 ha surface × 2.0 D (maximum)	800**	—	Sediment, macrophytes	Boyle et al. (1985)
Excavations	Aquatic and Vector Control Research Facility, University of California-Riverside, Oasis, CA	40*	0.004 ha surface × 0.3 D	10**	Well	Sediment, macrophytes	Mulla and Darwazeh (1976), Mulla et al. (1978, 1982)
Lined excavations	Oak Ridge National Laboratory, Oak Ridge, TN	8	0.002–0.003 ha surface × 0.7–0.8 D	11–14	Pond	Sediment, macrophytes	Cushman and Goyert (1984)
Excavations lined with Hypalon	Same as above	15	5 L × 5 W × 0.8–0.9 D	15	Pond	Sediment, macrophytes	Giddings et al. (1984)

(continued)

Table 10.2. (Continued)

Type/ Description	Location	Number of Units Available	Dimensions (m unless otherwise noted)	Volume (m^3)	Source of Water	Substratum	References
Excavations	Nelson Environmental Study Area, University of Kansas, Lawrence, KS	14	0.045 ha surface × 2.1 D (maximum)	500**	Reservoir and well	Sediment, macrophytes	deNoyelles et al. (1982, 1989), Dewey (1986), Kettle et al. (1987)
Excavations lined with polyethylene	Fifty Point Conservation Area, University of Guelph, Winona, ON	6	20 L × 10 W × 2.5 D	500**	Pond	Sediment, macrophytes	Stephenson and Mackie (1986)
Undescribed	Central Florida Research and Education Center, University of Florida, Sanford, FL	15*	6 L × 4 W × 0.45 D	11**	Well	—	Ali and Kok-Yokomi (1989)
Excavations lined with clay and topsoil	Southern Piedmont Agricultural Experiment Station, Virginia Polytechnic Institute and State University, Blackstone, VA	12	20.1 L × 20.1 W × 2.1 D (maximum)	517	Municipal, well, and reservoir	Sediment, emergent macrophytes	Layton (1989), Layton and Voshell (1991)
Diked marshes	Grizzly Island Wildlife Area, California Department of Fish and Game and University of California-Berkeley, Solano County, CA	12	22 L × 11 W × 0.6 D	145**	Canal	Pickleweed marsh	Batzer and Resh (1989)

*Probable number of units available; incomplete information provided by authors.
**Volume estimated from dimensions.

Properly constructed lentic mesocosms do not present as much of a problem with environmental safety because the test substance can be contained until it degrades. However, enclosure-type lentic mesocosms raise some important questions about environmental realism, particularly in long-term experiments. The enclosure of a part of a natural ecosystem also means the exclosure of other parts (Lundgren, 1985). For example, fish that are top predators are likely to be excluded, or at least underrepresented, and this could result in different community structure of lower trophic levels. Nutrients also are prevented from entering enclosures, possibly resulting in a shift towards oligotrophy with time. Nutrient recycling dynamics also may be altered, depending on nutrient input and predator-prey dynamics (Lehman, 1980), and algal and zooplankton communities may change as a result.

10.3.2.3. Wall Effects

Another potential problem with enclosures is that the sidewalls quickly become colonized with an Aufwuchs community that may achieve greater biomass than either the plankton or natural Aufwuchs. This could lead to an exceptionally high proportion of grazing benthic macroinvertebrates within an enclosure. Sidewall effects also can occur on sides of lotic channels and walls of some experimental ponds. The great amount of biological activity that can occur on walls forced Rodgers (1980, 1982) to include channel walls as a habitat for sampling. This is a somewhat dubious procedure, given that achieving environmental realism is a major reason for conducting a mesocosm study. Walls also must be taken into account when calculating effective concentrations of toxicants, and eventual fate of those substances within the enclosures, because some chemicals have a tendency to be adsorbed onto walls.

Sidewall effects can be minimized by making the ratio of water volume to wall area as great as possible. Lotic mesocosms can be constructed in a wide shallow profile with almost all of the channel covered by some type of natural substrate. Littoral enclosures will have at least one side with a natural boundary, and experimental ponds that require erected walls can be constructed with two sides having natural boundaries. Excavated experimental ponds do not have the complications of sidewall effects.

10.3.2.4. Size

This is an important consideration for all types of mesocosms. A test system must be large enough to support the biota of interest, but not so large as to be cost-prohibitive or unmanageable. The purpose of a mesocosm, however, is to model the whole ecosystem of which benthic macroinvertebrates are a part, and this requires that fish be present (Boyle, 1983). Recommendations for sizes of experimental systems necessary to maintain populations of fish range from 100–1,000 m^3 (Parsons, 1982; Uhlmann, 1985). Table 10.2 lists mesocosm studies that were conducted in systems ranging in size from 10–744 m^3. Some studies cited did not provide sufficient details to determine if populations of fish were supported, but reproduction of endemic fathead minnows was reported in littoral enclosures with volumes of 27–41 m^3 (Brazner et al., 1988; Siefert et al., 1989). Those littoral enclosures also supported young green sunfish that were introduced for a two-month study. Lotic mesocosms may have to be scaled larger to support fish populations. Experimental stream channels with a volume of $\approx$19 m^3 were incapable of supporting fish or crayfish (Bowling et al., 1980; Giesy and Allred, 1985). When crayfish were introduced into those channels in sufficient numbers for subsequent periodic sampling, they eliminated macrophytes and mussels.

10.3.2.5. Water Sources

The source of water is an important consideration for mesocosms used for toxicity testing. The large volumes required for some mesocosms, especially the open lotic type, make it necessary to use surface waters from nearby streams or reservoirs (Table 10.2). Wells can be used to supply water for some smaller lotic mesocosms and experimental ponds. The use of surface waters increases the likelihood of complications from other pollutants, but other characteristics of water quality and suspended particulates will approximate natural conditions more closely than using ground waters. However, in one type of experimental pond, a permanent reservoir of unpolluted natural surface waters was maintained with ground waters (Dewey, 1986; deNoyelles et al., 1989). In any type of mesocosm, water quality should be monitored very closely to ensure that it sustains aquatic life and is not a factor in the outcome of the toxicity tests (ASTM, 1988). For example, in one facility, it was necessary to treat the water from a deep well to remove CO_2 and to add O_2, nutrients, and hydrated lime so that inorganic water quality resembled that of surface water in the vicinity (Bowling et al., 1980).

10.3.2.6. Bottom Type

The substratum has been referred to as "... the stage upon which the drama of aquatic insect ecology is acted out" (Minshall, 1984, p. 358); hence, it will be a key component of any mesocosm (Table 10.2). In enclosure-type lentic mesocosms, natural sediment and perhaps macrophytes already will be present (Brazner et al., 1988; Siefert et al., 1989). A natural substratum will develop in excavated lentic mesocosms as a result of decomposition by aquatic microbes, deposition of dead plankton, and immigration of macrophytes. The process can be hastened by using sediments from an existing body of water and planting macrophytes (Cushman and Goyert, 1984;

Giddings et al., 1984), both of which also would serve to increase the homogeneity of individual units for benthic macroinvertebrates.

Most lotic mesocosms probably are intended to model lower-order streams that have riffle-pool sequences. Riffle habitat has been simulated (Table 10.2) by placing rocks with diameters from 2–15 cm ("pebble" and "cobble," according to the Wentworth classification of substratum particle sizes; Minshall, 1984); pool habitat has been simulated with a substratum consisting of a mixture of sand and silt. The materials used to construct riffles and pools should come from natural bodies of water to assure rapid colonization of the substratum by algae and microbes. However, the large size of some lotic mesocosms may make this logistically impossible. For example, in the 530-m^3 lotic mesocosms studied by Rodgers (1980, 1982, 1983), substratum for the pools came from a reservoir, but crushed limestone was used for the riffles. Some of the studies listed in Table 10.2 did not provide many details about the substratum, but toxicity tests should not be conducted in relatively new mesocosms unless the substratum has been transferred from mature environments or it has been well-colonized by an Aufwuchs community.

10.3.3. Experimental Design

Benthic macroinvertebrates have been studied in mesocosms to determine the effects of a wide variety of pollutants; the most frequent of these are insecticides and herbicides (Table 10.3). The purpose of this section is to review various aspects of the experimental designs that have been used, including statistical considerations, sampling, and measurements.

10.3.3.1. Replication

The number of replicate mesocosms assigned per treatment has ranged from one to four (Table 10.3). However, only two studies used four replicates, whereas one-fourth, one-third, and one-third of the studies used three, two, and one replicates, respectively. Studies in lotic mesocosms have used fewer replicates per treatment than those conducted in lentic mesocosms; perhaps lotic facilities have fewer separate units available for replication because the greater complexity necessary in constructing and operating some experimental streams makes them more expensive than their lentic counterparts (Warren and Davis, 1971). An alarming proportion of toxicity tests in lotic mesocosms have used only one replicate mesocosm per treatment. If the design of a toxicity test consists of only one mesocosm receiving each treatment (either a specified dose of the pollutant or a control), then the mesocosms are being subsampled rather than being replicated. This approach has been termed "pseudoreplication" (Hurlbert, 1984), and any statistical analysis ". . . must be viewed cautiously, at best" (Allan, 1984, p. 498; see also Cooper and Barmuta, Chapter 11). Among replicated meso-

Table 10.3. Summary of experimental designs for mesocosm studies involving benthic macroinvertebrates. Treatments: C = control, D = dose levels, R = replicates of each; sampling methods: RAS = representative artificial substrate, SAS = standardized artificial substrate; endpoints: div. = diversity, even. = evenness, rich. = taxa richness, tot. den. = total density, tot. bio. = total standing stock biomass, ind. den. = density of individual taxa, ind. bio. = standing stock biomass of individual taxa, ind. prod. = production of individual taxa, pres. abs. = presence or absence of taxa.

Type	Perturbation	Treatments/ Total Experimental Units	Duration	Sampling Frequency	Sampling Methods	Endpoints	References
I: Lotic	Cadmium	C, 2D, 2R/6	19 mo	1 mo	Multiplate SAS	Div., even., tot. den.	Bowling et al. (1980), Giesy and Allred (1985)
	Copper	C, 3D, 1R/4	5 yr	1 mo–1 yr	Portable invertebrate box sampler, drift net	Div., rich., tot. den., ind. den., ind. bio., ind. prod., functional feeding groups	Leland et al. (1986, 1989)
	Elevated temperature	C, 3D, 3R/12	4 mo	1 mo	Core, rock-tray RAS, scraper	Tot. den., tot. bio., ind. den., ind. bio., life cycles	Rodgers (1980)
	Elevated temperature	C, 3D, 3R/12	10 mo	1 mo	Same as above	Ind. den., ind. prod.	Rodgers (1982)
	Elevated temperature	C, 3D, 3R/12	4 mo	8 times*	Emergence trap, hand net	Fecundity	Rodgers (1983)
	Elevated temperature	C, 1D, 1R/2	5 mo	1 wk	Emergence trap	Ind. den., phenology	Nordlie and Arthur (1981)
	Elevated temperature	C, 1D, 1R/2	9 mo	1 mo	Substrate-filled RAS	Tot. den., ind. den., life cycle	Arthur et al. (1982)
	Acidification	C, 2D, 1R/3	17 wk	2 wk benthos, 1 wk emergence	Substrate-filled RAS, Ekman grab, drift net, emergence trap	Div., tot. den., ind. den.	Zischke et al. (1983)
	Diazinon	C, 2D, 1R/3	18 wk	2 wk	Substrate-filled RAS, drift net, emergence trap	Div., tot. den., ind. den.	Arthur et al. (1983)

	Chlorpyrifos	C, 3D, 1R/4	15 wk	2 wk	Substrate-filled RAS, drift net	Tot. den., ind. den.	Eaton et al. (1985)
	Pentachlorophenol	C, 3D, 1R/4	5 mo	1–2 wk	Substrate-filled RAS, drift net, emergence trap	Tot. den., ind. den., fecundity	Zischke et al. (1985), Hedtke and Arthur (1985)
	Chlorine	C, 3D, 1–2R/6	11 d	twice (4 d, 11 d)	Litter bags	Ind. den., leaf decomposition rate	Newman et al. (1987)
	p-Cresol	C, 1D, 1R/2	2 mo	1 wk	Artificial substrate*, macrophyte-clump RAS	Div., even., rich., tot. den., tot. bio., ind. den., ind. bio., life cycle	Cooper and Stout (1982, 1985)
	p-Cresol	C, 1D, 1R/2	10 d	once (10 d)	Leaf packs	Div., even., rich., tot. den., tot. bio., ind. den., leaf decomposition rate	Stout and Cooper (1983)
	Triphenyl-phosphate, sediment	C, 2D, 1R/3	7 mo	1 wk–1 mo	Hess sampler, drift net, emergence trap	Div., rich., tot. den.	Fairchild et al. (1987)
	Reduced discharge	C, 1D, 1R/2	2 mo	4 times drift, 4 times benthos	Drift net, Hess sampler	Tot. den., ind. den.	Corrarino and Brusven (1983)
II: Lentic	Nutrients, predation	C, 6D, 2–3R/20	3 yr	1 wk benthos, 2 per wk emergence (June–Sept.)	Ekman grab, emergence trap	Tot. bio., ind. den., ind. bio., ind. prod., life cycles	Hall et al. (1970)
	Chlorpyrifos	C, 4D, 2R/10	4 mo	1–14 d	Net on sled	Ind. den.	Hurlbert et al. (1970)
	Chlorpyrifos	C, 4D, 2R/10	3 mo	1–12 d	Net on sled	Ind. den., functional feeding groups	Hurlbert et al. (1972)
	Chlorpyrifos	C, 2D, 2R/6	2 mo	1–21 d	Multiplate SAS, emergence trap	Ind. den.	Macek et al. (1972)
	Chlorpyrifos	C, 3D, 1–2R/12	1 yr	4 d–9 mo	Multiplate SAS	Rich., tot. den., ind. den.	Brazner et al. (1988), Siefert et al. (1989)
	2 synthetic pyrethroids, 5 organophosphates	C, 2–4D, 2R/6–10	14 d	2–7 d	Mosquito dipper	Ind. den.	Mulla and Darwazeh (1976)
	6 synthetic pyrethroids	C, 3–4D, 2R/8–10	21 d	2–7 d	Mosquito dipper, mud scoop*	Ind. den.	Mulla et al. (1978)
	4 formulations of *Bacillus thuringiensis* var. *israelensis*	C, 2–4D, 2R/6–10	14 d	2–7 d	Mosquito dipper	Ind. den.	Mulla et al. (1982)

(*continued*)

Table 10.3. (Continued)

Type	Perturbation	Treatments/ Total Experimental Units	Duration	Sampling Frequency	Sampling Methods	Endpoints	References
	Methylparathion	C, 1D, 3R/6	4 mo	5 d–1 mo	Sweep net, air lift	Tot. den., ind. den.	Crossland (1984)
	Cypermethrin	C, 1D, 2R/4	1 yr	1 wk–2 mo	Benthic and surface artificial substrates*, emergence trap	Ind. den.	Hill (1985)
	Cypermethrin	C, 3D, 2R/8	1 yr	1 wk–2 mo	Same as above	Ind. den.	Hill (1985)
	2 benzoylphenylurea analogs	C, 1–3D, 3R/6–12	50 d	2–13 d	Zooplankton net, dip net	Ind. den.	Ali and Kok-Yokomi (1989)
	2,4-Dichlorophenoxy acetic acid (2,4-D)	C, 2D, 4R/12	5 mo	1 wk	Ekman grab	Tot. bio.	Boyle (1980)
	2,4-Dichlorophenoxy acetic acid (2,4-D)	C, 2D, 2R/6	10 mo	1–7 mo	Corer	Div., pres. abs., habitat, habits, functional feeding groups, pollution tolerance categories	Stephenson and Mackie (1986)
	Atrazine	C, 3D, 2R/8	5 mo	2 d–6 wk	Emergence trap	Div., even., rich., ind. den.	Dewey (1986)
	Pentachlorophenol	C, 1D, 3R/6	3 mo	2 wk sweep net, 3 times air lift (1–6 wk)	Sweep net, air lift	Tot. den., ind. den., fecundity	Crossland and Wolff (1985)
	Synthetic crude oil	C, 3D, 2R/8	11 mo	2 per mo	Ekman grab	Div., ind. den., ind. bio.	Cushman and Goyert (1984)
	Coal-derived oil	C, 5D, 2R/12	3 mo	1 wk	Emergence trap	Ind. den.	Giddings et al. (1984)
	3,4-Dichloroaniline	C, 2D, 3R/9	1 mo	once (1 mo)	Sweep net, air lift	Div., even., rich., ind. den.	Crossland and Hillaby (1985)
	Fluorene	C, 5D, 1–3R/14	5 mo	1 wk	Emergence trap	Div., rich., tot. den.	Boyle et al. (1985)

*Incomplete information provided by authors.

cosm toxicity tests, most have used only two replicates. These yield only one degree of freedom for statistical analyses, which increases the likelihood of a Type II error (i.e., concluding that no differences existed between dosed and control mesocosms when differences actually occurred; see discussion of Type I and Type II errors in Resh and McElravy, Chapter 5, Norris and Georges, Chapter 7, and Cooper and Barmuta, Chapter 11).

10.3.3.2. Duration

Most studies of benthic macroinvertebrates in both lotic and lentic mesocosms have lasted from three to six months; few studies were conducted for a year or longer (Table 10.3). The classical investigations by Hall et al. (1970) on the effects of nutrients and predation on community structure and production continued for three years in the Cornell experimental ponds (see Reice and Wohlenberg, Chapter 8), but benthic macroinvertebrates were followed only for about four months in each year. Most of the toxicity tests that have lasted for less than a month have been conducted primarily to determine the effectiveness of various insecticides for control of mosquito larvae, and effects on nontarget taxa have received only superficial consideration (e.g., Mulla and Darwazeh, 1976; Mulla et al., 1978, 1982).

Mesocosm toxicity tests should be designed to last at least long enough for the organisms of interest (usually those ecologically dominant) to complete a generation. However, the presence of fish may be the overriding factor that determines the appropriate length of a mesocosm toxicity test. Parsons (1982) indicated that studies of micronekton and nonreproducing fish in marine systems could extend from about one month to one year, but that mesocosm studies should continue for at least one year if reproducing fish populations are included. The following reasons indicate why mesocosm tests that include macroinvertebrates should last for at least one year if they hope to achieve a high degree of environmental realism:

1. Almost all benthic macroinvertebrates would pass through a complete life cycle. For semivoltine organisms (e.g., some dragonflies), one cohort would reach the adult stage and another cohort would begin its life cycle.
2. Taxa would not be omitted from study because of seasonal life cycles (e.g., winter stoneflies).
3. The biota would be exposed to a complete annual cycle of environmental factors such as temperature, light, and food, which serve as essential physiological and behavioral regulating mechanisms.
4. Ecosystem function (e.g., autotrophy vs. heterotrophy) is only apparent in temporal units of at least one year.

5. Toxic substances, such as heavy metals and pesticides, may persist for at least one complete annual cycle.

If studies last one year or longer, great care must be taken to ensure that mesocosms do not diverge significantly from the systems being modeled.

10.3.3.3. Frequency of Sampling

In most of the studies reported in Table 10.3, benthic macroinvertebrates have been sampled at intervals of one to two weeks. Apparently, scientists conducting mesocosm toxicity tests sample more frequently than scientists engaged in field evaluations of toxic substances (Voshell et al., 1989). In an experiment designed to follow the effects of a toxic substance on benthic macroinvertebrates, time intervals between sampling must be shorter than the life cycles of the organisms in the community. Chironomidae often exhibit the greatest diversity and abundance of any taxon in both lotic and lentic communities (Coffman and Ferrington, 1984); members of this family can complete their life cycles in one to three weeks under field conditions (Benke et al., 1984) and in as little as five days in laboratory studies (Mackey, 1977). In addition, some mayflies require only 19 days to complete development under field conditions (Benke and Jacobi, 1986). Frequent sampling also is required because the effective toxicity of some substances is brief; intervals will depend on the toxicant used.

10.3.3.4. Sampling Methods

The same sampling problems are encountered in mesocosms as in any field study of benthic macroinvertebrates; quantitative sampling is difficult because of the diversity of habits and habitats of the organisms. Table 10.3 indicates that four major categories of methods have been used about equally: artificial substrates, emergence traps, constant area samplers (core, grab, air lift), and devices that are moved through an unspecified amount of habitat to trap organisms (dipper, scoop, net). In addition, minor use of drift nets has been reported in lotic mesocosms.

Artificial substrates are used most frequently in lotic mesocosms. The reduction of variability by providing a constant habitat (Rosenberg and Resh, 1982) makes it easier to discern changes in the community resulting from treatment and is an advantage of these devices for mesocosm toxicity testing. A possible disadvantage is selective colonization by relatively few of the taxa that inhabit the mesocosms. This concern about using artificial substrates in mesocosms does not appear to have been considered in any of the studies listed in Table 10.3. Representative artificial substrates (RAS), which are containers filled with natural materials from the substratum, are likely to give more realistic results than can be obtained by using standardized artificial substrates (SAS), which are constructed of man-made materials (Rosenberg and Resh, 1982). Hardboard multiplate samplers (often referred to as Hester-Dendy samplers) are the most frequently used SASs.

Investigations in lentic mesocosms have relied mostly upon various devices that are moved through the water, mud, or vegetation to capture organisms. Results from this type of sampling are reported as catch per unit effort; sampling effort must be standardized carefully to assure valid comparisons of results between treatments. However, sampling methods that give results in relation to a constant area (e.g., grabs, corers, air-lifts, or RASs) may be better for comparing different treatments in mesocosm toxicity tests because greater standardization is assured. Emergence traps have been used extensively, particularly in lentic mesocosms, and in some cases have been the only sampling method. Emergence trapping can be quantitative, but it may not collect all of the taxa inhabiting a mesocosm (e.g., oligochaetes, molluscs, crustaceans, and insects such as stoneflies, odonates, and some mayflies that crawl on solid objects projecting out of the water to emerge).

Useful, innovative, sampling methods for benthic macroinvertebrates have been developed in response to needs for studying basic aquatic ecology (Merritt et al., 1984), but such methods have not been adopted in areas of applied aquatic ecology such as hazard assessment, although the same needs exist (Voshell et al., 1989). For example, at least three types of enclosures with nets (Hess, portable invertebrate box, T-sampler) would be effective for sampling riffle areas of lotic mesocosms, but the Hess sampler has been used only twice and the portable invertebrate box sampler only once in these habitats (Table 10.3). Available quantitative methods for using individual natural stones as sampling units, rather than net devices that would be too destructive, also appear to have been ignored in lotic mesocosms (Voshell et al., 1989). Finally, several devices would be better quantitative samplers for benthic macroinvertebrates on aquatic macrophytes than the ones listed in Table 10.3 (see Merritt et al., 1984).

10.3.3.5. Endpoints

Variables concerning community structure of benthic macroinvertebrates (diversity, taxa richness, total density, densities of individual taxa) have been reported most often in mesocosm toxicity tests (Table 10.3); however, functional variables such as fecundity, production, functional feeding groups, grazing rates, and leaf decomposition have received only scant attention. Standing stock biomass, both for the total community and individual taxa, is considered rarely but is a potentially useful measurement for analyzing structure or function.

10.3.4. Overview of Results

Mesocosm toxicity tests are needed to confirm environmental safety, but they are time-consuming and expensive. Therefore, it is prudent to consider what information is generated by their use that could not be obtained by conducting simpler tests (La Point and Perry, 1989). Because of the greater environmental realism provided by the mesocosm approach, several studies have reported results that would not have been expected based upon acute or chronic single species tests (e.g., Cooper and Stout, 1985; Crossland and Wolff, 1985; Eaton et al., 1985; Hedtke and Arthur, 1985; Dewey, 1986).

10.3.4.1. Secondary Effects

Toxic substances may exert either primary or secondary ecological effects (Hurlbert, 1975). Primary effects are caused by direct toxicological action on growth, survival, or reproduction, irrespective of the mechanisms of exposure or time intervals between exposure and expression of the ultimate effect. Secondary effects are ecosystem changes that may result from primary effects. A large number of secondary effects result from the complexity of ecosystems and the interrelatedness of their components. For example, changes could occur in predation, grazing, competition, energy flow, food and habitat availability, nutrient dynamics, and water quality. The most significant contributions of mesocosm studies have involved the identification of secondary effects (La Point and Perry, 1989); those involving benthic macroinvertebrates are summarized in Table 10.4.

10.3.4.2. Fate

The ability to analyze the fate of toxic substances in relation to ecological effects is another advantage of mesocosm toxicity tests over simpler laboratory tests. Effects of a toxic substance largely depend on the actual exposure that organisms receive in a natural environment. For example, the organophosphorus insecticide chlorpyrifos dissipated from the water in pond mesocosms within 24 hours, but it accumulated in the sediment in appreciable amounts for up to 14 days (Hurlbert et al., 1970, 1972). These authors concluded that benthic organisms would be at greater risk than nektonic or planktonic organisms. In contrast, the emergence of aquatic insects from pond mesocosms dosed with fluorene was unaffected even at the highest concentration of 10 mg/l (cf. the 48-hour EC50 for *Chironomus riparius* of 2.35 mg/l) (Boyle et al., 1985). However, fluorene never accumulated in the sediment, and it dissipated from the water before the majority of emergence took place.

10.3.4.3. Recovery

Documentation of recovery (resilience stability *sensu* Odum, 1983) is a third advantage of using mesocosm toxicity tests instead of laboratory tests (e.g., Hurlbert et al., 1970, 1972; Crossland, 1984; Giddings et al., 1984;

Table 10.4. Primary and secondary ecological effects involving benthic macroinvertebrates that have been reported in mesocosm toxicity tests.

Toxic Substance	Primary Effects	Secondary Effects	References
Insecticide	Lower numbers of macroinvertebrates	Lower fish growth because of less food	Crossland (1984), Brazner et al. (1988), Siefert et al. (1989)
Insecticide	Lower numbers of herbivorous macroinvertebrates and zooplankton	Blooms of algae because of less grazing	Hurlbert et al. (1972), Crossland (1984)
Insecticide	Lower numbers of predaceous macroinvertebrates and zooplankton	Eventual increase in abundance of some species of zooplankton because of less predation	Hurlbert et al. (1972), Crossland (1984)
Herbicide	Lower biomass of periphyton and macrophytes	Lower taxa richness and numbers of macroinvertebrates because of less food (periphyton, macrophytes) and habitat (macrophytes); lower fish growth because of less food (macro-invertebrates)	Dewey (1986), deNoyelles et al. (1989)
Herbicide	Lower biomass of macrophytes	Lower macroinvetebrate diversity because of reduced habitat complexity	Stephenson and Mackie (1986)
Coal-derived oil	Lower biomass of macrophytes	Higher numbers and biomass of one chironomid species because of increased detritus for food	Cushman and Goyert (1984)
Coal gasification by-product	Lower photosynthesis by filamentous algae	Lower numbers and biomass of macroinvertebrates because of low dissolved oxygen	Stout and Cooper (1983), Cooper and Stout (1985)
Metal	Lower numbers of crayfish in pools	Higher biomass of macrophytes in pools because of less grazing; lower biomass of Aufwuchs and macrophytes in flowing segments because of nutrient uptake in pools	Bowling et al. (1980), Giesy and Allred (1985)

Cooper and Stout, 1985; Hill, 1985; Brazner et al., 1988). Functional redundancy among ecosystem components enhances stability by giving ecosystems the capacity to recover from perturbation over time (Odum, 1983). The length of time required for recovery is of obvious interest to risk managers and only can be investigated in relatively natural, self-sustaining systems such as mesocosms. Most studies have reported recovery of macroinvertebrates within several weeks of episodic exposure(s) to low doses of various toxic substances (Hurlbert et al., 1970, 1972; Cooper and Stout, 1985; Brazner et al., 1988). For example, some recovery of chironomids occurred after 16 days in littoral enclosures treated with a low dose of chlorpyrifos, presumably because the enclosures could be recolonized readily by ovipositing adults from the adjacent, open pond (Brazner et al., 1988).

Mesocosms also can be used to simulate exposure of ecosystems to episodic pulses of chemicals at expected environmental concentrations (Touart, 1988; Touart and Slimak, 1989; see also Cooper and Barmuta, Chapter 11). Pulse-dosing is used rarely in laboratory toxicity studies, which may account for some of the problems in predicting ecosystem effects from laboratory data (Clark et al., 1986a,b). In some cases, episodic high doses of toxic substances, repeated episodic exposures, or repeated doses at lower concentrations have disrupted phytoplankton, zooplankton, insects, or fish; these disruptions lasted for more than a year (Giddings et al., 1984; Hill, 1985).

10.3.4.4. Long-Term Effects

Long-term effects of low-level pollution can be demonstrated in mesocosm studies (Lundgren, 1985). Although single-species acute tests remain the most frequently used toxicity tests, organisms may be exposed naturally to lower concentrations of toxic substances, and over longer time periods, than generally used in standard laboratory tests. Mesocosm tests would, therefore, provide more accurate information for risk managers. For example, a three-month exposure to coal-derived oil revealed that emergence of midges was reduced at only 3% of the 48-hour LC50 for *Chironomus tentans* (Giddings et al., 1984). Other mesocosm studies also have reported greater direct effects of a toxic substance than would have been predicted from laboratory acute tests alone (Arthur et al., 1983; Cushman and Goyert, 1984; Siefert et al., 1989).

10.3.5. Unresolved Issues

The advantages of mesocosm toxicity testing have been demonstrated, but their use is relatively new, and some areas will require further investigation. Five such areas are considered next.

10.3.5.1. Colonization and Succession

Ecosystems undergo changes in structure and function with time (i.e., ecosystem development or ecological succession); they tend to be more susceptible to perturbations during early stages of succession because they have less functional redundancy (Odum, 1983). Nevertheless, in only one lotic mesocosm toxicity test did an author mention that colonization was investigated prior to treatment (Rodgers, 1980, 1982, 1983); no other reports of successional dynamics are available. A similar situation exists for lentic mesocosms. In a study by Layton and Voshell (1991), typical lentic insects readily colonized 0.04 ha experimental ponds in the first year after construction, but nonflying macroinvertebrates were not well-established by the end of the first year. Clearly, the dynamics of colonization and succession need to be studied so that results of mesocosm toxicity tests can be interpreted accurately.

10.3.5.2. Environmental Realism

Future research should focus on examining realism and accuracy of mesocosms (Giesy and Allred, 1985). It has been suggested that experimental ponds are appropriate for mesocosm toxicity tests because they simulate shallow lakes, riverine embayments, and backwaters (Giddings and Franco, 1985; Touart, 1988; Touart and Slimak, 1989). Lotic mesocosms at the Monticello Ecological Research Station in Minnesota have been described as ". . . physically, chemically and biologically typical of warm-water, grassland, third- to fifth-order streams" (Cooper and Stout, 1985, p. 96); however, comparisons of these systems with natural bodies of water have not been reported, so their power to reproduce environmental effects has not been validated.

Mesocosms may have to be managed to obtain environmental realism. For example, two distinctive characteristics of low-order, lotic ecosystems are shading and allochthonous detritus inputs from riparian vegetation. Fairchild et al. (1987) simulated these features by placing shading structures over their lotic mesocosms from May to October and by distributing measured amounts of sugar maple leaves to each one for three weeks in November and one week in March.

10.3.5.3. Precision

A preliminary study is required in field research to determine the number of samples necessary to obtain a desired precision of estimates (Green, 1979; see also Resh and McElravy, Chapter 5, and Cooper and Barmuta, Chapter 11); however, these sorts of studies have never been done for mesocosm toxicity tests. Based upon published studies (Table 10.3), it appears that only a few replicates (one to four) are taken per treatment, probably because of the complexity of mesocosms (Fig. 10.1) and the resultant high costs of constructing and sampling them. For studies of benthic macroinvertebrates, a precision of ±40% often is considered reasonable (i.e., 95% confidence that the sample mean is within ±40% of the true population mean; Elliott, 1977). By analyzing 12 new, untreated pond mesocosms, Layton (1989) determined that three replicate ponds would provide reasonable precision for total density, but five or more replicates were necessary to determine densities of individual, dominant taxa. Using only two replicates would make it impossible to detect anything less than catastrophic changes in densities (100%–300%). Community structural measures such as number of taxa, diversity, and evenness could be estimated much more precisely (within 5%–15%), even by using only two replicates (Layton, 1989). This author also found that the variability of measures decreased the longer the ponds were allowed to colonize.

If the objectives of a toxicity study require greater precision than can possibly be achieved with mesocosms, and if the objectives do not place a high priority on accuracy, then a logical alternative would be to use microcosms (Giesy and Allred, 1985). However, environmental realism will be sacrificed in the process (Fig. 10.1).

10.3.5.4. Structure vs. Function

Some functional measurements should be included in all mesocosm toxicity tests because mesocosms are intended to mimic ecosystems. Although summary community metabolic measurements such as the photosynthesis/respiration ratio have been investigated in mesocosms (Giddings et al., 1984; Giesy and Allred, 1985), functional measurements of benthic macroinvertebrates have had relatively little attention (Table 10.3; Wallace, 1989). Many of the secondary ecological effects involving benthic macroinvertebrates listed in Table 10.4 could be more closely related to functional measurements, such as production, rather than structural measurements.

10.3.5.5. Integration of Effects

Although many secondary effects of toxic substances have been reported in mesocosm studies (Table 10.4), interactions between them have not been well-documented. It is assumed that simultaneous changes are correlated, but the appropriate empirical data to establish cause and effect relationships are missing. For example, reduced fish growth may occur simultaneously with a reduction of benthic macroinvertebrate populations. However, the impact may have been a direct effect on fish growth caused by long-term exposure to low levels of the toxic substance rather than an indirect one through diet. Much effort and expense go into mesocosm tests to make them quantitative, yet the most significant conclusions remain highly subjective.

Sufficient information exists on secondary effects of toxicants (Hurlbert, 1975) to develop *a priori* hypotheses and then test them by collecting appropriate data in mesocosm studies. Because many of the secondary effects that might involve benthic macroinvertebrates are related to trophic dynamics, quantitative analysis of the trophic basis of production (Benke, 1984) could be the approach to use. This involves measuring secondary production and determining the sources and eventual fates of the materials that support the observed level of production. Pathway analysis also may be useful for elucidating interactions that occur in mesocosm toxicity tests (Johnson et al., 1991). It is possible that mesocosms may be too complex for elucidating some cause and effect relationships (La Point and Perry, 1989); if so, it may be necessary to conduct studies in simpler microcosms (Bowling et al., 1980).

10.4. Conclusions

Toxicity research has been carried on for the last 100 years, but most of the activity connected with it occurred in the 1970s and resulted from changes in environmental legislation. However, the development of new toxicity test methods and the use of new organisms has been slow. This void will continue to exist because protection of the environment has centered around using existing, standardized tests, or refined versions of them, in order to rank effects of chemicals and effluents. However, environmental protection also requires protection of ecosystem integrity and maintenance of natural food webs. Collection of information on a range of species and subsequent use of these data in a decision-making process has not occurred in the United States, except possibly in the development of water quality criteria and standards.

Protection of ecosystem integrity will require an increase in the number of macroinvertebrate species used in toxicity tests. However, selection of test species and test methods will depend upon the management goals of regulatory agencies. If the objective is to regulate discharges or rank the toxicity of effluents or compounds, then almost any species, cultured or collected, can be used in acute tests. If the objective is to regulate discharges over time, then a species that can be cultured probably will be most useful because a constant supply of organisms of the same size or age can be obtained. If the objective is to understand chronic effects on growth or reproduction, to evaluate effects on different life stages, or to determine factors that modify toxicant sensitivity, then new test methods are needed because few macroinvertebrate species can be cultured in the laboratory. If the objective is to predict environmental effects, then not only are new methods needed but also much more information on the ecological roles and trophic interactions of dominant species is needed as well; this last objective has been the most difficult to achieve.

The results of Mayer and Ellersieck (1986) should be used as a guide to stimulate research. Although their database was limited, their survey indicated a global need to develop acute toxicity test methods for baetid and burrowing mayflies, caddisflies, and stoneflies. These organisms are sensitive to pollution and are good indicators of ecosystem health (e.g., Hilsenhoff, 1987), so they should be useful for calling attention to environmental problems. The development of water quality criteria in the United States demonstrates the need for additional test methods and the usefulness of using such information.

Initially, macroinvertebrates should be used in acute studies even if indigenous species have to be collected and acclimated to test conditions. Investigations should be directed toward identification of sensitive organisms or life stages and pertinent ecological variables that will modify organism sensitivity; the process should begin by using life-history information. In turn, this information will be useful for culturing test animals and developing defensible chronic test methods. Subsequently, this information also may provide a mechanism to rank potential ecosystem sensitivities to chemicals.

Despite all of the studies listed in Tables 10.2 and 10.3, the number of toxic substances that have been tested in mesocosms is a small fraction of those that could cause some degree of environmental damage. The complexity of mesocosm tests probably precludes their routine use (Bowling et al., 1980), but they will be required in *risk assessment* for pollutants, given the following criteria: (1) a major environmental concern exists because of widespread exposure and likely effects based upon laboratory studies; (2) chemical persistence and fate concerns are indicated from laboratory and field-monitoring studies; and (3) exposure and effects data from mesocosm studies can be used to determine principles of ecosystem-level effects and, thereby, yield predictive capabilities. Pesticides probably will be the focus of much of the mesocosm testing done in the near future (Touart, 1988; Touart and Slimak, 1989).

In the final analysis, the most important reason for the slow development of new toxicity test methods and slow adoption of new test organisms is the neglect that basic ecologists have shown to the applied discipline of environmental toxicology (Buikema and Benfield, 1979). Basic ecology will benefit from seeking answers to practical questions (Slobodkin, 1988), and a unified multidisciplinary approach that includes innovations from basic ecology is needed to address the impacts of toxicants on ecosystems (Maciorowski, 1988). By introducing toxic substances into realistic, replicable, model ecosystems, aquatic ecologists and environmental toxicologists can work together to test theories, develop explanations, and simultaneously address practical regulatory concerns that will provide immediate benefits to society.

Acknowledgments

We are grateful to V.D. Christman, R.J. Layton, and S.W. Hiner for their assistance with the preparation of this manuscript. We are also grateful to various personnel at the USEPA for their conversations and help in determining data requirements under various aspects of the U.S. Federal Code. A.L. Buikema, Jr. was supported in part on an assignment with the Office of Pesticide Programs, USEPA, under the Intergovernmental Personnel Act. J.R. Voshell, Jr. was supported by research funds from the Virginia Agricultural Experiment Station.

References

Adams, W.J., R.A. Kimerle, B.B. Heidolph, and P.R. Michael. 1983. Field comparison of laboratory-derived acute and chronic toxicity data. In *Aquatic Toxicology and Hazard Assessment: Sixth Symposium,* eds. W.E. Bishop, R.D. Cardwell, and B.B. Heidolph, pp. 367–85. American Society for Testing and Materials Special Technical Publication 802. American Society for Testing and Materials, Philadelphia, PA.

Ali, A. and M.L. Kok-Yokomi. 1989. Field studies on the impact of a new benzoylphenylurea insect growth regulator (UC-84572) on selected aquatic nontarget invertebrates. *Bulletin of Environmental Contamination and Toxicology* 42:134–41.

Allan, J.D. 1984. Hypothesis testing in ecological studies of aquatic insects. In *The Ecology of Aquatic Insects,* eds. V.H. Resh and D.M. Rosenberg, pp. 484–507. Praeger Pubs., New York.

Anderson, B.G. 1980. Aquatic invertebrates in tolerance investigations from Aristotle to Naumann. In *Aquatic Invertebrate Bioassays,* eds. A.L. Buikema, Jr. and J. Cairns, Jr., pp. 3–35. American Society for Testing and Materials Special Technical Publication 715. American Society for Testing and Materials, Philadelphia, PA.

Anderson, R.L. 1980. Chironomidae toxicity tests—biological background and procedures. In *Aquatic Invertebrate Bioassays,* eds. A.L. Buikema, Jr. and J. Cairns, Jr., pp. 70–80. American Society for Testing and Materials Special Technical Publication 715. American Society for Testing and Materials, Philadelphia, PA.

APHA (American Public Health Association). 1985. *Standard Methods for the Examination of Water and Waste Water,* 16th ed. American Public Health Association, Washington, DC.

Armitage, B.J. 1980. Effects of temperature on periphyton biomass and community composition in the Browns Ferry experimental channels. In *Microcosms in Ecological Research,* ed. J.P. Giesy, Jr., pp. 668–83. Symposium Series 52, Technical Information Center, U.S. Department of Energy, Washington, DC. (Available as Conference-781101, National Technical Information Service, Springfield, VA.)

Arthur, J.W., J.A. Zischke, K.N. Allen, and R.O. Hermanutz. 1983. Effects of diazinon on macroinvertebrates and insect emergence in outdoor experimental channels. *Aquatic Toxicology* 4:283–302.

Arthur, J.W., J.A. Zischke, and G.L. Ericksen. 1982. Effect of elevated water temperature on macroinvertebrate communities in outdoor experimental channels. *Water Research* 16:1465–77.

ASTM (American Society for Testing and Materials). 1988. *Annual Book of ASTM Standards*. Section 11, Water and Environmental Technology. Vol. 11.04, Pesticides; resource recovery; hazardous substances and oil spill responses; waste disposal; biological effects. American Society for Testing and Materials, Philadelphia, PA.

Baker, J.M. and G.B. Crapp. 1974. Toxicity tests for predicting the ecological effects of oil and emulsifier pollution on littoral communities. In *Ecological Aspects of Toxicity Testing of Oils and Dispersants,* eds. L.R. Beynon and E.B. Cowell, pp. 23–40. John Wiley, New York.

Banse, K. 1982. Experimental marine ecosystem enclosures in a historical perspective. In *Marine Mesocosms: Biological and Chemical Research in Experimental Ecosystems,* eds. G.D. Grice and M.R. Reeve, pp. 11–24. Springer-Verlag, New York.

Batzer, D.P. and V.H. Resh. 1989. Waterfowl management and mosquito production in diked salt marshes: preliminary considerations and mesocosm design. *Proceedings of the California Mosquito and Vector Control Association* 56:153–7.

Benke, A.C. 1984. Secondary production of aquatic insects. In *The Ecology of Aquatic Insects,* eds. V.H. Resh and D.M. Rosenberg, pp. 289–322. Praeger Pubs., New York.

Benke, A.C. and D.I. Jacobi. 1986. Growth rates of mayflies in a subtropical river and their implications for secondary production. *Journal of the North American Benthological Society* 5:107–14.

Benke, A.C., T.C. Van Arsdall, Jr., D.M. Gillespie, and F.K. Parrish. 1984. Invertebrate productivity in a subtropical blackwater river: the importance of habitat and life history. *Ecological Monographs* 54:25–63.

Birge, W.J. and J.A. Black. 1982. Statement on surrogate species clusters concept. In *Surrogate Species Workshop,* Anonymous, pp. A6–5 to A6–7. Report TR–507–36B. U.S. Environmental Protection Agency, Washington, DC.

Blanck, H., G. Wallin, and S.-Å. Wängberg. 1984. Species-dependent variation in algal sensitivity to chemical compounds. *Ecotoxicology and Environmental Safety* 8:339–51.

Bowling, J.W., J.P. Giesy, H.J. Kania, and R.L. Knight. 1980. Large-scale microcosms for assessing fates and effects of trace contaminants. In *Microcosms in Ecological Research,* ed. J.P. Giesy, Jr., pp. 224–47. Symposium Series 52, Technical Information Center, U.S. Department of Energy, Washington, DC. (Available as Conference-781101, National Technical Information Service, Springfield, VA.)

Boyle, T.P. 1980. Effects of the aquatic herbicide 2,4-D DMA on the ecology of experimental ponds. *Environmental Pollution* (Series A) 21:35–49.

Boyle, T.P. 1983. Role and application of semicontrolled ecosystem research in the assessment of environmental contaminants. In *Aquatic Toxicology and Hazard Assessment: Sixth Symposium,* eds. W.E. Bishop, R.D. Cardwell, and B.B. Heidolph, pp. 406–13. American Society for Testing and Materials Special Technical Publication 802. American Society for Testing and Materials, Philadelphia, PA.

Boyle, T.P., S.E. Finger, R.L. Paulson, and C.F. Rabeni. 1985. Comparison of laboratory and field assessment of fluorene—Part II: Effects on the ecological structure and function of experimental pond ecosystems. In *Validation and Predictability of Laboratory Methods for Assessing the Fate and Effects of Contaminants in Aquatic Ecosystems,* ed. T.P. Boyle, pp. 134–51. American Society for Testing and Materials Special Technical Publication 865. American Society for Testing and Materials, Philadelphia, PA.

Brazner, J.C., S.J. Lozano, M.L. Knuth, S.L. Bertelsen, L.J. Heinis, D.A. Jensen, E.R. Kline, S.L. O'Halloran, K.W. Sargent, D.K. Tanner, and R.E. Siefert. 1988. *The Effects of Chlorpyrifos on a Natural Aquatic System: a Research Design for Littoral Enclosure Studies and Final Research Report.* U.S. Environmental Protection Agency, Duluth, MN.

Brungs, W.A. and D.I. Mount. 1978. Introduction to a discussion of the use of aquatic toxicity tests for evaluation of the effects of toxic substances. In *Estimating the Hazard of Chemical Substances to Aquatic Life,* eds. J. Cairns, Jr., K.L. Dickson, and A.W. Maki, pp. 15–26. American Society for Testing and Materials Special Technical Publication 657. American Society for Testing and Materials, Philadelphia, PA.

Buikema, A.L., Jr. 1983. Variation in static acute toxicity test results with *Daphnia magna* exposed to refinery effluents and reference toxicants. *Oil and Petrochemical Pollution* 1:189–98.

Buikema, A.L., Jr. and E.F. Benfield. 1979. Use of macroinvertebrate life history information in toxicity tests. *Journal of the Fisheries Research Board of Canada* 36:321–8.

Buikema, A.L., Jr., B.R. Niederlehner, and J. Cairns, Jr. 1982. Biological monitoring. Part IV—Toxicity testing. *Water Research* 16:239–62.

Burks, S.L. and J.L. Wilhm. 1977. Bioassays with a natural assemblage of benthic macroinvertebrates. In *Aquatic Toxicology and Hazard Evaluation,* eds. F.L. Mayer and J.L. Hamelink, pp. 127–36. American Society for Testing and Materials Special Technical Publication 634. American Society for Testing and Materials, Philadelphia, PA.

Cairns, J., Jr. 1957. Environment and time in fish toxicity. *Industrial Wastes* 2:1–5.

Cairns, J., Jr. 1983. Are single species toxicity tests alone adequate for estimating environmental hazard? *Hydrobiologia* 100:47–57.

Cairns, J., Jr., ed. 1985. *Multispecies Toxicity Testing*. Pergamon Press, New York.

Cairns, J., Jr. 1986. The myth of the most sensitive species. *BioScience* 36:670–2.

Cairns, M.A., A.V. Nebeker, J.H. Gakstatter, and W.L. Griffis. 1984. Toxicity of copper-spiked sediments to freshwater invertebrates. *Environmental Toxicology and Chemistry* 3:435–46.

Chapman, P.M. 1986. Sediment quality criteria from the sediment quality triad: an example. *Environmental Toxicology and Chemistry* 5:957–64.

Clark, J.R., P.W. Borthwick, L.R. Goodman, J.M. Patrick, Jr., E.M. Lores, and J.C. Moore. 1986a. Comparison of laboratory toxicity test results with responses of estuarine animals exposed to fenthion in the field. *Environmental Toxicology and Chemistry* 6:151–60.

Clark, J.R., P.W. Borthwick, L.R. Goodman, J.M. Patrick, Jr., E.M. Lores, and J.C. Moore. 1986b. Field and laboratory toxicity tests with shrimp, mysids and sheepshead minnows exposed to fenthion. In *Aquatic Toxicology and Environmental Fate: 9th Volume,* eds. T.M. Poston and R. Purdy, pp. 161–76. American Society for Testing and Materials Special Technical Publication 921. American Society for Testing and Materials, Philadelphia, PA.

Clements, W.H., D.S. Cherry, and J. Cairns, Jr. 1988a. Structural alterations in aquatic insect communities exposed to copper in laboratory streams. *Environmental Toxicology and Chemistry* 7:715–22.

Clements, W.H., D.S. Cherry, and J. Cairns, Jr. 1988b. Impact of heavy metals on insect communities in streams: a comparison of observational and experimental results. *Canadian Journal of Fisheries and Aquatic Sciences* 45:2017–25.

Clements, W.H., D.S. Cherry, and J. Cairns, Jr. 1989. The influence of copper exposure on predator-prey interactions in aquatic insect communities. *Freshwater Biology* 21:483–8.

Coffman, W.P. and L.C. Ferrington, Jr. 1984. Chironomidae. In *An Introduction to the Aquatic Insects of North America,* 2nd ed., eds. R.W. Merritt and K.W. Cummins, pp. 551–652. Kendall/Hunt Publishing, Dubuque, IA.

Cooper, W.E. and R.J. Stout. 1982. Assessment of transport and fate of toxic materials in an experimental stream ecosystem. In *Modeling the Fate of Chemicals in the Aquatic Environment,* eds. K.L. Dickson, A.W. Maki, and J. Cairns, Jr., pp. 347–78. Ann Arbor Science Pubs., Ann Arbor, MI.

Cooper, W.E. and R.J. Stout. 1985. The Monticello experiment: a case study. In *Multispecies Toxicity Testing,* ed. J. Cairns, Jr., pp. 96–116. Pergamon Press, New York.

Corrarino, C.A. and M.A. Brusven. 1983. The effects of reduced stream discharge on insect drift and stranding of near shore insects. *Freshwater Invertebrate Biology* 2:88–98.

Crossland, N.O. 1984. Fate and biological effects of methyl parathion in outdoor ponds and laboratory aquaria. II. Effects. *Ecotoxicology and Environmental Safety* 8:482–95.

Crossland, N.O. and D. Bennett. 1984. Fate and biological effects of methyl parathion in outdoor ponds and laboratory aquaria. I. Fate. *Ecotoxicology and Environmental Safety* 8:471–81.

Crossland, N.O. and J.M. Hillaby. 1985. Fate and effects of 3,4-dichloroaniline in the laboratory and in outdoor ponds: II. Chronic toxicity to *Daphnia* spp. and other invertebrates. *Environmental Toxicology and Chemistry* 4:489–99.

Crossland, N.O. and C.J.M. Wolff. 1985. Fate and biological effects of pentachlorophenol in outdoor ponds. *Environmental Toxicology and Chemistry* 4:73–86.

Cushman, R.M. 1984. Chironomid deformities as indicators of pollution from a synthetic, coal-derived oil. *Freshwater Biology* 14:179–82.

Cushman, R.M. and J.C. Goyert. 1984. Effects of a synthetic crude oil on pond benthic insects. *Environmental Pollution* (Series A) 33:163–86.

Davis, J.C. 1977. Standardization and protocols of bioassays—their role and significance for monitoring, research and regulatory usage. In *Proceedings of the 3rd Aquatic Toxicity Workshop, Halifax, NS, November 2–3, 1976,* eds. W.R. Parker, E. Pessah, P.G. Wells, and G.F. Westlake, pp. 1–14. Surveillance Report EPS–5–AR–77–1. Environmental Protection Service, Environment Canada, Halifax, N.S.

deNoyelles, F., Jr. and W.D. Kettle. 1985. Experimental ponds for evaluating bioassay predictions. In *Validation and Predictability of Laboratory Methods for Assessing the Fate and Effects of Contaminants in Aquatic Ecosystems,* ed. T.P. Boyle, pp. 91–103. American Society for Testing and Materials Special Technical Publication 865. American Society for Testing and Materials, Philadelphia, PA.

deNoyelles, F., Jr., W.D. Kettle, C.H. Fromm, M.F. Moffett, and S.L. Dewey. 1989. Use of experimental ponds to assess the effects of a pesticide on the aquatic environment. In *Using Mesocosms to Assess the Aquatic Ecological Risk of Pesticides: Theory and Practice,* ed. J.R. Voshell, Jr., pp. 41–56. Miscellaneous Publication No. 75, Entomological Society of America, Lanham, MD.

deNoyelles, F., W.D. Kettle, and D.E. Sinn. 1982. The responses of plankton communities in experimental ponds to atrazine, the most heavily used pesticide in the United States. *Ecology* 63:1285–93.

Dewey, S.L. 1986. Effects of the herbicide atrazine on aquatic insect community structure and emergence. *Ecology* 67:148–62.

Doudoroff, P., B.G. Anderson, G.E. Burdick, P.S. Galtsoff, W.B. Hart, R. Patrick, E.R. Strong, E.W. Surber, and W.M. Van Horn. 1951. Bio-assay methods for the evaluation of acute toxicity of industrial wastes to fish. *Sewage and Industrial Wastes* 23:1380–97.

Eaton, J., J. Arthur, R. Hermanutz, R. Kiefer, L. Mueller, R. Anderson, R. Erickson, B. Nordling, J. Rogers, and H. Pritchard. 1985. Biological effects of continuous and intermittent dosing of outdoor experimental streams with chlorpyrifos. In *Aquatic Toxicology and Hazard Assessment: Eighth Symposium,* eds. R.C. Bahner and D.J. Hansen, pp. 85–118. American Society for Testing and Materials Special Technical Publication 891. American Society for Testing and Materials, Philadelphia, PA.

Elliott, J.M. 1977. Some methods for the statistical analysis of samples of benthic macroinvertebrates, 2nd ed. *Freshwater Biological Association Scientific Publication* No. 25:1–156.

Ewell, W.S., J.W. Gorsuch, R.O. Kringle, K.A. Robillard, and R.C. Spiegel. 1986. Simultaneous evaluation of the acute effects of chemicals on seven aquatic species. *Environmental Toxicology and Chemistry* 5:831–40.

Fairchild, J.F., T. Boyle, W.R. English, and C. Rabeni. 1987. Effects of sediment and contaminated sediment on structural and functional components of experimental stream ecosystems. *Water, Air, and Soil Pollution* 36:271–93.

Fremling, C.R. and W.L. Mauck. 1980. Methods for using nymphs of burrowing mayflies (Ephemeroptera, *Hexagenia*) as toxicity test organisms. In *Aquatic Invertebrate Bioassays,* eds. A.L. Buikema, Jr. and J. Cairns, Jr., pp. 81–97. American Society for Testing and Materials Special Technical Publication 715. American Society for Testing and Materials, Philadelphia, PA.

Giddings, J.M. 1980. Types of aquatic microcosms and their research applications. In *Microcosms in Ecological Research,* ed. J.P. Giesy, Jr., pp. 248–66. Symposium Series 52, Technical Information Center, U.S. Department of Energy, Washington, DC. (Available as Conference-781101, National Technical Information Service, Springfield, VA.)

Giddings, J.M. 1981. Laboratory tests for chemical effects on aquatic population interactions and ecosystem properties. In *Methods for Ecological Toxicology: a Critical Review of Laboratory Multispecies Tests,* ed. A.S. Hammons, pp. 23–92. Ann Arbor Science Pubs., Ann Arbor, MI.

Giddings, J.M. and P.J. Franco. 1985. Calibration of laboratory bioassays with results from microcosms and ponds. In *Validation and Predictability of Laboratory Methods for Assessing the Fate and Effects of Contaminants in Aquatic Ecosystems,* ed. T.P. Boyle, pp. 104–19. American Society for Testing and Materials Special Technical Publication 865. American Society for Testing and Materials, Philadelphia, PA.

Giddings, J.M., P.J. Franco, R.M. Cushman, L.A. Hook, G.R. Southworth, and A.J. Stewart. 1984. Effects of chronic exposure to coal-derived oil on freshwater ecosystems: II. Experimental ponds. *Environmental Toxicology and Chemistry* 3:465–88.

Giesy, J.P., Jr., ed. 1980. *Microcosms in Ecological Research.* Symposium Series 52, Technical Information Center, U.S. Department of Energy, Washington, DC. (Available as Conference-781101, National Technical Information Service, Springfield, VA.)

Giesy, J.P. and P.M. Allred. 1985. Replicability of aquatic multispecies test systems. In *Multispecies Toxicity Testing,* ed. J. Cairns, Jr., pp. 187–247. Pergamon Press, New York.

Giesy, J.P., R.L. Graney, J.L. Newsted, C.J. Rosiu, A. Benda, R.G. Kreis, Jr., and F.J. Horvath. 1988. Comparison of three sediment bioassay methods using Detroit River sediments. *Environmental Toxicology and Chemistry* 7:483–98.

Giesy, J.P., Jr. and E.P. Odum. 1980. Microcosmology: introductory comments. In *Microcosms in Ecological Research,* ed. J.P. Giesy, Jr., pp. 1–13. Symposium Series 52, Technical Information Center, U.S. Department of Energy, Washington, DC. (Available as Conference-781101, National Technical Information Service, Springfield, VA.)

Green, R.H. 1979. *Sampling Design and Statistical Methods for Environmental Biologists.* John Wiley, New York.

Grice, G.D. and M.R. Reeve. 1982. Introduction and description of experimental ecosystems. In *Marine Mesocosms: Biological and Chemical Research in Experimental Ecosystems,* eds. G.D. Grice and M.R. Reeve, pp. 1–9. Springer-Verlag, New York.

Grothe, D.R. and R.A. Kimerle. 1985. Inter- and intralaboratory variability in *Daphnia magna* effluent toxicity test results. *Environmental Toxicology and Chemistry* 4:189–92.

Hall, D.J., W.E. Cooper, and E.E. Werner. 1970. An experimental approach to the production dynamics and structure of freshwater animal communities. *Limnology and Oceanography* 15:839–928.

Hammons, A.S., ed. 1981. *Methods for Ecological Toxicology: a Critical Review of Laboratory Multispecies Tests*. Ann Arbor Science Pubs., Ann Arbor, MI.

Hansen, S.R. and R.R. Garton. 1982a. The effects of diflubenzuron on a complex laboratory stream community. *Archives of Environmental Contamination and Toxicology* 11:1–10.

Hansen, S.R. and R.R. Garton. 1982b. Ability of standard toxicity tests to predict the effects of the insecticide diflubenzuron on laboratory stream communities. *Canadian Journal of Fisheries and Aquatic Sciences* 39:1273–88.

Harass, M.C. and F.B. Taub. 1985. Comparison of laboratory microcosms and field responses to copper. In *Validation and Predictability of Laboratory Methods for Assessing the Fate and Effects of Contaminants in Aquatic Ecosystems,* ed. T.P. Boyle, pp. 57–74. American Society for Testing and Materials Special Technical Publication 865. American Society for Testing and Materials, Philadelphia, PA.

Hart, W.B., P. Doudoroff, and J. Greenbank. 1945. *The Evaluation of the Toxicity of Industrial Wastes, Chemicals and Other Substances to Fresh-Water Fishes*. Waste Control Laboratory, Atlantic Refining Co., Philadelphia, PA.

Heath, R.T. 1980. Are microcosms useful for ecosystem analysis? In *Microcosms in Ecological Research,* ed. J.P. Giesy, Jr., pp. 333–47. Symposium Series 52, Technical Information Center, U.S. Department of Energy, Washington, DC. (Available as Conference-781101, National Technical Information Service, Springfield, VA.)

Hedtke, S.F. and J.W. Arthur. 1985. Evaluation of a site-specific water quality criterion for pentachlorophenol using outdoor experimental streams. In *Aquatic Toxicology and Hazard Assessment: Seventh Symposium,* eds. R.D. Cardwell, R. Purdy, and R.C. Bahner, pp. 551–64. American Society for Testing and Materials Special Technical Publication 854. American Society for Testing and Materials, Philadelphia, PA.

Henderson, C. 1957. Application factors to be applied to bioassays for the safe disposal of toxic wastes. In *Biological Problems in Water Pollution,* ed. C.M. Tarzwell, pp. 31–7. Public Health Service, U.S. Department of Health, Education and Welfare, Washington, DC.

Hershey, A.E. 1987. Tubes and foraging behavior in larval Chironomidae: implications for predator avoidance. *Oecologia* (Berlin) 73:236–41.

Hill, I.R. 1985. Effects on non-target organisms in terrestrial and aquatic environments. In *The Pyrethroid Insecticides,* ed. J.P. Leahey, pp. 151–262. Taylor and Francis, Philadelphia, PA.

Hilsenhoff, W.L. 1987. An improved biotic index of organic stream pollution. *Great Lakes Entomologist* 20:31–9.

Horning, W.B., II. and C.I. Weber. 1985. *Short-Term Methods for Estimating the Chronic Toxicity of Effluents and Receiving Waters to Freshwater Organisms*. EPA/600/4–85/014. U.S. Environmental Protection Agency, Cincinnati, OH.

Hurlbert, S.H. 1975. Secondary effects of pesticides on aquatic ecosystems. *Residue Reviews* 57:81–148.
Hurlbert, S.H. 1984. Pseudoreplication and the design of ecological field experiments. *Ecological Monographs* 54:187–211.
Hurlbert, S.H., M.S. Mulla, J.O. Keith, W.E. Westlake, and M.E. Dusch. 1970. Biological effects and persistence of Dursban in freshwater ponds. *Journal of Economic Entomology* 63:43–52.
Hurlbert, S.H., M.S. Mulla, and H.R. Willson. 1972. Effects of an organophosphorus insecticide on the phytoplankton, zooplankton, and insect populations of fresh-water ponds. *Ecological Monographs* 42:269–99.
Jenkins, D.G., R.J. Layton, and A.L. Buikema, Jr. 1989. State of the art in aquatic ecological risk assessment. In *Using Mesocosms to Assess the Aquatic Ecological Risk of Pesticides: Theory and Practice,* ed. J.R. Voshell, Jr., pp. 18–32. Miscellaneous Publication No. 75, Entomological Society of America, Lanham, MD.
Johnson, M.L., D.G. Huggins, and F. deNoyelles, Jr. 1991. Ecosystem modeling with LISREL: a new approach for measuring direct and indirect effects. *Ecological Applications.* 1:383–98.
Kettle, W.D., F. deNoyelles, Jr., B.D. Heacock, and A.M. Kadoum. 1987. Diet and reproductive success of bluegill recovered from experimental ponds treated with atrazine. *Bulletin of Environmental Contamination and Toxicology* 38:47–52.
Kimball, K.D. and S.A. Levin. 1985. Limitations of laboratory bioassays: the need for ecosystem-level testing. *BioScience* 35:165–71.
Krueger, H.O., J.P. Ward, and S.H. Anderson. 1988. *A Resource Manager's Guide for Using Aquatic Organisms to Assess Water Quality for Evaluation of Contaminants.* Biological Report 88 (20). Fish and Wildlife Service, U.S. Department of Interior, Washington, DC.
La Point, T.W. and J.A. Perry. 1989. Use of experimental ecosystems in regulatory decision making. *Environmental Management* 13:539–44.
Larson, L. and J. Hyland. 1987. *Ambient Aquatic Life Water Quality Criteria for Zinc—1987.* EPA/440/5–87/003. U.S. Environmental Protection Agency, Washington, DC.
Layton, R.J. 1989. Macroinvertebrate colonization and production in new experimental ponds. Ph.D. dissertation, Virginia Polytechnic Institute and State Univ., Blacksburg, VA.
Layton, R.J. and J.R. Voshell, Jr. 1991. Colonization of new experimental ponds by benthic macroinvertebrates. *Environmental Entomology* 20:110–7.
LeBlanc, G.A. 1984. Interspecies relationships in acute toxicity of chemicals to aquatic organisms. *Environmental Toxicology and Chemistry* 3:47–60.
Lehman, J.T. 1980. Nutrient recycling as an interface between algae and grazers in freshwater communities. In *Evolution and Ecology of Zooplankton Communities,* ed. W.C. Kerfoot, pp. 251–63. Special Symposium Vol. 3, American Society of Limnology and Oceanography. Univ. Press of New England, Hanover, NH.
Leland, H.V., S.V. Fend, J.L. Carter, and A.D. Mahood. 1986. Composition and abundance of periphyton and aquatic insects in a Sierra Nevada, California, stream. *Great Basin Naturalist* 46:595–611.

Leland, H.V., S.V. Fend, T.L. Dudley, and J.L. Carter. 1989. Effects of copper on species composition of benthic insects in a Sierra Nevada, California, stream. *Freshwater Biology* 21:163–79.

Levin, S.A., K.D. Kimball, W.H. McDowell, and S.F. Kimball, eds. 1984. New perspectives in ecotoxicology. *Environmental Management* 8:375–442.

Little, G. 1968. Induced winter breeding and larval development in the shrimp, *Palaemonetes pugio* Holthuis (Caridea, Palaemonidae). *Crustaceana Supplement* 2:19–26.

Livingston, R.J. 1988. Use of freshwater macroinvertebrate microcosms in the impact evaluation of toxic wastes. In *Functional Testing of Aquatic Biota for Estimating Hazards of Chemicals,* eds. J. Cairns, Jr. and J.R. Pratt, pp. 166–218. American Society for Testing and Materials Special Technical Publication 988. American Society for Testing and Materials, Philadelphia, PA.

Livingston, R.J. and D.A. Meeter. 1985. Correspondence of laboratory and field results: what are the criteria for verification? In *Multispecies Toxicity Testing,* ed. J. Cairns, Jr., pp. 76–88. Pergamon Press, New York.

Loewengart, G. and A.W. Maki. 1985. Multispecies toxicity tests in safety assessment of chemicals: necessity or curiosity? In *Multispecies Toxicity Testing,* ed. J. Cairns, Jr., pp. 1–12. Pergamon Press, New York.

Lundgren, A. 1985. Model ecosystems as a tool in freshwater and marine research. *Archiv für Hydrobiologie Supplement* 70:157–96.

Lynch, T.R., H.E. Johnson, and W.J. Adams. 1982. The fate of atrazine and a hexachlorobiphenyl isomer in naturally-derived model stream ecosystems. *Environmental Toxicology and Chemistry* 1:179–92.

Lynch, T.R., H.E. Johnson, and W.J. Adams. 1985. Impact of atrazine and hexachlorobiphenyl on the structure and function of model stream ecosystems. *Environmental Toxicology and Chemistry* 4:399–413.

Macek, K.J., D.F. Walsh, J.W. Hogan, and D.D. Holz. 1972. Toxicity of the insecticide Dursban to fish and aquatic invertebrates in ponds. *Transactions of the American Fisheries Society* 101:420–7.

Maciorowski, A.F. 1988. Populations and communities: linking toxicology and ecology in a new synthesis. *Environmental Toxicology and Chemistry* 7:677–8.

Maciorowski, H.D. and R.McV. Clarke. 1980. Advantages and disadvantages of using invertebrates in toxicity testing. In *Aquatic Invertebrate Bioassays,* eds. A.L. Buikema, Jr. and J. Cairns, Jr., pp. 36–47. American Society for Testing and Materials Special Technical Publication 715. American Society for Testing and Materials, Philadelphia, PA.

Mackey, A.P. 1977. Growth and development of larval Chironomidae. *Oikos* 28:270–5.

Maki, A.W. 1980. Evaluation of toxicant effects on structure and function of model stream communities: correlations with natural stream effects. In *Microcosms in Ecological Research,* ed. J.P. Giesy, Jr., pp. 583–609. Symposium Series 52, Technical Information Center, U.S. Department of Energy, Washington, DC. (Available as Conference-781101, National Technical Information Service, Springfield, VA.)

Malueg, K.W., G.S. Schuytema, J.H. Gakstatter, and D.F. Krawczyk. 1984. Toxicity of sediments from three metal-contaminated areas. *Environmental Toxicology and Chemistry* 3:279–91.

Mauck, W.L., F.L. Mayer, Jr., and D.D. Holz. 1976. Simazine residue dynamics in small ponds. *Bulletin of Environmental Contamination and Toxicology* 16:1–8.

Mauck, W.L. and L.E. Olson. 1977. Polychlorinated biphenyls in adult mayflies (*Hexagenia bilineata*) from the upper Mississippi River. *Bulletin of Environmental Contamination and Toxicology* 17:387–90.

Mayer, F.L., Jr. and M.R. Ellersieck. 1986. *Manual of Acute Toxicity: Interpretation and Data Base of 410 Chemicals and 66 Species of Freshwater Animals.* Resource Publication 160. Fish and Wildlife Service, U.S. Department of Interior, Washington, DC.

Merritt, R.W. and K.W. Cummins, eds. 1984. *An Introduction to the Aquatic Insects of North America,* 2nd ed. Kendall/Hunt Publishing, Dubuque, IA.

Merritt, R.W., K.W. Cummins, and V.H. Resh. 1984. Collecting, sampling, and rearing methods for aquatic insects. In *An Introduction to the Aquatic Insects of North America,* 2nd ed., eds. R.W. Merritt and K.W. Cummins, pp. 11–26. Kendall/Hunt Publishing, Dubuque, IA.

Minshall, G.W. 1984. Aquatic insect-substratum relationships. In *The Ecology of Aquatic Insects,* eds. V.H. Resh and D.M. Rosenberg, pp. 358–400. Praeger Pubs., New York.

Mount, D.I. 1982. Aquatic surrogates. In *Surrogate Species Workshop,* Anonymous, pp. A6–2 to A6–4. Report TR–507–36B. U.S. Environmental Protection Agency, Washington, DC.

Mount, D.I. 1985. Scientific problems in using multispecies toxicity tests for regulatory purposes. In *Multispecies Toxicity Testing,* ed. J. Cairns, Jr., pp. 13–8. Pergamon Press, New York.

Mulla, M.S. and H.A. Darwazeh. 1976. Field evaluation of new mosquito larvicides and their impact on some nontarget insects. *Mosquito News* 36:251–6.

Mulla, M.S., B.A. Federici, and H.A. Darwazeh. 1982. Larvicidal efficacy of *Bacillus thuringiensis* serotype H-14 against stagnant-water mosquitoes and its effects on nontarget organisms. *Environmental Entomology* 11:788–95.

Mulla, M.S., H.A. Navvab-Gojrati, and H.A. Darwazeh. 1978. Biological activity and longevity of new synthetic pyrethroids against mosquitoes and some nontarget insects. *Mosquito News* 38:90–6.

Newman, R.M., J.A. Perry, E. Tam, and R.L. Crawford. 1987. Effects of chronic chlorine exposure on litter processing in outdoor experimental streams. *Freshwater Biology* 18:415–28.

Nordlie, K.J. and J.W. Arthur. 1981. Effect of elevated water temperature on insect emergence in outdoor experimental channels. *Environmental Pollution* (Series A) 25:53–65.

Odum, E.P. 1983. *Basic Ecology.* Saunders College Publishing, Philadelphia, PA.

Odum, E.P. 1984. The mesocosm. *BioScience* 34:558–62.

Oseid, D.M. and L.L. Smith, Jr. 1974. Factors influencing acute toxicity estimates of hydrogen sulfide to freshwater invertebrates. *Water Research* 8:739–46.

Parrish, P.R. 1985. Acute toxicity tests. In *Fundamentals of Aquatic Toxicology,* eds. G.M. Rand and S.R. Petrocelli, pp. 31–57. Hemisphere Publishing, Washington, DC.

Parsons, T.R. 1982. The future of controlled ecosystem enclosure experiments. In *Marine Mesocosms: Biological and Chemical Research in Experimental Ecosystems,* eds. G.D. Grice and M.R. Reeve, pp. 411–8. Springer-Verlag, New York.

Patrick, R. 1949. A proposed biological measure of stream conditions, based on a survey of the Conestoga Basin, Lancaster County, Pennsylvania. *Proceedings of the Academy of Natural Sciences of Philadelphia* 101:277–341.

Patrick, R., J. Cairns, Jr., and A. Scheier. 1968. The relative sensitivity of diatoms, snails, and fish to twenty common constituents of industrial wastes. *Progressive Fish-Culturist* 30:137–40.

Peltier, W.H. and C.I. Weber. 1985. *Methods for Measuring the Acute Toxicity of Effluents to Freshwater and Marine Organisms.* EPA/600/4–85/013. U.S. Environmental Protection Agency, Washington, DC.

Petersen, R.C., Jr. and L.B.M. Petersen. 1985. Net anomalies as an indicator of stress in a freshwater invertebrate. Paper presented at the 6th Annual Meeting of the Society of Environmental Toxicology and Chemistry, St. Louis, MO, November 10–13, 1985. (Abstract, p. 46)

Rodgers, E.B. 1980. Effects of elevated temperatures on macroinvertebrate populations in the Browns Ferry experimental ecosystems. In *Microcosms in Ecological Research,* ed. J.P. Giesy, Jr., pp. 684–702. Symposium Series 52, Technical Information Center, U.S. Department of Energy, Washington, DC. (Available as Conference-781101, National Technical Information Service, Springfield, VA.)

Rodgers, E.B. 1982. Production of *Caenis* (Ephemeroptera: Caenidae) in elevated water temperatures. *Freshwater Invertebrate Biology* 1:2–16.

Rodgers, E.B. 1983. Fecundity of *Caenis* (Ephemeroptera: Caenidae) in elevated water temperatures. *Journal of Freshwater Ecology* 2:213–8.

Rosenberg, D.M. and V.H. Resh. 1982. The use of artificial substrates in the study of freshwater benthic macroinvertebrates. In *Artificial Substrates,* ed. J. Cairns, Jr., pp. 175–235. Ann Arbor Science Pubs., Ann Arbor, MI.

Rosenberger, D.R., E. Long, R. Bogardus, E. Farbenbloon, R. Hitch, and S. Hitch. 1978. *Considerations in Conducting Bioassays.* Technical Report D–78–23. U.S. Army Engineers Waterways Experiment Station, Vicksburg, MS.

Schuytema, G.S., D.F. Krawczyk, W.L. Griffis, A.V. Nebeker, M.L. Robideaux, B.J. Brownawell, and J.C. Westall. 1988. Comparative uptake of hexachlorobenzene by fathead minnows, amphipods and oligochaete worms from water and sediment. *Environmental Toxicology and Chemistry* 7:1035–46.

Selby, D.A., J.M. Ihnat, and J.J. Messer. 1985. Effects of subacute cadmium exposure on a hardwater mountain stream microcosm. *Water Research* 19:645–55.

Sheehan, P.J. 1980. The ecotoxicology of copper and zinc: studies on a stream macroinvertebrate community. Ph.D. dissertation, Univ. of California, Davis, CA.

Siefert, R.E., S.J. Lozano, J.C. Brazner, and M.L. Knuth. 1989. Littoral enclosures for aquatic field testing of pesticides: effects of chlorpyrifos on a natural system. In *Using Mesocosms to Assess the Aquatic Ecological Risk of Pesticides: Theory and Practice,* ed. J.R. Voshell, Jr., pp. 57–73. Miscellaneous Publication No. 75, Entomological Society of America, Lanham, MD.

Slobodkin, L.B. 1988. Intellectual problems of applied ecology. *BioScience* 38:337–42.

Slooff, W. 1983. Benthic macroinvertebrates and water quality assessment: some toxicological considerations. *Aquatic Toxicology* 4:73–82.

Slooff, W., J.A.M. van Oers, and D. de Zwart. 1986. Margins of uncertainty in ecotoxicological hazard assessment. *Environmental Toxicology and Chemistry* 5:841–52.

Solomon, K.R., Panel Chairman. 1986. *Pyrethroids: Their Effects on Aquatic and Terrestrial Ecosystems.* NRCC No. 24376. Associate Committee on Scientific Criteria for Environmental Quality, National Research Council of Canada, Ottawa, ON.

Sprague, J.B. 1969. Measurement of pollutant toxicity to fish. I. Bioassay methods for acute toxicity. *Water Research* 3:793–821.

Sprague, J.B. 1970. Measurement of pollutant toxicity to fish. II. Utilizing and applying bioassay results. *Water Research* 4:3–32.

Sprague, J.B. 1971. Measurement of pollutant toxicity to fish. III. Sublethal effects and "safe" concentrations. *Water Research* 5:245–66.

Sprague, J.B. 1973. The ABC's of pollutant bioassay using fish. In *Biological Methods for the Assessment of Water Quality,* eds. J. Cairns, Jr. and K.L. Dickson, pp. 6–30. American Society for Testing and Materials Special Technical Publication 528. American Society for Testing and Materials, Philadelphia, PA.

Stay, F.S., A. Katko, C.M. Rohm, M.A. Fix, and D.P. Larsen. 1988. Effects of fluorene on microcosms developed from four natural communities. *Environmental Toxicology and Chemistry* 7:635–44.

Stephan, C.E., D.I. Mount, D.J. Hansen, J.H. Gentile, and G.A. Chapman. 1985. *Guidelines for Deriving Numerical National Water Quality Criteria for the Protection of Aquatic Organisms and Their Uses.* PB85–227049/REB. U.S. Environmental Protection Agency, Washington, DC.

Stephenson, M. and G.L. Mackie. 1986. Effects of 2,4-D treatment on natural benthic macroinvertebrate communities in replicate artificial ponds. *Aquatic Toxicology* 9:243–51.

Stephenson, R.R. and D.F. Kane. 1984. Persistence and effects of chemicals in small enclosures in ponds. *Archives of Environmental Contamination and Toxicology* 13:313–26.

Stout, R.J. and W.E. Cooper. 1983. Effect of p-cresol on leaf decomposition and invertebrate colonization in experimental outdoor streams. *Canadian Journal of Fisheries and Aquatic Sciences* 40:1647–57.

Suter, G.W., II. and D.S. Vaughn. 1985. Extrapolation of ecotoxicity data: choosing tests to suit the assessment. In *Synthetic Fossil Fuel Technologies—Results of Health and Environmental Studies,* ed. K.E. Cowser, pp. 387–99. Butterworth Pubs., Boston.

Tarzwell, C.M. 1958. The use of bioassays in the safe disposal of electroplating wastes. *Annual Technical Proceedings, American Electroplaters Society* 45:60–2.

Taub, F.B. 1985. Toward interlaboratory (round-robin) testing of a standardized aquatic microcosm. In *Multispecies Toxicity Testing,* ed. J. Cairns, Jr., pp. 165–86. Pergamon Press, New York.

Taub, F.B., A.C. Kindig, and L.L. Conquest. 1986. Preliminary results of interlaboratory testing of a standardized aquatic microcosm. In *Community Toxicity Testing*, ed. J. Cairns, Jr., pp. 93–120. American Society for Testing and Materials Special Technical Publication 920. American Society for Testing and Materials, Philadelphia, PA.

Touart, L.W. 1988. *Hazard Evaluation Division Technical Guidance Document—Aquatic Mesocosm Tests to Support Pesticide Registrations*. EPA-540/9–88/035. Hazard Evaluation Division, U.S. Environmental Protection Agency, Arlington, VA.

Touart, L.W. and M.W. Slimak. 1989. Mesocosm approach for assessing the ecological risk of pesticides. In *Using Mesocosms to Assess the Aquatic Ecological Risk of Pesticides: Theory and Practice,* ed. J.R. Voshell, Jr., pp. 33–40. Miscellaneous Publication No. 75, Entomological Society of America, Lanham, MD.

Uhlmann, D. 1985. Scaling of microcosms and the dimensional analysis of lakes. *Internationale Revue der gesamten Hydrobiologie* 70:47–62.

Urban, D.J. and N.J. Cook. 1986. *Hazard Evaluation Division Standard Evaluation Procedure—Ecological Risk Assessment*. EPA 540/9–86–167. Hazard Evaluation Division, U.S. Environmental Protection Agency, Arlington, VA.

USEPA (U.S. Environmental Protection Agency). 1979. Guidance for the premanufacture testing: discussion of policy issues, alternative approaches and test methods. *Federal Register* 44:16240.

Voshell, J.R., Jr., ed. 1989. *Using Mesocosms to Assess the Aquatic Ecological Risk of Pesticides: Theory and Practice*. Miscellaneous Publication No. 75, Entomological Society of America, Lanham, MD.

Voshell, J.R., Jr., R.J. Layton, and S.W. Hiner. 1989. Field techniques for determining the effects of toxic substances on benthic macroinvertebrates in rocky-bottomed streams. In *Aquatic Toxicology and Hazard Assessment: 12th Volume,* eds. U.M. Cowgill and L.R. Williams, pp. 134–55. American Society for Testing and Materials Special Technical Publication 1027. American Society for Testing and Materials, Philadelphia, PA.

Wallace, J.B. 1989. Structure and function of freshwater ecosystems: assessing the potential impact of pesticides. In *Using Mesocosms to Assess the Aquatic Ecological Risk of Pesticides: Theory and Practice,* ed. J.R. Voshell, Jr., pp. 4–17. Miscellaneous Publication No. 75, Entomological Society of America, Lanham, MD.

Wallace, R.R., H.B.N. Hynes, and N.K. Kaushik. 1975. Laboratory experiments on the factors affecting the activity of *Gammarus pseudolimnaeus* Bousfield. *Freshwater Biology* 5:533–46.

Warren, C.E. and G.E. Davis. 1971. Laboratory stream research: objectives, possibilities, and constraints. *Annual Review of Ecology and Systematics* 2:111–44.

Wiley, M. and S.L. Kohler. 1984. Behavioral adaptations of aquatic insects. In *The Ecology of Aquatic Insects,* eds. V.H. Resh and D.M. Rosenberg, pp. 101–33. Praeger Pubs., New York.

Wilson, R.S. and J.D. McGill. 1977. A new method of monitoring water quality in a stream receiving sewage effluent, using chironomid pupal exuviae. *Water Research* 11:959–62.

Winner, R.W., M.W. Boesel, and M.P. Farrell. 1980. Insect community structure as an index of heavy-metal pollution in lotic ecosystems. *Canadian Journal of Fisheries and Aquatic Sciences* 37:647–55.

Woltering, D.M. 1985. Population responses to chemical exposure in aquatic multispecies systems. In *Multispecies Toxicity Testing,* ed. J. Cairns, Jr., pp. 61–75. Pergamon Press, New York.

Zapp, J.A., Jr. 1980. Historical consideration of interspecies relationships in toxicity assessment. In *Aquatic Toxicology,* eds. J.G. Eaton, P.R. Parrish, and A.C. Hendricks, pp. 2–10. American Society for Testing and Materials Special Technical Publication 707. American Society for Testing and Materials, Philadelphia, PA.

Zischke, J.A., J.W. Arthur, R.O. Hermanutz, S.F. Hedtke, and J.C. Helgen. 1985. Effects of pentachlorophenol on invertebrates and fish in outdoor experimental channels. *Aquatic Toxicology* 7:37–58.

Zischke, J.A., J.W. Arthur, K.J. Nordlie, R.O. Hermanutz, D.A. Standen, and T.P. Henry. 1983. Acidification effects on macroinvertebrates and fathead minnows (*Pimephales promelas*) in outdoor experimental channels. *Water Research* 17:47–63.

11

Field Experiments in Biomonitoring

Scott D. Cooper and Leon A. Barmuta

11.1. Introduction

The use of well-replicated, controlled field experiments has resulted in important insights into the structure and functioning of terrestrial, marine, and freshwater ecological systems. Although this powerful approach often has been directed at basic ecological questions, scientists and managers increasingly have recognized that field experiments are an integral part of many biomonitoring programs. In this chapter, we first examine the relative merits of various approaches to applied ecological problems, including surveys, laboratory studies, and field experiments, and discuss the role of field experiments in environmental monitoring. A discussion of the elements of good experimental design follows, including the need for replication, concurrent controls, and the precise manipulation of target variables. We then review common methods for (and problems with) the statistical analysis of ecological data and emphasize the importance of scale in the design and interpretation of field experiments. Throughout this discussion, we use literature examples to illustrate our points, but we do not present a comprehensive review of field experiments. The chapter concludes with recommendations for conducting field experiments.

11.2. Approaches to Biomonitoring Programs

A variety of approaches has been used to determine relationships between environmental variables, such as anthropogenic pollutants, and the abundances and distributions of benthic macroinvertebrates. These include field surveys, and laboratory and field experiments.

11.2.1. Field Surveys

Surveys examine relationships among biotic and abiotic factors across a number of sites or times, some of which differ in specific environmental attributes. Differences in, for example, the abundances of macroinvertebrates at two or more sites or times with different environmental conditions are then attributed to these differences in conditions. Common examples of such surveys include examinations of relationships between the abundances of macroinvertebrate species and physicochemical factors (e.g., pH, heavy metals) among lakes (e.g., Almer et al., 1974; Roff and Kwiatkowski, 1977; Yan and Strus, 1980); comparisons of unaffected upstream areas to impacted downstream areas (e.g., Norris et al., 1982; Marchant et al., 1984; Gore, 1980); contrasts between streams draining basins with different land-use practices (e.g., Dance and Hynes, 1980; Gurtz and Wallace, 1984; Brown and King, 1987); and monitoring programs that compare different points in time (e.g., before vs. after a disturbance) within a system (e.g., Pearson 1984).

Because the environmental variables of interest are rarely manipulated directly by the investigator, this approach generates correlational data that may be difficult to interpret for several reasons (Green, 1979; Cooper and Dudley, 1988). First, compared sites or times usually differ in a variety of ways so it is unclear which factors are responsible for any perceived patterns. Second, two variables may be correlated because both are related to a third, perhaps unmeasured and incidental, variable. Third, relationships between environmental and response variables only can be examined over the range of values of the environmental variable found in the surveyed sites. If this range of values is small, sampling variation may preclude detection of a relationship between the environmental and response variables. More importantly, it is not clear whether the survey results can be applied to sites or times with environmental values lying outside the surveyed range. This is especially problematic for managers who need to extrapolate survey results to predict the effects of planned or possible impacts on ecological systems. Fourth, correlational analyses rely on static measurements and are, therefore, poor at elucidating dynamic relationships between factors, that is, those involving feedbacks and time lags (Carpenter and Kitchell, 1987, 1988). Fifth, because surveys often are based on measurements taken at one point in time, they may miss important extreme events. For example, pH depressions associated with snow melt may have long-lasting effects on macroinvertebrate assemblages; however, surveys completed in late summer, after pH depressions have been ameliorated by buffering processes, may indicate little regarding the relationship between acidification and faunal assemblages.

Sometimes the results of surveys are subjected to sophisticated regression or multivariate analyses in an attempt to overcome some of these problems (e.g., Marchant et al., 1984; Wright et al., 1984). The stepwise variable selection techniques sometimes used in these analyses can yield erroneous results if some of the variables are highly intercorrelated (see Green, 1979).

Some of the more popular multivariate techniques (e.g., Principal Components Analysis) are useful for describing intercorrelations in the data set, provided the assumptions are met, but such analyses are best for generating hypotheses rather than testing them, simply because the investigator has no direct control over the environmental variables of interest (Green, 1979).

Field surveys are considered in detail by Resh and McElravy (Chapter 5), Resh and Jackson (Chapter 6), and Norris and Georges (Chapter 7).

11.2.2. Laboratory Experiments

Laboratory experiments have been used commonly to determine the effects of different concentrations of toxins on aquatic macroinvertebrates, but often have been criticized because of their artificiality (Schindler, 1977; Kimball and Levin, 1985; see also Buikema and Voshell, Chapter 10). Laboratory experiments usually separate an organism from its physical, chemical, and biological environment. For example, a pollutant may affect a given organism because of toxic effects on its competitors, predators, or its food base rather than the organism itself; however, such effects would not be uncovered in single-species bioassays (Clements et al., 1989). The organisms commonly used in laboratory studies are usually a few species that can be readily cultured. In many cases, it is apparent that results using these species have little relevance to more sensitive species present in the natural environment (Cairns, 1983).

Some laboratory experiments attempt to simulate natural conditions by transferring natural assemblages, substrates, and water to laboratory containers (e.g., Tonnessen 1984; Clements et al., 1989). Apart from the difficulties inherent in attempting such transfers, certain environmental conditions, such as those associated with weather, cannot be simulated in the laboratory. Water chemistry in recirculating or stagnant laboratory systems can diverge from initial natural values, and certain large-scale features, such as stream meanders or lake stratification, cannot be recreated. Because these factors can influence an organism's response to pollutants, results from laboratory studies only should be extrapolated to the field with extreme caution.

11.2.3. Field Experiments

The usual procedure for circumventing the problems inherent in surveys or laboratory bioassays is to perform controlled experiments *in situ*. Researchers doing field experiments manipulate environmental variables of interest and then examine the responses of aquatic organisms to those manipulations. Field experiments range in spatial scale from small enclosures to whole-lake or whole-stream manipulations (Fig. 11.1) and in temporal scale from "pulse" experiments, where responses to an impact applied at a single point in time are measured, to "press" experiments, where treatments are

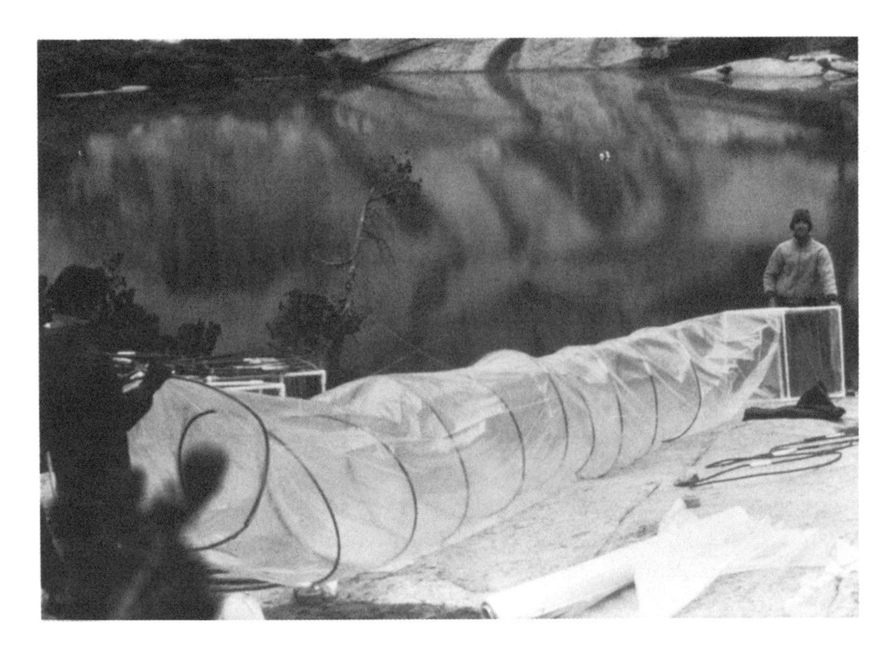

Figure 11.1. Field microcosms and mesocosms, and whole-system experiments. A and B: Bags used in testing the effects of acidification on the plankton and benthos of a small lake (Melack et al., 1987; Barmuta et al., 1990; see Fig. 11.6). A: Enclosures were 1 m in diameter, ~10 m deep, and contained 7.6 m^3 of water. PVC frames at the bottom of the bags were pushed into the lake bed thereby enclosing the whole water column.

maintained, sometimes over many years (see Buikema and Voshell, Chapter 10).

We believe that field experiments are a necessary part of biomonitoring efforts for the following reasons:

1. *To calibrate biomonitoring programs.* In many cases, the effects of perturbations on natural systems are inferred by examining relationships between changes in macroinvertebrate assemblages and the onset, duration, and magnitude of perturbations. The responses of different taxa to perturbations usually are not known, so field experiments are needed to determine directly the responses of intact assemblages to specific environmental changes. Congruence between the results of field experiments and observed changes in natural systems greatly increases our confidence in attributing changes to observed environmental perturbations. Field experiments also can show which species are sensitive to particular pollutants (indicator

Figure 11.1. (Continued) B: Enclosures were 1 m in diameter, ~5 m deep, and contained ~3.5 m^3 of epilimnetic water.

species), allowing monitoring efforts to focus on these sensitive species.

2. *Greater control over relevant variables*. Field experiments can be used to examine the direct effects of a given impact or proposed course of action. Levels of the environmental variables can be raised or lowered deliberately to levels not currently experienced by the system to allow prediction of effects (e.g., Barmuta et al., 1990). In addition, anthropogenic stresses on lakes and streams are often complex mixtures of physical and chemical factors whose effects only can be unraveled by good factorial experimental designs. Such designs not only allow the effects of individual pollutants to be

Figure 11.1. (Continued) C: Tubs (each with bottom area of ~0.13 m^3) used in examining the effects of predators on prey insects in stream pools. Tubs were gravity fed, flow was adjusted to reflect natural discharge, and all macroinvertebrate inputs and losses were measured using drift nets at inlets and outlets (Cooper et al., 1988a).

determined, but also show how two or more pollutants interact to produce synergistic, or antagonistic effects on aquatic organisms.

3. *Determining proximal mechanisms.* Together with detailed knowledge of the natural history of aquatic organisms, the results of field experiments can indicate the pathways of interaction among biotic and abiotic factors that produce the observed responses. Subsequent experiments can explicitly determine organisms' responses to factors that, in turn, are directly affected by environmental change. For example, experimental acidification may increase numbers of some taxa owing to a release from competition with other taxa that are reduced by acid inputs. Subsequent direct manipulations of competitors that are reduced by acidification can allow the delineation of indirect effects of acid inputs mediated through effects on competitive relationships. Furthermore, determination of mechanisms controlling responses to environmental change can allow the construction of mechanistic models that predict the responses

Figure 11.1. (Continued) D: Stream channels used in examining the responses of benthic insects and drift to pulsed acidification (Cooper et al., 1988b; Hopkins et al., 1989). Water flowed downhill through a pipe to the plywood reservoir and then was distributed to each of 12 replicate channels (2 m long × 0.2 m wide × 0.2 m high) by auxiliary pipes.

of organisms to novel situations. Because mechanisms are likely to be more general than relationships among static parameters, these experimentally derived models will be more powerful than predictions based on correlational data.

Although the need for field experiments in biomonitoring is clear, it must be emphasized that they should be viewed as another, albeit powerful, tool available to the applied ecologist (Schindler, 1987). Field experiments can be complemented with information derived from laboratory studies and field surveys, and some examples are discussed below. Nevertheless, it is essential that experiments be rigorously conducted so that their results can be interpreted unambiguously. Poor design and execution can result in erroneous conclusions about the impacts of environmental perturbations that can, in turn, mislead managers. In the following section, we describe the design and analysis of field experiments and examine some of the problems of applying these techniques.

Figure 11.1. (Continued) E: *In situ* stream channels (8 m × 0.5 m × 0.6 m) used to examine the responses of macroinvertebrates to continuous acidification (Griffiths, 1987).

11.3. Design, Analysis, and Interpretation of Field Experiments

All good research begins with well-formulated questions, although distressingly few environmental impact assessments bother to formulate them (Rosenberg et al., 1981). Each ecological question should be framed by a series of hypotheses that address the posed questions. The null hypothesis of no effect or impact of the independent (criterion) variable(s) on the dependent (response) variable(s) is tested against the alternative; the most powerful alternative test stipulates the expected direction of deviation from the null hypothesis rather than the more usual "two-sided" test (Green, 1979).

Figure 11.1. (Continued) F: Littoral enclosures (21 m long × 12 m wide) used to examine the effects of pumpkinseed sunfish on snails (Osenberg, 1988). The sides of the cages were nylon netting (1.3 cm mesh) and extended from the surface, where they were buoyed by floats (visible here), to the bottom. Cages extended from the shoreline to the profundal zone (depth ~5.0–5.5 m) and enclosed both littoral vegetation (76% of the area) and profundal sediments.

The first step is to select the dependent and independent variables to be used in the experiments. The results of field surveys and laboratory studies can be useful in identifying important questions, guiding the selection of variables, and suggesting the appropriate intensities of manipulations to be used in a field experiment.

Response variables, of course, will depend on the exact questions being asked, and will range from biochemical or physiological to ecosystem-level responses (Fig. 11.2). These responses range in spatial and temporal scales from 10^{-8} m and 10^{-3} s to 10^{6} m and 10^{9} s. Because we are dealing here with a variety of macroinvertebrate species and life stages, different hierarchies of response variables overlap in space and time, which is depicted by symbol overlaps in Fig. 11.2. Individual responses range from the behavior of sedentary individuals over short time spans to the movement of mobile individuals over their lifetimes. Population responses range through

Figure 11.1. (Continued) G and H: Acid addition experiments in Norris Creek, Hubbard Brook Experimental Forest, New Hampshire, a third-order stream with a mean width of 1.8 m (Hall et al., 1980). Sulfuric acid from the carboy was added to Norris Creek, lowering pH to ~4 for 4–5 mo; pH in the upstream reference section was 5.7–6.4 during the experiment.

the population dynamics of individual species in individual patches, linked patches, and whole systems at time scales up to a number of generations (Wiens et al., 1986). Community scales include the spatial ranges through several generations of the largest consumers that affect macroinvertebrates (usually fish; Carpenter, 1988), whereas ecosystem scales include long-term catchment processes.

Although Fig. 11.2 was constructed for medium-sized lakes and rivers,

Figure 11.1. (Continued)

individual through ecosystem responses can be constrained by the size of the natural units (e.g., patches, ponds, local drainage areas), and community and ecosystem responses can be constrained by the size, mobility, and abundance of top predators and the nature and extent of catchment processes. Similar scale diagrams can be constructed for other groups of organisms; actual scales for individual and population responses, for example, will depend on the size, mobility, and generation times of target organisms as well as habitat constraints (such as the size and duration of patches). Biogeographical patterns and evolutionary time would, of course, encompass larger spatial and temporal scales than shown here. Key issues in ecology include the extent to which low hierarchical levels of response translate into macroscopic responses, the nature and extent of constraints imposed by higher

Figure 11.1. (Continued) I: Response of Lake 226 to nutrient addition, Experimental Lakes Area, Ontario (Schindler, 1974; Schindler and Fee, 1974). This double-basin lake was divided by installation of a vinyl sea curtain. Nitrogen and carbon were added to the lower basin, which remained clear. Nitrogen, carbon, and phosphorus were added to the upper basin, which developed an algal bloom.

on lower hierarchical levels, and feedbacks between different response levels.

In the following discussion, we will concentrate on population and community responses because these have been the principal foci of past field-based biomonitoring efforts. In addition, many management questions are framed in terms of these variables. This does not negate the importance of biochemical and physiological variables; they promise to be valuable in pro-

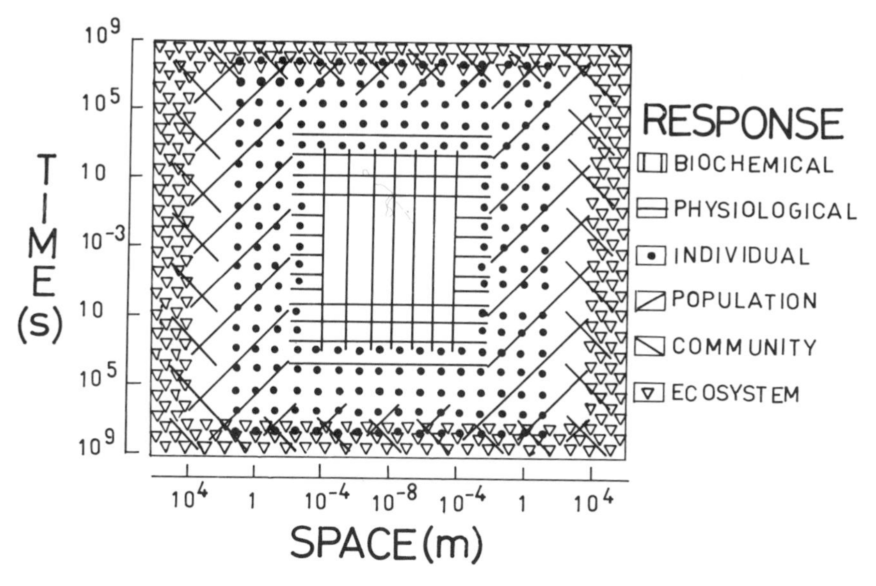

Figure 11.2. Spatial and temporal scales of different response variables for benthic macroinvertebrates. Derived from concepts and data in Allen and Starr (1982), Harris (1986), Legendre and Demers (1984), and Minshall (1988).

viding both sensitive indicators of response to environmental change and mechanistic explanations of the effects of environmental insults (e.g., see Johnson et al., Chapter 4). However, the expression of changes in these variables at the population or community level remains to be determined.

For population and community level studies, the appropriate level of taxonomic discrimination depends very much on the questions being asked and the magnitude and scope of the anticipated impacts. Severe impacts may be detectable using generic or even higher taxonomic levels (Green, 1979), whereas other studies will need taxa identified to the species level (Resh and Unzicker, 1975; see also Resh and McElravy, Chapter 5).

The selection of the identity and appropriate levels of the independent variables will depend on the nature of anthropogenic perturbations and observed, predicted, or planned magnitudes and durations of perturbations. When manipulating levels of the independent variables, it is usually best to use a graded series with replication at each level, including control replicates. The range over which the variables are manipulated should encompass their expected or measured intensities and durations. For preexisting problems, it is relatively easy to examine routinely collected baseline data; in contrast, experiments to predict the impact of levels of a novel pollutant may require input from many disciplines and the use of sophisticated models

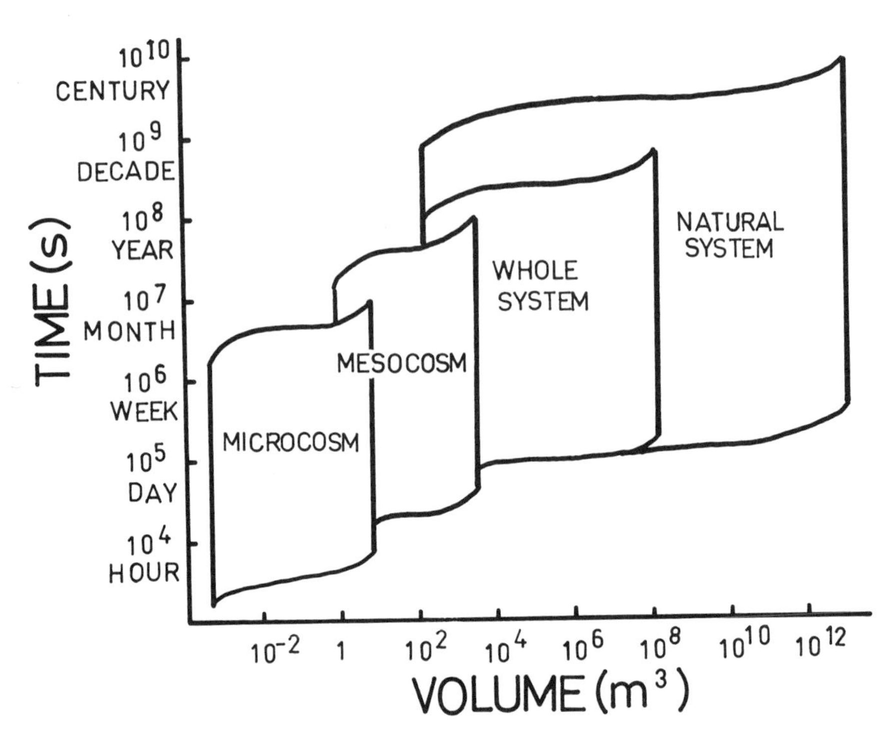

Figure 11.3. Spatial and temporal scales of microcosm, mesocosm, and whole-system experiments, and scales for natural community and ecosystem responses. Derived from concepts and data in Diamond (1986), Frost et al. (1988), and Kitchell et al. (1988).

to estimate the likely range and duration of the putative impact (Graney et al., 1989; Jenkins et al., 1989).

11.3.1. Importance of Scale

The spatial scale of field experiments ranges from microcosms through mesocosms to whole-lake or stream manipulations (Fig. 11.3). Buikema and Voshell (Chapter 10) define mesocosms as experimental containers holding >10 m^3; however, considerable variation exists in the use of the terms "microcosm" and "mesocosm" in the literature, which is depicted by overlap in the spatial scales of these experimental units in Fig. 11.3. Comparisons of response (Fig. 11.2) and experimental (Fig. 11.3) scales show that field microcosm experiments, for example, may be appropriate for examining the short-term behavior of populations in local patches, but may indicate little about the responses of communities to perturbations of whole systems. Although community and population responses in very small natural systems, such as container habitats (Barton and Smith, 1984), may be simulated accurately by small field or laboratory containers, community and population responses in natural systems ranging from small streams and ponds to large rivers and lakes are of interest here. Temporal scales for whole systems are truncated at the lower and higher ends of Fig. 11.3 because of the logistical difficulties of sampling whole natural systems over very short time periods and because attention here is restricted to recent ecological time. Long-term (> decades) reconstructions of the history of natural systems require recourse to paleoecological methods (see Walker, Chapter 9). Overlap between experimental and natural scales (Fig. 11.3) reflects the degree to which experimental systems accurately simulate the heterogeneity and size of natural systems and the duration of natural ecological responses. As defined here, microcosm experiments are "unrealistic," whereas the largest mesocosms (constructed ponds, channels) can accurately mimic small, natural systems (small ponds and streams). Results of even whole-system experiments (10^5 to 10^7 m^3; e.g., Carpenter et al., 1987; Schindler, 1977; Eriksson and Tengelin, 1987; Evans, 1989) cannot be extrapolated with confidence to even larger natural systems. Manipulation of large natural systems ($>10^7$ m^3) presents formidable logistical difficulties; however, effects of perturbation can be inferred from long-term data sets that encompass periods of time before and after disturbance or from paleoecological reconstructions.

In designing experiments, ecologists often are faced with difficult choices: small enclosures are easy to replicate and manipulate but can be environmentally unrealistic, whereas whole lakes and streams represent the ultimate in realism but are difficult to manipulate and replicate (Carpenter, 1989). Certainly, all of these experimental units are useful, but it is extremely important to define the natural scales at which the results have relevance and understand the limitations imposed by the technique adopted (Wise, 1984; Wiens et al., 1986).

Large-scale stream experiments commonly use an upstream, unimpacted area as the control and areas downstream of an impact as the experimental sites (e.g., Spence and Hynes, 1971; Yasuno et al., 1982; Culp et al., 1986; Hall et al., 1980; Hall, 1989). This design has the following problems: (1) environmental conditions upstream and downstream are likely to differ; (2) upstream processes may affect downstream results; and (3) no true replication of treatments is possible.

The other extreme is the small, *in situ* enclosure. These devices quickly diverge from natural conditions: edge effects are exacerbated, mixing is prevented or current and sediment conditions are modified, and the behavior of organisms may be altered by confinement (e.g., restriction of immigration, accidental exclusion of predators or competitors) (Cooper et al., 1990; Peckarsky and Penton, 1990). The results from small-scale enclosures may be inapplicable to natural situations not only because of enclosure artifacts but also because of the mismatch between the scale addressed by the questions (e.g., what are the effects of a perturbation on this lake or stream?) and the scale represented by the enclosure (often only part of a habitat) (Figs. 11.1–11.3).

The problems of replication, ease of manipulation, size, and fidelity to natural conditions may be solved by using *in situ* or bankside microcosms or mesocosms such as large bags in lakes (Schindler et al., 1980; Marmorek, 1984; Barmuta et al., 1990), and channels within or alongside streams (e.g., see summary of ORSTOM Onchocerciasis Control Programme experiments in Muirhead-Thomson, 1987; Hopkins et al., 1989; Hart and Robinson, 1990). Such enclosures may allow researchers to set up replicated study systems under nearly natural conditions. In addition, similar conditions among replicates can be created at the outset, minimizing problems associated with sampling or interreplicate variation. This feature has its strengths and weaknesses. On one hand, such uniformity will increase the precision of final density estimates; on the other hand, such systems do not incorporate the heterogeneity found in natural, larger systems.

If enclosures satisfactorily mimic natural conditions, then the results can be extrapolated, with some confidence, to natural systems. The most straightforward way to assess this is to compare values of chemical, physical, and biological variables within control enclosures to those in the natural environment. If these values are congruent over the course of the experiment, then it is likely that no "cage effects" exist. Even if unmanipulated enclosures are similar to the larger, natural system, it is still possible that the responses of organisms to treatment in enclosures will differ from treatment effects in lakes or streams. For example, addition of pollutants may result in direct toxic effects on mobile organisms confined within enclosures. In the natural system, however, such mobile organisms may be able to avoid these lethal effects by moving into refuges or out of the area completely. The only way to assess this is to conduct experiments at a variety of scales. If such experiments produce similar results, then mesocosm experiments may suffice. If the results differ, however, it may be that manipulations of larger, open systems (e.g., ponds, lakes, streams) are the only viable alternatives, although mesocosms still may be used for determining the direct and interactive effects of a variety of factors (Schindler, 1987). Because factorial experimental designs require extensive replication, it is unlikely that such experiments can be conducted on whole streams or lakes.

Small natural or large artificial systems, such as littoral enclosures, experimental or natural ponds, and large, artificial stream channels obviate some of the problems associated with lake bag or small channel experiments (Buikema and Voshell, Chapter 10). Because of their size, these systems incorporate much of the heterogeneity found in larger, natural systems. Thus, refuges may be included and sufficient space may be available for large, mobile consumers (e.g., fish) to complete their life cycles. However, such systems may prove difficult and expensive to construct, manipulate, and replicate, and some artifacts may still remain. For example, littoral enclosures may inhibit the movement of organisms and materials from limnetic zones, resulting in conditions unrepresentative of natural littoral zones; small ponds may not stratify, thus restricting their utility for extrapolation to larger lakes. Similar concerns apply to large stream channels: they rarely incorporate meanders, and they usually differ from natural streams in the ratios of depths and areas of pool and riffle sections (Leopold, et al., 1964; Warren and Davis, 1971; Rodgers, 1983; Zischke et al., 1983; Buikema and Voshell, Chapter 10). Consequently, the natural range of variation in substrate, depth, and velocity may be restricted. Recirculating systems may have additional problems with nutrient limitation and "recycling" of emigrants. Furthermore, natural or experimental ponds and large stream channels often vary considerably among replicates prior to manipulation, and some of the effects of the manipulations may be lost in the "noise" of pond-to-pond or channel-to-channel variation (Hall et al., 1970). Nevertheless, some questions are addressed appropriately by such systems, and they have been used effectively (Hurlbert et al., 1972; Rodgers, 1983; Crossland and Wolff, 1985; Dewey, 1986). Certainly, results from large mesocosms can be applied to natural systems of similar size having similar conditions. The distinction between microcosms, mesocosms, and natural systems is, therefore, somewhat arbitrary, and the critical concern is the degree to which experimental systems faithfully mimic the natural system of interest (Fig. 11.3).

Two aspects of time in experiments merit attention: (1) the duration of manipulations, and (2) the duration of monitoring to assess the impacts of those manipulations. In pulse experiments, manipulations are applied for short periods of time, generally as single events, and responses to these pulses are followed. In press experiments, in contrast, treatment levels are maintained throughout experiments (Bender et al., 1984). Both pulse and press experiments measure a community's resistance to change in the face of a temporary or permanent disturbance. Pulse experiments also measure the ability of a community to recover from a perturbation, whereas press experiments measure the response to new, altered conditions. The duration of monitoring will depend on the responses of the dependent variables. If sublethal effects are suspected, then monitoring should continue through several life cycles of common animals in order to describe demographic repercussions.

The choice of press or pulse experiments will be determined, to a large extent, by the duration of anthropogenic disturbances recorded or anticipated in the natural environment. For example, pulsed additions of acid to experimental treatments may simulate pulsed acidification events, such as those associated with snow melt or rainstorms (e.g., Hall et al., 1985; Hopkins et al., 1989), whereas continual addition of acid to maintain constant, lowered pH levels explores the effects of chronic acidification on experimental assemblages (e.g., Schindler et al., 1985). However, short-term press experiments may not effectively model responses to long-term or repeated exposure to the perturbation; longer time spans may allow colonization by resistant species or even selection of resistant strains in local populations.

Although seldom addressed experimentally, the frequency of pulsed perturbations is an important consideration when designing experiments to evaluate the impacts of a series of perturbations on natural systems. Similarly, the timing and rapidity of environmental manipulations should closely mimic those observed or predicted in the natural environment. Whether manipulations should be done abruptly or gradually also will depend on the nature of real-world disturbances.

Ecologists intending to use field experiments have to decide whether to use replicated mesocosms, which may diverge in important ways from the natural system they seek to model during the course of the experiment, or whether to manipulate entire systems and risk ambiguous results that are not amenable to traditional, rigorous statistical analysis. A potential solution is to conduct small-scale experiments that record responses to graded treatment levels and to calibrate these results simultaneously with the results of simpler, whole-system manipulations. However, few analyses explicitly have compared the results of experiments conducted at different spatial and temporal scales. In evaluating the roles of various nutrients in controlling phytoplankton growth, early experiments done in small jars consistently showed that carbon was limiting to algal growth, because the jars inhibited mixing of atmospheric carbon dioxide. In contrast, large-scale field experiments in Canadian Shield lakes showed that phosphorus was the major nutrient limiting growth (summarized in Schindler, 1977, 1987).

Schindler et al. (1985) reported that the relative sensitivities of groups such as benthic crustaceans and cladocerans in a whole-lake acidification experiment were similar to those predicted from earlier survey and small-scale field and laboratory studies, but many other whole-lake results were different (e.g., primary production, decomposition rates, species composition of zooplankton and forage-fish assemblages, fish condition). Carpenter and Kitchell (1988) reported that the qualitative responses of phytoplankton species to direct manipulation of zooplankton in small *in situ* bags often differed from phytoplankton responses to whole-lake manipulation of zooplankton grazers effected by manipulating top fish predators. In evaluating the effects of pollutants on stream invertebrates, Clements et al. (1988) compared the results from outdoor stream microcosms to correlational data collected above and below a pollutant outfall and found a considerable degree of congruence. Hall (1989) experimentally acidified a small lake-outlet stream in the Experimental Lakes Area of northwestern Ontario for four days. Concurrently, he compared the macroinvertebrate assemblage in the outlet stream of a lake that had been experimentally acidified for three to five years (Schindler et al., 1985) to macroinvertebrate assemblages in similar lake-outlet streams nearby. Community responses to short-term acidification often were different from those to long-term acidification. For example, the relative abundance of collector-filterer taxa decreased in response to short-term acidification, but the relative abundance of collector-filterers was higher in the long-term acidified site than in reference sites. The distribution of individuals among taxa also was different in the long-term vs. short-term acidification site, relative to reference areas.

These admittedly mixed results indicate that evaluating the influence of scale at which experiments are done has only begun. It is unlikely that results obtained from 1-m^3 enclosures can be extrapolated to Lake Superior, and some questions only can be addressed by whole-ecosystem manipulations (Likens, 1985; Schindler, 1987). It is probable, however, that the degree to which results from mesocosms can be extrapolated to large, natural systems will depend on how closely experimental scales approach those of the natural systems under study. In addition, the results cited above indicate that micro- or mesocosm experiments may be useful in delineating direct, often toxic effects of pollutants on organisms. However, long-term, large-scale effects of pollutants on systems probably only can be determined through whole-system experiments because of complexities introduced by indirect and multiplicative effects. For example, the complexity of trophic interactions, time lags in responses of organisms to perturbations, and pollutant-induced physicochemical alterations to environmental structure all can affect community responses to anthropogenic inputs (Schindler et al., 1985).

A final caveat concerns opportunities for performing whole-system experiments. The number of these done is small because of time and expense, but whole-system alterations occur continuously as a result of human activities. By cooperating with industry and governmental agencies, researchers could place proposed perturbations within the context of a properly planned experimental design. Costly environmental impact assessments presently are required for many planned developments. These assessments usually include an inventory of potentially affected aquatic resources, followed by educated guesses of the impact of proposed developments on these resources. However, rigorous analyses of the impacts of various kinds of developments or pollutant inputs on the aquatic biota are sorely needed, preferably using well-replicated field experiments. Diversion of some environmental impact funds to field experiments at least will provide governmental and consulting personnel with a firmer basis for evaluating environmental impacts of various developments, rather than the "survey-and-guess" procedure that predominates at present.

11.3.2. Controls, Presampling, and Replication

11.3.2.1. Controls and Presampling

Field experiments require adequate concurrent controls to evaluate the effects of manipulations on aquatic communities. Samples also should be taken from the control and treatment sites before the manipulations are applied ("presampling"), to assess similarity of the sites and hence adequacy of the controls (e.g., Hall et al., 1980; Ormerod et al., 1987; Leland et al., 1989; Barmuta et al., 1990). Stratifying samples among various microhabitats can be useful in minimizing sampling costs and ensuring that samples are taken from habitats that are as similar as possible in treatment and control areas (Green, 1979; Lamberti and Resh, 1979). Given any preexisting differences, it may be possible to frame the experimental question in terms of a *change* in the response variables, or attempt to "factor out" the differences using analysis of covariance.

11.3.2.2. Replication, Pseudoreplication, and Some Solutions

Two areas likely will diverge over time, even without manipulating one of them. Comparisons of single control with single treatment units may yield ambiguous results because it is unclear whether the observed changes would have occurred in the absence of the impact. Eberhardt (1976) called this a "pseudodesign." Hurlbert (1984) called it "pseudoreplication" and showed that it has been a common flaw in designing ecological experiments (see also Norris and Georges, Chapter 7).

The most common form of pseudoreplication is spatial (Fig. 11.4A) and usually involves the comparison of a single control site with a single site for each manipulation using a number of samples taken at each site. This is not a test of the null hypothesis because the replicates are nested within each site; it merely shows whether the sites are similar or different at a given intensity of sampling. When single true replicates are assigned to each treatment, statistically testing the effects of the treatment is invalid because no true estimate exists of the within-treatment variance against which between-treatment effects can be compared. For a rigorous test of the null hypothesis, it is necessary that more than one experimental unit or site be assigned to each treatment, and that these units, not samples from within the units, be used in statistical analyses of the effects of the manipulations or treatments. It is important to ensure that replicates assigned to different treatments are not spatially segregated, so that effects of treatment are not confounded with effects of location. When the number of experimental units is low, Hurlbert (1984) argues that interspersion of treatments among replicates may be more effective at preventing treatment segregation than simple random assignment of replicates to treatments.

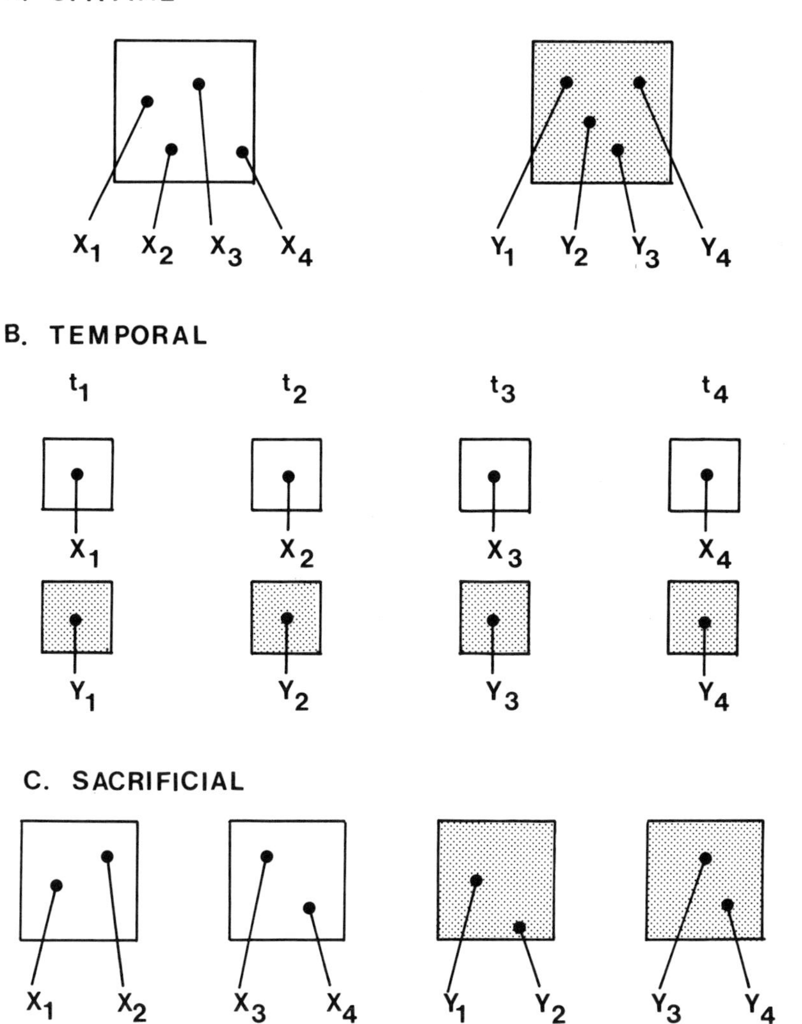

Figure 11.4. Examples of three kinds of pseudoreplicated experimental designs. Pseudoreplication results from treating the sample units ($x_1, \ldots, x_4$; $y_1, \ldots, y_4$) as independent samples taken from each treatment (represented by shaded and unshaded boxes). In A, the sample units are only nested within each experimental unit; in B, the same experimental units are sampled repeatedly at consecutive times in such a way that successive observations are not independent of previous observations; in C, the sample units either are regarded incorrectly as independent replicates or are pooled prior to analysis. After Hurlbert (1984).

Temporal pseudoreplication (Fig. 11.4B) is similar, but in this case the samples are taken at different points through time, rather than through space, within a replicate (Hurlbert, 1984). Samples taken through time from the same replicate are likely to be correlated and are, therefore, not independent (Underwood, 1981). Again, by treating each sample rather than each replicate unit as a statistical replicate, researchers artificially inflate the degrees of freedom of the test and invalidly test hypotheses with pseudoreplicates rather than true ones. As in spatial pseudoreplication, the solution to temporal pseudoreplication is simply to assign a number of replicate units to each treatment and to treat these units as the replicates in statistical analyses. A mean value can be calculated over the experimental period for each replicate by averaging the samples taken through time, and these mean values then can be used in standard statistical analyses. If the interval between samples varies, then means may need to be time-weighted when calculating replicate values for analyses. Alternatively, investigators can use statistical techniques that are specifically designed for time-series data, such as repeated measures analysis of variance (ANOVA) (Winer, 1971) or the more robust multivariate equivalent, profile analysis (Harris, 1985; see Holomuzki, 1989, and Barmuta et al., 1990, for examples). Provided assumptions are met, these procedures permit valid testing of the differences between times and the description of the shape of the response over time in terms of a polynomial trend. Norris and Georges (Chapter 7) consider time-series analysis in detail.

"Sacrificial" pseudoreplication occurs when researchers have a number of replicate units per treatment (Fig. 11.4C), but then inappropriately analyze units nested within experimental units as independent replicates or ignore the replicates by pooling the data (Hurlbert, 1984). This form of pseudoreplication often is manifested in the inappropriate use of contingency tables. For example, an investigator may have five containers assigned to a control treatment and five to an experimental treatment and be interested in the numbers of dead vs. alive animals in each treatment. In performing statistical analyses, the researcher may improperly set up a two by two contingency table, control vs. experimental x alive vs. dead, using total numbers of dead vs. alive macroinvertebrates added across all replicates within a treatment as the cell values. As Hurlbert clearly has shown, however, large differences in a single replicate may have a large effect on the results of the χ^2 analysis even if all of the other replicates are similar.

None of the foregoing should argue against taking nested samples within treatment units or sites, provided they are analyzed properly. Often such nested designs provide valuable additional information about sources of variation within an experiment (Green, 1979; Hurlbert, 1984). Similarly, some studies will be impossible to replicate in the true sense of the word, but if they are executed well enough, they will yield unequivocal results. Hurlbert

(1984) cites the work at Hubbard Brook (e.g., Likens, 1985) and the Experimental Lakes Area (e.g., Schindler, 1987) as good examples of unreplicated experiments that still yielded good results. Replication of control systems, the use of closely paired catchments, or "staircase" experimental designs in which treatment is initiated at different times in different experimental units (Walters et al., 1988) are all approaches that can increase confidence in the results of such studies (Carpenter, 1989). What is principally at issue in pseudoreplication, however, is the incorrect application of conventional statistical analyses (Hurlbert, 1984).

The appropriate methods to analyze unreplicated experiments have provoked some controversy. Although the common solution to most pseudoreplication problems is to assign a number of replicate units to each treatment and to treat these units as statistical replicates, sometimes situations exist in which it is logistically impossible to set up more than one or a few replicate units for each treatment, particularly when experimental units are large (whole streams and lakes). This is a particularly awkward problem because managers usually are interested in the impacts of human perturbations on whole systems, and the processes at work may operate at scales outside the scope of mesocosms (Schindler, 1987). Carpenter (1989) recently performed a power analysis on simulations of whole-lake experiments to determine the amount of replication needed to detect statistically the indirect effects of tenfold changes in piscivore biomass on phytoplankton biomass. His analyses indicated that ten replicate control and experimental lakes were needed to detect responses to this perturbation, a level of replication that would require tremendous amounts of money and effort. Trautmann et al. (1982) also found that extensive premanipulation data sets on many lakes were necessary to achieve adequate power to detect changes in phytoplankton biomass resulting from the banning of phosphate detergents in New York State. As a potential solution, Carpenter (1989) suggested examining the consistency of results obtained from pseudoreplicated whole-lake experiments conducted in a variety of places at a variety of times.

Often, however, not only are the opportunities for replication restricted, but the question itself is site-specific (e.g., what is the effect of *this* outfall on *this* river?). For such circumstances, Stewart-Oaten et al. (1986) suggested a statistical design that they termed Before-After-Control-Impact (BACI), and similar techniques have been proposed under the general heading of Random Intervention Analyses (RIA) (Carpenter et al., 1989). The general features of these designs include monitoring control and experimental sites before and after perturbations have occurred, and using the difference between control and experimental areas as the response variable. The samples at each time are treated as replicates, and mean differences between the two areas before and after the impact are compared by a t-test or a U-test (Fig. 11.5). The assumptions that need to be met are explained fully by

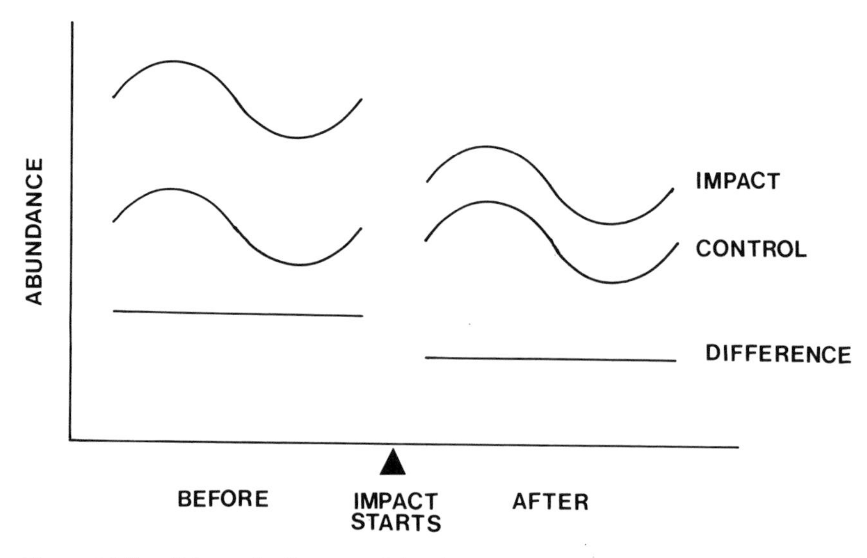

Figure 11.5. Schematic diagram of the abundance of a selected species before and after the start of some impact in a control and an impacted site. The appropriate statistical test compares the *difference* in abundance between the sites before and after the start of the impact. After Stewart-Oaten et al. (1986).

Stewart-Oaten et al. (1986). However, it must be emphasized that for this procedure to be valid, a long time series of data needs to be collected both before and after the impact.

Millard et al. (1985) discuss further the problem of using changes in abundance (rather than changes in the difference of abundance) in such situations, particularly when the data are likely to be correlated in space or time. Such autocorrelated data violate assumptions of independence in ANOVA tests and may inflate the Type I error rate (see Section 11.3.2.3, "Power," below). They suggest using multivariate time-series analysis for data in which frequent sampling results in temporal autocorrelation, whereas spatial autocorrelations may be treated by using multivariate analysis of variance (MANOVA).

Another approach is to use Monte Carlo simulations to generate a distribution of differences between control and experimental areas based on data taken before perturbations occur. The null hypothesis of no change in the difference between experimental and control areas is rejected if observed differences between control and experimental sites fall into the $P < 0.05$ tail of the generated distribution after perturbation. Carpenter et al. (1989) used a slightly different approach. Under the null hypothesis of no change in the difference between control and experimental systems before and after manipulation, they generated a null distribution based on random assignment of interecosystem differences between pre- and postmanipulation periods, hence the term Random Intervention Analysis. Their test statistic was the observed difference between average interecosystem differences before vs. after manipulation, and its distribution was estimated by random permutation of the differences throughout the monitored period. If the observed difference between average pre- and postmanipulation differences fell into the $P < 0.05$ tail of the distribution, then the null hypothesis was rejected (i.e., a statistically significant change in interecosystem differences had been detected). RIA has the advantages of not requiring assumptions as restrictive as parametric analyses and not being greatly affected by serial autocorrelation, at least for the data examined by Carpenter et al. (1989; but see Stewart-Oaten et al. 1992). Severe autocorrelation would necessitate use of more conservative Type I error levels.

Although these techniques will contribute greatly to the rigorous analysis of changes in control vs. experimental areas after perturbations, they should not be regarded as a panacea to the ills of pseudoreplication. These techniques demonstrate that a change has occurred, but cannot unequivocally attribute the change to the applied manipulation. The experimental area still may be affected differentially by something other than the manipulation being studied. This will be most problematic if the impact is localized to one of the control or experimental areas, is large, and results in long-lasting effects (Stewart-Oaten et al., 1986). An ability to detect such disruptions requires thorough knowledge of the study area. Alternatively, two systems may respond differently to a common change (e.g., in the weather) because of differences in their morphometries, chemistries, or biotas (Magnuson et al., 1990). Premanipulation sampling of a control and an experimental system in one year, and postmanipulation sampling in the next, could result in ambiguous results because it would be unclear if any changes in the differences between these systems were due to treatment or to the differential responses of the systems to interannual differences in environmental factors such as the weather. The problem, of course, is that systems must be sampled for many years to establish patterns of interannual variability to act as a baseline for comparison to postmanipulation changes, especially if inferences are to be made exclusively from RIA. Given current funding protocols and the rapidity of development, it is unlikely that scientists will be able to begin a monitoring program many years before a perturbation occurs.

Critical needs exist for analyses that determine relationships between sampling frequency and serial correlation for ecological data and for analyses that partition total variation in such data into components of different temporal scale. Such analyses would help researchers determine the appropriate frequency of sampling and provide a guide for framing null hypotheses. Carpenter (1989) examined a data set derived from plankton samples taken every two weeks and found little evidence for serial correlation. R. Griffiths (Ontario Ministry of the Environment, London, ON, personal communication) partitioned total variation in a benthic data set taken from a Canadian lake into different temporal scales. Duplicate samples were taken at each of several stations at monthly intervals over seven years. He found that the relative importances of seasonal and interannual variation varied from site to site and that interannual variation could account for 20% to 45% of the variation in the premanipulation data set. Goldman et al. (1989) found considerable interannual variation in mean primary productivity in a California lake over 28 years and a marginally significant autocorrelation between primary production values for one year and the next. After removing autocorrelational artifacts by using autoregressive integrated moving average (ARIMA) models (Box and Jenkins, 1976), they found a clear linkage between mean primary production and meteorological conditions. At present, insufficient data are available in most studies to evaluate appropriate scales of temporal sampling, although it is apparent that sampling must continue for several years to provide an adequate baseline for comparison of conditions before and after an impact.

11.3.2.3. Power

Statistical tests determine the probability that observed results can be attributed to chance. The Type I error (α) is the probability of rejecting the null hypothesis of no effect, even when it is true; the Type II error (β) is the probability of accepting the null hypothesis when it is false (see also Resh and McElravy, Chapter 5, and Norris and Georges, Chapter 7). A powerful analysis will minimize Type II error. The power of a test depends on three factors (Cohen, 1977): (1) the Type I error rate (typically set at $\alpha = 0.05$); (2) the number of replicates collected; and (3) the size of the effect one wishes to detect.

Ecologists historically have been concerned primarily with Type I error because they want to be certain that observed differences are real. In biomonitoring, however, the costs of committing a Type II error, that is, accepting the null hypothesis when it is false, may be as great or greater than those of committing a Type I error (Toft and Shea, 1983; Peterman, 1990). For a given variable and level of replication, an inverse relationship exists between Type I and Type II errors; selection of increasingly conservative values of α will increase the probability (β) of committing a Type II error. Consequently, circumstances may arise when more liberal levels of α should be adopted (e.g., Dewey, 1986) if the costs of committing a Type II error are excessive (Toft and Shea, 1983).

Power increases with more replication within treatments. If the within-treatment variation is large, then the additional replication required will in-

Table 11.1. Number of replicates required to detect a 33%, 50%, or 90% decrease of the mean abundance shown in the first column for lotic benthos. Values are for a t-test for two treatments setting $\alpha = 0.05$, power = 0.80 (see Cohen, 1977) and using Morin's (1985) regression to calculate variances assuming sampler size is held constant at 0.1 m^2.

	Number of Replicates per Treatment		
Mean (number per m^2)	For 33% Decrease	For 50% Decrease	For 90% Decrease
5	306	125	31
10	204	87	22
50	100	40	11
100	74	30	8
500	40	16	5
1,000	31	13	4
5,000	20	9	4
10,000	17	8	3
50,000	12	6	3
100,000	11	5	3
500,000	10	5	3

crease for a given α and size of effect one wishes to detect. The value of doing pilot experiments to determine appropriate sample sizes should be obvious. However, for highly variable data, reasonable power will be achieved at reasonable cost only if it is sufficient to detect relatively large differences between treatments (Allan, 1984). Morin (1985) demonstrated that more replicate samples were necessary to achieve a desired level of precision when estimating sample means for rare species than for common ones. A corollary of this is that more replicates are needed to demonstrate an effect at a given power for rare organisms (Table 11.1), so it is probably better to select relatively abundant organisms as dependent variables for quantitative studies. Cohen (1977) is a good introductory text on power analyses, and Gerrodette (1987) presents power analyses for detecting trends.

11.3.3. Analysis of Experimental Results

The choice of statistical analyses should be part of the design phase of experimentation. To a large degree, the design should be dictated by practical considerations, features of the system such as potentially confounding covariates, and the chosen statistical analyses. Number of replicates per treatment, cross-classification of treatments, and assignment of treatments may depend critically on the statistical analysis that will be used to analyze data. For example, many nonparametric tests for paired data require at least five replicate pairs, and some orthogonal comparisons may require equal cell sizes (Winer, 1971; Conover, 1980).

Before embarking upon the analysis it is imperative to look at the data first and check for errors, outliers, and impossible or nonsensical values (Green, 1979; Underwood, 1981; Norris and Georges, Chapter 7). The next step is to assess whether the data meet the assumptions of the analysis. ANOVA, for example, requires that the error terms be normally distributed, that treatments be independent, and that variances be homogeneous. Violations of these assumptions can have large effects on the statistical results. Fortunately, some of these problems may be overcome by transforming the data, although violation of the independence assumption cannot be corrected after the data have been collected (Winer, 1971; Green, 1979).

Different transformations are appropriate for different types of data, sometimes on *a priori* grounds, and these are fully discussed in several basic texts (Elliott, 1977; Green, 1979; Sokal and Rohlf, 1981). Traditionally, logarithmic transformations have been applied routinely to abundance data of benthic macroinvertebrates, but some controversy exists about whether fractional power transformations may provide a better general purpose method (Downing, 1979; Chang and Winnell, 1981). It appears that the choice of the best transformation will depend on the data themselves (Downing, 1981; Vézina, 1988). The adequacy of whichever method is chosen should be checked (Downing, 1981; Underwood, 1981). Norris and Georges (Chapter 7) consider transformations in detail.

If the data still violate the critical assumptions, nonparametric methods could be used instead, although they are generally less powerful than their parametric counterparts (Green, 1979). These methods do make some assumptions, although they are less restrictive than those for parametric tests. For example, ordinal tests assume that data are continuous, that is, that they can assume any value within the precision of the measuring instrument. Counts violate this assumption, but these tests appear to be fairly robust to this violation. In addition, it should be remembered that some data (e.g., those expressed as ranks) imply use of nonparametric methods *a priori*. Remember, nonparametric techniques are not simply a "backup" for when parametric methods fail (Conover, 1980).

The most powerful ANOVA analyses involve preplanned comparisons between treatments (Winer, 1971; Sokal and Rohlf, 1981); this requires some previous knowledge of the system to determine which comparisons will be of interest. Although planned comparisons probably could be used more frequently in ecology, many experiments still will be exploratory and ecologists are more likely to need *a posteriori* methods to test significance among treatment means. Day and Quinn (1989) exhaustively review all the methods available and provide recommendations for the most appropriate techniques under various circumstances.

If a number of treatment levels (say > 5) are used in experiments, then responses can be analyzed using regression techniques. Analyses spanning a wide range of treatment levels are useful, both because they provide response curves for the full range of observed or predicted perturbation levels and because they provide precise equations for predicting environmental impact over the manipulated range. In addition, the advent of inexpensive computers and useful statistical software is making sophisticated nonlinear regression techniques easier to apply in situations where simple linear or polynomial techniques are inadequate (Fig. 11.6). Graney et al. (1989) argue convincingly that application of regression analyses to designs using a number of treatment levels, with limited replication at each level, provides more useful and sensitive results than ANOVAs applied to designs using fewer levels but more replicates per treatment.

In many cases, analysis of covariance (ANCOVA), can be used to factor out the influences of potential confounding variables and to indicate possible relationships between the manipulated variable and various unmanipulated variables. For example, in many field experiments a certain amount of variation occurs among replicates within a treatment for target dependent variables. Initial or premanipulation levels of a response variable in individual replicates can be used as covariates in an ANCOVA to adjust values in treatments to account for these initial variations, provided that the required assumptions are met (Underwood, 1981). Alternatively, a researcher may suspect that variation among replicates within a treatment is caused by variation in a previously measured environmental factor among replicates. An ANCOVA that treats this environmental factor as a covariate will indicate the effects of treatment on an organism's response to the environmental factor, allow an examination of treatment effects after confounding environmental influences have been considered, and suggest possible interactions between the manipulated variable and the environmental factor. Because manipulation of more than three variables (with sufficient replication) often becomes logistically impossible, these hybrid correlational/manipulative approaches may be the only alternative for indicating interactions among a variety of factors in affecting response variables, and also suggest the strong influence of environmental factors that should be considered in future experimental designs.

Where possible, ANOVA or regression analyses should be followed by examination of the residuals to check the adequacy of the statistical model. Systematic patterns in plots of residuals against predicted values indicate inadequate or wrong transformation, and plotting residuals against such variables as time of collection or order of sample processing sometimes can reveal unsuspected correlations or biases in the data (Daniel and Wood, 1980; Draper and Smith, 1981).

In practice, many experiments collect data on several taxa rather than just

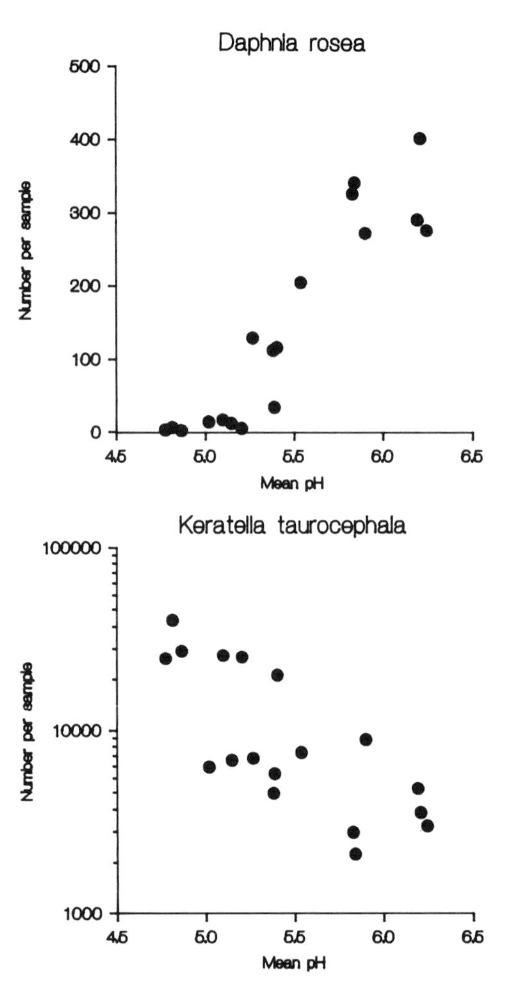

Figure 11.6. Responses of two zooplankton species to experimental acidification of *in situ* enclosures in a small lake elucidated by nonlinear regression techniques. Mean abundance of *Daphnia rosea* and *Keratella taurocephala* in each replicate enclosure is plotted vs. mean pH of each enclosure over the experimental period. *Daphnia*'s response to pH was described using a simplex, least-squares model-fitting procedure to estimate thresholds and slopes. *Daphnia* abundance = X_0 [299 (pH − 5.01)]. 5.01 is the threshold pH below which *Daphnia* abundance = 0, and X_0 is an "indicator variable" which = 0 below the threshold, and = 1 above the threshold pH. All thresholds and slopes differed significantly from zero, and the equation accounted for 87% of the variation in *Daphnia* abundance ($F = 118.0$, $P < 0.001$). *Keratella*'s abundance was log-linearly related to pH: Log_{10} *Keratella* abundance = 16.8 − 1.4 pH ($r^2 = 0.57$, $F = 21.3$, $P < 0.001$). The volume of a sample was 0.11 m^3. From Barmuta et al. (1990).

one. If researchers are interested in finding an "indicator species," then a series of univariate tests (ANOVAs, t-tests) will indicate the responses of each taxon to experimental manipulations, although a large number of such analyses will inflate the Type I error. Where firm hypotheses are being tested, rather than examining the data for promising patterns requiring further experimentation, inflation of the Type I error can be minimized by applying the sequential Bonferroni's correction (Harris, 1985; Rice, 1989). Alternatively, researchers may be interested in whole-community responses, preferring to analyze the simultaneous responses of all taxa to manipulation by performing multivariate equivalents of the univariate tests (e.g., MANOVA). Green (1979) presents a hypothetical worked example and also emphasizes that including too many dependent variables is problematic. If reducing the number of dependent variables is necessary, he recommends using techniques such as Principal Components Analysis (PCA) to derive a new set of transformed variables which are then used as the dependent variables (see Morin, 1987, for an example). MANOVAs can be followed by appropriate discriminant analyses to determine which species are best used to distinguish between treatments. Chatfield and Collins (1980), Harris (1985), and Manly (1986) provide good introductions to multivariate hypothesis testing.

A variety of multivariate methods also are available for exploring and displaying data with many dependent variables (see Norris and Georges, Chapter 7). Such techniques as clustering, ordination, and network trees are useful for generating hypotheses. Recent work by Faith et al. (1987) and Minchin (1987) suggest the most robust techniques for ecological purposes, ter Braak and Prentice (1988) and James and McCulloch (1990) provide comprehensive reviews, and progress is being made towards providing significance tests for analyses based on these methods (e.g., Faith and Norris, 1989; Faith, 1990).

11.3.4. Interpretation of Analyses

The degree to which experimental results can be extrapolated to natural environments will depend on the degree to which experimental replicates are open to biotic processes such as colonization and emigration (Cooper et al., 1990; see also Buikema and Voshell, Chapter 10), as well as abiotic processes such as mixing, atmospheric inputs, etc. (Schindler, 1987).

For example, many investigators begin experiments with cages or channels provided with natural substrates devoid of macroinvertebrates and then monitor the response of the colonizing macroinvertebrates to manipulations. Without comparisons to natural assemblages, the results of such experiments should not be extrapolated to natural conditions for two reasons. First, macroinvertebrate assemblages colonizing cages or channels are often different from those resident in natural benthic substrates because macroinvertebrate species differ in their relative mobilities (Townsend and Hildrew, 1976; Williams and Hynes, 1976). Second, the responses of individuals to environmental perturbations may differ if they are moving into or through the system compared to being resident in the substrate.

To ensure that the results of manipulations can be extrapolated to natural communities, researchers either should compare assemblages in unmanipulated enclosures to those in natural systems to show that they are similar, or they should initially stock experimental units with natural assemblages. Even when experimental enclosures contain assemblages similar to those found in the natural environment at the beginning of experiments, alteration of macroinvertebrate colonization rates can have a large effect on the experimental results. For example, Hopkins et al. (1989) found that most baetid mayfly nymphs drifted out of experimental channels in response to a pulsed addition of acid; however, baetid densities in control and experimental channels were similar two days after acid inputs ended, owing to the rapid recolonization of open experimental channels by drifting nymphs. In contrast, Cooper et al. (1988b) found severe, lasting effects of pulsed acidification on baetid densities when drift into the channels was blocked by nets.

Decisions on whether or not to allow colonization will hinge on the scale to which experimental results are to be extrapolated. If researchers are interested primarily in the effects of a localized perturbation on a small patch of habitat, then colonization of small enclosures should simulate what happens at that scale in natural situations (i.e., enclosures should be open to natural colonization). Alternatively, if lake bags or stream channels represent "miniature" lakes and streams, and the results are to be extrapolated to larger natural systems, then experimental colonization rates should simulate natural intersystem (lake-to-lake, stream-to-stream) dispersal. Scale considerations can have profound effects on the interpretation of experimental results for a number of other reasons (see Section 11.3.1, "Importance of Scale," above).

If statistical significance is being used as the criterion for a treatment effect, it always should be remembered that our capability to detect statistical significance is as much a function of the power of the statistical test as it is of a real treatment effect. Statistical tests of null hypotheses should be accompanied by power analyses, which indicate the probability of making a Type II error. Statements of probabilities of Type I and II errors are especially critical to managers because both errors carry attendant resource and economic costs. Some statement of test power will guide managers in assessing the probability that acceptance of the null hypothesis of no treatment effect will be erroneous, particularly when treatment effects are not statistically significant.

11.4. Conclusions

Field experiments constitute an extremely powerful approach for determining the effects of anthropogenic perturbations on ecological systems. They should be an essential part of biomonitoring programs because they can be used in the following ways: (1) to interpret observed responses to environmental change; (2) to calibrate monitoring programs; (3) to identify sensitive species; (4) to predict responses to possible perturbations; (5) to disentangle direct and indirect effects of perturbations; and (6) to determine the direct and interactive effects of a variety of variables on ecological systems. The results of properly designed, executed, and analyzed field experiments can be more unambiguously interpreted than the results of surveys, natural experiments, or laboratory bioassays. The rigor with which manipulative experiments are designed and analyzed can affect the results and their interpretation. Too often, the interpretation of past field experiments has been compromised by experimental artifacts, the misidentification of experimental replicates, the low power and poor choice of statistical tests, the scale of manipulations, and inadequate controls. By keeping well-defined questions and hypotheses clearly in mind, and by following such questions and hypotheses with rigorous designs and statistical analyses, a researcher should be able to avoid most of the pitfalls outlined above.

In the interests of clarity in the design, execution, analysis, and interpretation of manipulative experiments, we offer the following recommendations:

1. Clearly state the ecological question, including a precise identification of dependent and independent variables.
2. Clearly outline a null hypothesis and a series of alternative hypotheses, which represent possible outcomes to experimental manipulations.
3. In designing experiments to test hypotheses, investigators should set up unmanipulated, control sites or arenas that are treated exactly as manipulated areas, except for alterations of the target independent variable.
4. Each of the control and manipulated treatments should be well-replicated so that rigorous statistical analyses can be applied to the results. Replicates assigned to different treatments should not be spatially segregated. The design of the experiment should clearly define experimental replicates, that is, the sites or arenas that are manipulated or used as controls, and these units, *not* samples or individuals within them, should be used as replicates in statistical analyses. For time-series data, these analyses would include repeated measures ANOVAs or profile analyses.

5. All experimental units should be sampled prior to manipulations to ensure that experimental replicates are similar at the outset. Where it is logistically impossible to have more than one to a few replicates per treatment, it is especially important that a long pre-manipulation data set be acquired to act as a baseline for evaluating the effects of manipulations.
6. The researcher should recognize explicitly the spatial and temporal scale of experiments relative to the spatial and temporal scale of questions. Ideally, experiments should be performed at the same scale to which the results are to be extrapolated. If this is logistically impossible and experimental units are smaller than natural systems (lakes, streams), then investigators should compare response values inside experimental units to those in the natural system to evaluate "cage" artifacts.
7. The spatial scale of experiments, the duration and frequency of manipulations (press vs. pulse), the choice of dependent and independent variables, and the range of treatment levels should be tailored to the scale, duration, magnitude, and frequency of observed, predicted, or potential perturbations and their postulated effects on response variables. Scale and the choice of response variables are intimately related: physiological responses are quick and temporary, whereas ecosystem responses often are delayed and long-lasting.
8. Rigorous statistical analyses are needed to ascertain that observed responses could not have occurred by chance. Use of statistical significance as the criterion for treatment effects, however, is as much a function of the power of statistical tests as of a real treatment effect. Statistical analyses of data should not only include a statement of the probability of making a Type I error, but should also assess the power of tests applied.
9. Data should be examined to ascertain that they meet the assumptions of statistical tests (e.g., in the case of parametric statistics) by examining residuals, variance to mean relationships, and testing for homogeneity of variances. If assumptions are not met, then data may require transformation; adequacy of the transformation should be checked before applying the test. Alternatively, statistics that do not make such restrictive assumptions, such as nonparametric or distribution-free methods, may be used.
10. In situations where it is impossible to set up more than one replicate for each treatment, then control and manipulated sites should be measured for some time before and after manipulations are started. Investigators can then apply BACI or RIA methodologies

to the data to see if differences between control and manipulated sites change after manipulation. However, without additional information, such as logical mechanisms for such effects based on other descriptive or experimental data, it is difficult to attribute any observed changes to experimental manipulations.

11. Regression analyses can be used to provide precise, predictive equations for relationships between dependent and independent variables in experiments where a number of treatment levels are used. ANCOVA can be used to factor out the influence of potentially confounding variables on treatment effects and indicate possible relationships between manipulated and unmanipulated variables.
12. The choice of univariate vs. multivariate tests often will depend on whether an investigator is interested in population- or community-level responses to perturbations. In the former case, univariate tests may indicate sensitive indicator species; however, the results of numerous univariate tests must be approached with caution because of the probability of committing a Type I error by chance. Alternatively, multivariate hypothesis-testing analyses (e.g., MANOVA) can indicate the effects of treatment on whole assemblages. Descriptive multivariate techniques (e.g., PCA) can indicate relationships among dependent or among independent variables, and can reduce and transform such data to a small number of variables that then can be used in standard statistical analyses.
13. Researchers should consider potential experimental artifacts, colonization conditions, scale of manipulations, the nature of controls, and the power of tests in interpreting the results of experiments, particularly in precisely delineating the natural conditions under which the experimental results have relevance.

It is our contention that field experiments for directly assessing the effects of anthropogenic perturbations on aquatic resources should be more widely used, thereby providing managers with a rigorous base for evaluating or predicting the effects of observed or potential disturbances on freshwater systems. Such experiments provide a more powerful and less ambiguous database, compared with other approaches, for evaluating the effects of environmental change. A critical need in field experiments is careful delineation of the degree to which experimental results can be extrapolated to other situations or scales. For example, it is important to know the degree to which small-scale, short-term experiments can be extrapolated accurately to whole systems. The most straightforward way to examine the scale-dependence of ecological results is to perform similar experimental manipulations at a variety of spatial and temporal scales.

Because of the high cost of whole-system manipulations, researchers from academic, governmental, and private institutions should coordinate their efforts to evaluate the environmental effects of possible or planned perturbations within the context of a rigorous experimental design. The use of experiments that are rigorously designed, executed, analyzed, and interpreted can transform the art of environmental impact assessment into a science.

Acknowledgments

We are grateful to R. Griffiths, R. Hall, D. Hart, V.G. Mite, and G. Quinn for discussions or making unpublished data available to us. J.D. Allan and R. Voshell, Jr., commented constructively on the manuscript. Thanks also to S. Wiseman and especially K. Kratz for preparation of the figures. D. Divins, T. Dudley, R. Griffiths, R. Hall, K. Kratz, K. Mills, C. Osenberg, J. Sickman, and K. Tonnessen kindly provided slides and photographs. J. Melack generously provided funds. Our greatest debt, however, is to R.H. Green, S.H. Hurlbert, J.D. Allan, and A.J. Underwood for their excellent reviews of experimental design and analysis in ecology.

References

Allan, J.D. 1984. Hypothesis testing in ecological studies of aquatic insects. In *The Ecology of Aquatic Insects,* eds. V.H. Resh and D.M. Rosenberg, pp. 484–507. Praeger Pubs., New York.

Allen, T.F.H. and T.B. Starr. 1982. *Hierarchy: Perspectives for Ecological Complexity*. Univ. of Chicago Press, Chicago.

Almer, B., W. Dickson, C. Ekstrom, E. Hornstrom, and U. Miller. 1974. Effects of acidification on Swedish lakes. *Ambio* 3:30–6.

Barmuta, L.A., S.D. Cooper, S.K. Hamilton, K.W. Kratz, and J.M. Melack. 1990. Responses of zooplankton and zoobenthos to experimental acidification in a high-elevation lake (Sierra Nevada, California, U.S.A.). *Freshwater Biology* 23:571–86.

Barton, D.R. and S.M. Smith. 1984. Insects of extremely small and extremely large aquatic habitats. In *The Ecology of Aquatic Insects,* eds. V.H. Resh and D.M. Rosenberg, pp. 456–83. Praeger Pubs., New York.

Bender, E.A., T.J. Case, and M.E. Gilpin. 1984. Perturbation experiments in community ecology: theory and practice. *Ecology* 65:1–13.

Box, G.E.P. and G.M. Jenkins. 1976. *Time Series Analysis, Forecasting and Control*. Holden-Day, San Francisco.

Brown, S.S. and D.K. King. 1987. Community metabolism in natural and agriculturally disturbed riffle sections of the Chippewa River, Isabella County, Michigan. *Journal of Freshwater Ecology* 4:39–51.

Cairns, J., Jr. 1983. Are single species toxicity tests alone adequate for estimating environmental hazard? *Hydrobiologia* 100:47–57.

Carpenter, S.R. 1988. Transmission of variance through lake food webs. In *Complex Interactions in Lake Communities,* ed. S.R. Carpenter, pp. 119–35. Springer-Verlag, New York.

Carpenter, S.R. 1989. Replication and treatment strength in whole-lake experiments. *Ecology* 70:453–63.

Carpenter, S.R., T.M. Frost, D. Heisey, and T.K. Kratz. 1989. Randomized intervention analysis and the interpretation of whole-ecosystem experiments. *Ecology* 70:1142–52.

Carpenter, S.R. and J.F. Kitchell. 1987. The temporal scale of variance in limnetic primary production. *American Naturalist* 129:417–33.

Carpenter, S.R. and J.F. Kitchell. 1988. Consumer control of lake productivity. *BioScience* 38:764–9.

Carpenter, S.R., J.F. Kitchell, J.R. Hodgson, P.A. Cochran, J.J. Elser, M.M. Elser, D.M. Lodge, D. Kretchmer, X. He, and C.N. von Ende. 1987. Regulation of lake primary productivity by food web structure. *Ecology* 68:1863–76.

Chang, W.Y.B. and M.H. Winnell. 1981. Comment on the fourth-root transformation. *Canadian Journal of Fisheries and Aquatic Sciences* 38:126–7.

Chatfield, C. and A.J. Collins. 1980. *Introduction to Multivariate Analysis*. Chapman and Hall, London.

Clements, W.H., D.S. Cherry, and J. Cairns, Jr. 1988. Impact of heavy metals on insect communities in streams: a comparison of observational and experimental results. *Canadian Journal of Fisheries and Aquatic Sciences* 45:2017–25.

Clements, W.H., D.S. Cherry, and J. Cairns, Jr. 1989. The influence of copper exposure on predator-prey interactions in aquatic insect communities. *Freshwater Biology* 21:483–8.

Cohen, J. 1977. *Statistical Power Analysis for the Behavioral Sciences*. Academic Press, New York.

Conover, W.J. 1980. *Practical Nonparametric Statistics,* 2nd ed. John Wiley, New York.

Cooper, S.D. and T.L. Dudley. 1988. The interpretation of "controlled" vs "natural" experiments in streams. *Oikos* 52:357–61.

Cooper, S.D., T.L. Dudley, and M.R. Shaw. 1988a. Effects of interstitial and cryptic predators on stream pool assemblages. Paper presented at the 36th Annual Meeting of the North American Benthological Society, Tuscaloosa, AL, May 17–20, 1988. *Bulletin of the North American Benthological Society* 5:85. Abstract.

Cooper, S.D., K. Kratz, R.W. Holmes, and J.M. Melack. 1988b. *An Integrated Watershed Study: an Investigation of the Biota in the Emerald Lake System and Stream Channel Experiments*. Report for California Air Resources Board. Marine Science Institute and Department of Biological Sciences, University of California, Santa Barbara, CA.

Cooper, S.D., S.J. Walde, and B.L. Peckarsky. 1990. Prey exchange rates and the impact of predators on prey populations in streams. *Ecology* 71:1503–14.

Crossland, N.O. and C.J.M. Wolff. 1985. Fate and biological effects of pentachlorophenol in outdoor ponds. *Environmental Toxicology and Chemistry* 4:73–86.

Culp, J.M., F.J. Wrona, and R.W. Davies. 1986. Response of stream benthos and drift to fine sediment deposition versus transport. *Canadian Journal of Zoology* 64:1345–51.

Dance, K.W. and H.B.N. Hynes. 1980. Some effects of agricultural land use on stream insect communities. *Environmental Pollution* (Series A) 22:19–28.

Daniel, C. and F.S. Wood. 1980. *Fitting Equations to Data,* 2nd ed. John Wiley, New York.

Day, R.W. and G.P. Quinn. 1989. Comparisons of treatments after an analysis of variance in ecology. *Ecological Monographs* 59:433–63.

Dewey, S.L. 1986. Effects of the herbicide atrazine on aquatic insect community structure and emergence. *Ecology* 67:148–62.

Diamond, J. 1986. Overview: laboratory experiments, field experiments, and natural experiments. In *Community Ecology,* eds. J. Diamond and T.J. Case, pp. 3–22. Harper and Row, New York.

Downing, J.A. 1979. Aggregation, transformation, and the design of benthos sampling programs. *Journal of the Fisheries Research Board of Canada* 36:1454–63.

Downing, J.A. 1981. How well does the fourth-root transformation work? *Canadian Journal of Fisheries and Aquatic Sciences* 38:127–9.

Draper, N.R. and H. Smith. 1981. *Applied Regression Analysis,* 2nd ed. John Wiley, New York.

Eberhardt, L.L. 1976. Quantitative ecology and impact assessment. *Journal of Environmental Management* 4:27–70.

Elliott, J.M. 1977. Some methods for the statistical analysis of samples of benthic invertebrates, 2nd ed. *Freshwater Biological Association Scientific Publication* No. 25:1–156.

Eriksson, M.O.G. and B. Tengelin. 1987. Short-term effects of liming on perch *Perca fluviatalis* populations in acidified lakes in south-west Sweden. *Hydrobiologia* 146:187–91.

Evans, R.A. 1989. Response of limnetic insect populations of two acidic, fishless lakes to liming and brook trout *(Salvelinus fontinalis)* introduction. *Canadian Journal of Fisheries and Aquatic Sciences* 46:342–51.

Faith, D.P. 1990. Benthic macroinvertebrates in biological surveillance: Monte Carlo significance tests on functional groups' responses to environmental gradients. *Environmental Monitoring and Assessment* 14:247–64.

Faith, D.P., P.R. Minchin, and L. Belbin. 1987. Compositional dissimilarity as a robust measure of ecological distance. *Vegetatio* 69:57–68.

Faith, D.P. and R.H. Norris. 1989. Correlation of environmental variables with patterns of distribution and abundance of common and rare freshwater macroinvertebrates. *Biological Conservation* 50:77–98.

Frost, T.M., D.L. DeAngelis, S.M. Bartell, D.J. Hall, and S.H. Hurlbert. 1988. Scale in the design and interpretation of aquatic community research. In *Complex Interactions in Lake Communities,* ed. S.R. Carpenter, pp. 229–58. Springer-Verlag, New York.

Gerrodette, T. 1987. A power analysis for detecting trends. *Ecology* 68:1364–72.

Goldman, C.R., A. Jassby, and T. Powell. 1989. Interannual fluctuations in primary production: meteorological forcing at two subalpine lakes. *Limnology and Oceanography* 34:310–23.

Gore, J.A. 1980. Ordinational analysis of benthic communities upstream and downstream of a prairie storage reservoir. *Hydrobiologia* 69:33–44.

Graney, R.L., J.P. Giesy, Jr., and D. DiToro. 1989. Mesocosm experimental design strategies: advantages and disadvantages in ecological risk assessment. In *Using Mesocosms to Assess the Aquatic Ecological Risk of Pesticides: Theory and Practice,* ed. J.R. Voshell, Jr., pp. 74–88. Miscellaneous Publication No. 75, Entomological Society of America, Lanham, MD.

Green, R.H. 1979. *Sampling Design and Statistical Methods for Environmental Biologists*. John Wiley, New York.

Griffiths, R.W. 1987. Acidification as a model for the response of aquatic communities to stress. Ph.D. dissertation, Univ. of Waterloo, Waterloo, ON.

Gurtz, M.E. and J.B. Wallace. 1984. Substrate-mediated response of stream invertebrates to disturbance. *Ecology* 65:1556–69.

Hall, D.J., W.E. Cooper, and E.E. Werner. 1970. An experimental approach to the production dynamics and structure of freshwater animal communities. *Limnology and Oceanography* 15:839–928.

Hall, R.J. 1989. Structural organization of benthic communities in lake outflows exposed to short- (4-days) and long-term (3–5 years) pH disturbance. Paper presented at the 37th Annual Meeting of the North American Benthological Society, Guelph, ON, May 16–19, 1989. *Bulletin of the North American Benthological Society* 6:96. Abstract.

Hall, R.J., C.T. Driscoll, G.E. Likens, and J.M. Pratt. 1985. Physical, chemical, and biological consequences of episodic aluminum additions to a stream. *Limnology and Oceanography* 30:212–20.

Hall, R.J., G.E. Likens, S.B. Fiance, and G.R. Hendrey. 1980. Experimental acidification of a stream in the Hubbard Brook Experimental Forest, New Hampshire. *Ecology* 61:976–89.

Harris, G.P. 1986. *Phytoplankton Ecology: Structure, Function and Fluctuation.* Chapman and Hall, London.

Harris, R.J. 1985. *A Primer of Multivariate Statistics,* 2nd ed. Academic Press, New York.

Hart, D.D. and C.T. Robinson. 1990. Resource limitation in a stream community: phosphorus enrichment effects on periphyton and grazers. *Ecology* 71:1494–1502.

Holomuzki, J.R. 1989. Salamander predation and vertical distributions of zooplankton. *Freshwater Biology* 21:461–72.

Hopkins, P.S., K.W. Kratz, and S.D. Cooper. 1989. Effects of an experimental acid pulse on invertebrates in a high altitude Sierra Nevada stream. *Hydrobiologia* 171:45–58.

Hurlbert, S.H. 1984. Pseudoreplication and the design of ecological field experiments. *Ecological Monographs* 54:187–211.

Hurlbert, S.H., M.S. Mulla, and H.R. Willson. 1972. Effects of an organophosphorus insecticide on the phytoplankton, zooplankton, and insect populations of fresh-water ponds. *Ecological Monographs* 42:269–99.

James, F.C. and C.E. McCulloch. 1990. Multivariate analysis in ecology and systematics: panacea or Pandora's box? *Annual Review of Ecology and Systematics* 21:129–66.

Jenkins, D.G., R.J. Layton, and A.L. Buikema, Jr. 1989. State of the art in aquatic ecological risk assessment. In *Using Mesocosms to Assess the Aquatic Ecological Risk of Pesticides: Theory and Practice,* ed. J.R. Voshell, Jr., pp. 18–32. Miscellaneous Publication No. 75, Entomological Society of America, Lanham, MD.

Kimball, K.D. and S.A. Levin. 1985. Limitations of laboratory bioassays: the need for ecosystem-level testing. *BioScience* 35:165–71.

Kitchell, J.F., S.M. Bartell, S.R. Carpenter, D.J. Hall, D.J. McQueen, W.E. Neill, D. Scavia, and E.E. Werner. 1988. Epistemology, experiments, and pragmatism. In *Complex Interactions in Lake Communities,* ed. S.R. Carpenter, pp. 263–80. Springer-Verlag, New York.

Lamberti, G.A. and V.H. Resh. 1979. Substrate relationships, spatial distribution patterns, and sampling variability in a stream caddisfly population. *Environmental Entomology* 8:561–7.

Legendre, L. and S. Demers. 1984. Towards dynamic biological oceanography and limnology. *Canadian Journal of Fisheries and Aquatic Sciences* 41:2–19.

Leland, H.V., S.V. Fend, T.L. Dudley, and J.L. Carter. 1989. Effects of copper on species composition of benthic insects in a Sierra Nevada, California stream. *Freshwater Biology* 21:163–79.

Leopold, L.B., M.G. Wolman, and J.P. Miller. 1964. *Fluvial Processes in Geomorphology*. W.H. Freeman, San Francisco.

Likens, G.E. 1985. An experimental approach for the study of ecosystems. *Journal of Ecology* 73:381–96.

Magnuson, J.J., B.J. Benson, and T.K. Kratz. 1990. Temporal coherence in the limnology of a suite of lakes in Wisconsin, U.S.A. *Freshwater Biology* 23:145–59.

Manly, B.F.J. 1986. *Multivariate Statistical Methods. A Primer*. Chapman and Hall, London.

Marchant, R., P. Mitchell, and R. Norris. 1984. Distribution of benthic invertebrates along a disturbed section of the La Trobe River, Victoria: an analysis based on numerical classification. *Australian Journal of Marine and Freshwater Research* 35:355–74.

Marmorek, D. 1984. Changes in the temporal behavior and size structure of plankton systems in acid lakes. In *Early Biotic Responses to Advancing Lake Acidification,* ed. G.R. Hendrey, pp. 23–41. Butterworth Pubs., Boston.

Melack, J.M., S.D. Cooper, R.W. Holmes, J.O. Sickman, K. Kratz, P. Hopkins, H. Hardenbergh, M. Thieme, and L. Meeker. 1987. *Chemical and Biological Survey of Lakes and Streams Located in the Emerald Lake Watershed, Sequoia National Park*. Report for California Air Resources Board. Marine Science Institute and Department of Biological Sciences, University of California, Santa Barbara, CA.

Minchin, P.R. 1987. An evaluation of the relative robustness of techniques for ecological ordination. *Vegetatio* 69:89–108.

Minshall, G.W. 1988. Stream ecosystem theory: a global perspective. *Journal of the North American Benthological Society* 7:263–88.

Millard, S.P., J.R. Yearsley, and D.P. Lettenmaier. 1985. Space-time correlation and its effects on methods for detecting aquatic ecological change. *Canadian Journal of Fisheries and Aquatic Sciences* 42:1391–1400.

Morin, A. 1985. Variability of density estimates and the optimization of sampling programs for stream benthos. *Canadian Journal of Fisheries and Aquatic Sciences* 42:1530–4.

Morin, P.J. 1987. Salamander predation, prey facilitation and seasonal succession in microcrustacean communities. In *Predation: Direct and Indirect Impacts on Aquatic Communities,* eds. W.C. Kerfoot and A. Sih, pp. 174–87. Univ. Press of New England, Hanover, NH.

Muirhead-Thomson, R.C. 1987. *Pesticide Impact on Stream Fauna: With Special Reference to Macroinvertebrates*. Cambridge Univ. Press, Cambridge, England.

Norris, R.H., P.S. Lake, and R. Swain. 1982. Ecological effects of mine effluents on the South Esk River, north-eastern Tasmania. III. Benthic macroinvertebrates. *Australian Journal of Marine and Freshwater Research* 33:789–809.

Ormerod, S.J., P. Boole, C.P. McCahon, N.S. Weatherley, D. Pascoe, and R.W. Edwards. 1987. Short-term experimental acidification of a Welsh stream: comparing the biological effects of hydrogen ions and aluminium. *Freshwater Biology* 17:341–56.

Osenberg, C.W. 1988. Body size and the interaction of fish predation and food limitation in a freshwater snail community. Ph.D. dissertation, Michigan State Univ., East Lansing, MI.

Pearson, R.G. 1984. Temporal changes in the composition and abundance of the macro-invertebrate communities of the River Hull. *Archiv für Hydrobiologie* 100:273–98.

Peckarsky, B.L. and M.A. Penton. 1990. Effects of enclosures on stream microhabitat and invertebrate community structure. *Journal of the North American Benthological Society* 9:249–61.

Peterman, R.M. 1990. The importance of reporting statistical power: the forest decline and acidic deposition example. *Ecology* 71:2024–7.

Resh, V.H. and J.D. Unzicker. 1975. Water quality monitoring and aquatic organisms: the importance of species identification. *Journal of the Water Pollution Control Federation* 47:9–19.

Rice, W.R. 1989. Analyzing tables of statistical tests. *Evolution* 43:223–5.

Rodgers, E.B. 1983. Fecundity of *Caenis* (Ephemeroptera: Caenidae) in elevated water temperatures. *Journal of Freshwater Ecology* 2:213–8.

Roff, J.C. and R.E. Kwiatkowski. 1977. Zooplankton and zoobenthos communities of selected northern Ontario lakes of different acidities. *Canadian Journal of Zoology* 55:899–911.

Rosenberg, D.M., V.H. Resh, S.S. Balling, M.A. Barnby, J.N. Collins, D.V. Durbin, T.S. Flynn, D.D. Hart, G.A. Lamberti, E.P. McElravy, J.R. Wood, T.E. Blank, D.M. Schultz, D.L. Marrin, and D.G. Price. 1981. Recent trends in environmental impact assessment. *Canadian Journal of Fisheries and Aquatic Sciences* 38:591–624.

Schindler, D.W. 1974. Eutrophication and recovery in experimental lakes: implications for lake management. *Science* 184:897–9.

Schindler, D.W. 1977. Evolution of phosphorous limitation in lakes. *Science* 195:260–2.

Schindler, D.W. 1987. Detecting ecosystem responses to anthropogenic stress. *Canadian Journal of Fisheries and Aquatic Sciences* 44 (Supplement 1):6–25.

Schindler, D.W. and E.J. Fee. 1974. Experimental Lakes Area: whole-lake experiments in eutrophication. *Journal of the Fisheries Research Board of Canada* 31:937–53.

Schindler, D.W., R.H. Hesslein, R. Wagemann, and W.S. Broecker. 1980. Effects of acidification on mobilization of heavy metals and radionuclides from the sediments of a freshwater lake. *Canadian Journal of Fisheries and Aquatic Sciences* 37:373–7.

Schindler, D.W., K.H. Mills, D.F. Malley, D.L. Findlay, J.A. Shearer, I.J. Davies, M.A. Turner, G.A. Linsey, and D.R. Cruikshank. 1985. Long-term ecosystem stress: the effects of years of experimental acidification on a small lake. *Science* 228:1395–1401.

Sokal, R.R. and F.J. Rohlf. 1981. *Biometry,* 2nd ed. W.H. Freeman, San Francisco.

Spence, J.A. and H.B.N. Hynes. 1971. Differences in benthos upstream and downstream of an impoundment. *Journal of the Fisheries Research Board of Canada* 28:35–43.

Stewart-Oaten, A., J.R. Bence, and C.W. Osenberg. 1992. Assessing effects of unreplicated perturbations: no simple solutions. *Ecology* 73:1396–1404.

Stewart-Oaten, A., W.W. Murdoch, and K.R. Parker. 1986. Environmental impact assessment: "pseudoreplication" in time? *Ecology* 67:929–40.

ter Braak, C.J.F. and I.C. Prentice. 1988. A theory of gradient analysis. *Advances in Ecological Research* 18:271–317.

Toft, C.A. and P.J. Shea. 1983. Detecting community-wide patterns: estimating power strengthens statistical inference. *American Naturalist* 122:618–25.

Tonnessen, K.A. 1984. Potential for aquatic ecosystem acidification in the Sierra Nevada, California. In *Early Biotic Responses to Advancing Lake Acidification,* ed. G.R. Hendrey, pp. 147–69. Butterworth Pubs., Boston.

Townsend, C.R. and A.G. Hildrew. 1976. Field experiments on the drifting, colonization and continuous redistribution of stream benthos. *Journal of Animal Ecology* 45:759–72.

Trautmann, N.M., C.E. McCulloch, and R.T. Oglesby. 1982. Statistical determination of data requirements for assessment of lake restoration programs. *Canadian Journal of Fisheries and Aquatic Sciences* 39:607–10.

Underwood, A.J. 1981. Techniques of analysis of variance in experimental marine biology and ecology. *Oceanography and Marine Biology: an Annual Review* 19:513–605.

Vézina, A.F. 1988. Sampling variance and the design of quantitative surveys of the marine benthos. *Marine Biology* 97:151–5.

Walters, C.J., J.S. Collie, and T. Webb. 1988. Experimental designs for estimating transient responses to management disturbances. *Canadian Journal of Fisheries and Aquatic Sciences* 45:530–8.

Warren, C.E. and G.E. Davis. 1971. Laboratory stream research: objectives, possibilities, and constraints. *Annual Review of Ecology and Systematics* 2:111–44.

Wiens, J.A., J.F. Addicott, T.J. Case, and J. Diamond. 1986. Overview: the importance of spatial and temporal scale in ecological investigations. In *Community Ecology,* eds. J. Diamond and T.J. Case, pp. 145–53. Harper and Row, New York.

Williams, D.D. and H.B.N. Hynes. 1976. The recolonization mechanisms of stream benthos. *Oikos* 27:265–72.

Winer, B.J. 1971. *Statistical Principles in Experimental Design,* 2nd ed. McGraw-Hill, New York.

Wise, D.H. 1984. The role of competition in spider communities: insights from field experiments with a model organism. In *Ecological Communities: Conceptual Issues and the Evidence,* eds. D.R. Strong, Jr., D. Simberloff, L.G. Abele, and A.B. Thistle, pp. 42–53. Princeton Univ. Press, Princeton, NJ.

Wright, J.F., D. Moss, P.D. Armitage, and M.T. Furse. 1984. A preliminary classification of running-water sites in Great Britain based on macro-invertebrate species and the prediction of community type using environmental data. *Freshwater Biology* 14:221–56.

Yan, N.D. and R. Strus. 1980. Crustacean zooplankton communities of acidic, metal-contaminated lakes near Sudbury, Ontario. *Canadian Journal of Fisheries and Aquatic Sciences* 37:2282–93.

Yasuno, M., J. Ohkita, and S. Hatakeyama. 1982. Effects of temephos on macrobenthos in a stream of Mt. Tsukuba. *Japanese Journal of Ecology* 32:29–38.

Zischke, J.A., J.W. Arthur, K.J. Nordlie, R.O. Hermanutz, D.A. Standen, and T.P. Henry. 1983. Acidification effects on macroinvertebrates and fathead minnows (*Pimephales promelas*) in outdoor experimental channels. *Water Research* 17:47–63.

12

Future Directions in Freshwater Biomonitoring Using Benthic Macroinvertebrates

R.O. Brinkhurst

12.1. Introduction

As I read through the various contributions to this book, I could see that many of the immediate future developments in biomonitoring have been mentioned. Simple extrapolation from the present situation is quite easy to do, and a catalog of new approaches would read like a recipe book. I believe that almost any biologist could work out a way in which his or her specialty could be applied to biomonitoring. Physiologists have shown already that simple organ-system preparations, such as heart-muscle or nerve-muscle isolates, can be used to test toxicity. Almost any taxonomic group can be exploited by an "expert" to produce field diagnoses of ecosystem health. A salmon biologist might tell you that studying benthos is a waste of time. A microbiologist might tell you to use a microcosm instead of studying a natural system. Likewise, sediment respiration can be used as a toxicology "test bed." Recent novel toxicological approaches examine the effects of contaminants directly on chromosomes, on larval metamorphoses, and on changes in behavior; they also document the appearance of genetic monsters or malformed offspring. Researchers have examined histological changes and chemical burdens in various animal tissues. Sentinel organisms, such as marine mussels, are the modern equivalent of the miner's canary (see Johnson et al., Chapter 4). The list of potentially new methods is legion.

The big problem is not so much the creation of lively new methods (or, cynically speaking, supporting our research field by suggesting it become the next panacea for environmental science); rather, the problem is getting these new methods accepted by environmental managers, politicians, and the courts. The slow, painful progress of multivariate statistical methods in the applied realm, and within government agencies in particular, serves as a good example of the reluctance of administrators to change procedures. I have heard senior managers reply to a series of presentations on why diversity indices should be replaced by incorporating taxonomic information within the available data (discussed later in this chapter) by saying: "We find diversity indices useful and will continue to use them." Researchers become impatient with the glacial rate at which new information becomes incorporated into statutes and regulations, but working within a government agency soon teaches one some of the reasons for that. It is far easier to manage and, if need be, prosecute with a small number of precise regulations than it is with a complex law requiring expert interpretation. Both our political and legal processes are slow and deliberate in a conscious attempt to avoid hasty mistakes. Legislation, once written, is difficult to rewrite, given the checks and balances in the system under which we live. Despite apparent central power, authoritarian regimes seem not to have done any better with environmental issues, but this may have more to do with their priorities than with the slow adoption of new science.

All this means that we have already an extensive array of methods available that we do not utilize fully. The future may require better education of both biologists and their managers so that these methods are understood and exploited. A set of simple, inexpensive, foolproof tests that can be applied by cheap labor with a minimum of thought will never exist despite the wishes of index-makers. Nevertheless, a relatively small fraction of our efforts still should be focused on creating and developing new methods. Such methods must be proposed on the basis of careful parallel trials against existing techniques in order to justify their promotion. Because I do not wish to suggest many radical changes in our methodology, I have thought it best to adopt a philosophical approach to this essay, rather than describing methods that are already in limited use or that could be dreamed up without careful testing.

12.2. The New Challenges

12.2.1. The Post-Industrial Society

Nine hundred fifty-one Superfund priority landfill sites have been identified in the USA, having an estimated clean-up cost of $300 billion (US). Apparently, more have been identified but not acknowledged, and presumably more are unknown. We are now aware of humans as a geophysical-scale force in the environment. We cannot imagine that we are setting out on a clean-up program that will fix things for all time, and so our biomonitoring scientists may find themselves running just to keep up.

We know about ozone layers, global warming, and acid rain, yet our current management regimes, based heavily on routine toxicology, are set in a point source scenario. Our future clearly should be much more concerned with general perturbations and non-point source issues. Although we have made substantial progress in the control of water pollution in Europe as well as in North America, ugly problems with malformed waterfowl and fish heavily contaminated with pesticides and PCBs and covered with pathogenic lesions hint at more general problems still to be faced. Recovery of the sport fishery in the Laurentian Great Lakes, after its near destruction by pollution, overfishing, and sea lampreys in the 1950s, has cost more than $8 billion in the United States. This recovery is marred by the suggestion that any woman intending to have a child at any time, or persons under 16 years of age, should not eat salmonids, carp, or catfish from many areas in the lakes, or that other people should not eat these fish more than once per week.

12.2.2. The Rest of the World Plays Catch-Up

It is hard to imagine what horrors are associated with the gross levels of water pollution that an aquatic biologist encounters in the Third World. I have visited the People's Republic of China twice attempting to collect aquatic oligochaetes from clean water. The existence of clean-water species in a group of so-called indicator organisms such as the oligochaetes will be mentioned below, but it is sufficient to say here that they were nowhere to be found. Reports exist of normal communities of benthos along the Sino-Soviet border from work done during a previous era of cooperation between China and the U.S.S.R. Otherwise, the wide rivers, shown on maps of the middle regions of China, usually are reduced to a thin trickle in the middle of a wide valley. The Gobi Desert surges toward Beijing at an accelerating pace. Only the widths of the bridges and the large areas of dry river bed remain as testimony to the former size of these rivers. The water is sometimes jet black, sometimes rainbow-hued with chemicals, or it is steaming with heat. Even springs among the limestone hills of Guilin contain worldwide "indicator" species, reflecting upstream contamination not alleviated by a spell underground. This must be a view of conditions that affected places like Ironbridge, England, immediately after the industrial revolution. The same scenes are repeated in Eastern Europe, to judge from recent media coverage of places where environment has been sacrificed to exploitation and exported profit. Our future might involve exporting expertise to such places and a boom in pollution-related industries, including the provision of consultants.

12.2.3. Achievements of the Industrial World

Despite the growing list of global problems mentioned above, we can be proud of the results of our efforts in accomplishing three essential tasks: (1) deciding on relevant information to acquire; (2) rendering that information understandable to the public, our managers, our politicians, and our lawyers; and (3) translating that understanding into laws that can be applied and policed without totally disrupting the economy of the nation. As a result, the worst point source abuses of aquatic ecosystems in the western world have been controlled. We have achieved this despite controlling only a minuscule share of the resources (in terms of money and brains) that are applied to areas such as space, defense, and fundamental physics.

However, appallingly low levels of support are severely hampering our efforts. For example, our field work depends on taxonomy, often said to be tedious, time-consuming, and expensive. This need not be true if complete, well-designed, clear taxonomic guides existed that were updated frequently. But how can these be produced when the staffs of many organizations devoted to taxonomic research compete for funds on a scale of only hundreds of dollars a year? How can we attract capable students into such a field when little or no prospect of employment exists, and managers and granting agencies suggest that taxonomists never complete their projects? It is a self-fulfilling prophesy to give someone a large, underfunded task and then complain that they fail to produce. We can hope, then, that as we come to realize the importance of checking on the health of natural systems, in addition to managing effluents by simple means such as toxicology (see below), that the provision of tools for the job will be recognized and that considerable support for ongoing national faunal surveys will materialize. The first step in management is to know what resources are available. I do not believe that we can manage ecosystems on the basis of processes without knowing their contents.

12.3. Field vs. Laboratory: Applying What We Already Know

I would like now to consider the debate between toxicologists and field ecologists and see what the future might hold for both. The toxicological approach to pollution control, coupled with concepts such as water quality standards and priority pollutant lists, seems to spring from a kind of linear and reductionist thinking that I term essentially "urban." By this I mean the approach of engineers and architects who conceive of cities with no farm fields to provide food and no balance between waste treatment and the absorptive capacity of the environment (Odum, 1989, and see below). This kind of utopian, future world can be found in many magazine articles based on urban thinking. Food is assumed to be imported into, and wastes exported from, such urban constructions at the expense of less-endowed societies in the same way that the Japanese, and before them the British, expanded their own tiny bits of real estate by importing raw materials to support their economic dominance of others. Waste management is by output, wherein the potential impact of wastes is reduced to an acceptable level by filtering. The material passing through the filter utilizes the absorptive capacity of the ecosystem. The material collected on the filter still requires disposal in some other form with or without treatment.

I believe that an element of this "urban" mind set exists in those who try to extrapolate a modified toxicology into an environmental concept by using multispecies tests, microcosms, and mesocosms. These are expensive (apart from the economy of multispecies tests of the type where targets are simply combined into single tests instead of a series) and are no more ecological than the original, much more tightly controlled, approach. Take, for instance, comments in this book (e.g., Chapters 6 and 10) to the effect that chironomids are pollution-tolerant. Any field biologist knows that such a statement applies to a few but not to all species of chironomids. The same claim could be made for tubificid oligochaetes, which most of the index-makers classify as pollution-tolerant. Some time ago, P. M. Chapman and I endeavored to create a little havoc in this tidy world by doing a series of very simple bioassays with tubificids (see references cited in Chapman and Brinkhurst, 1984). Little work has been done with these animals, perhaps because people think they are all tolerant or because people still think that they are hard to identify. We tested a wide range of tubificid species from oligotrophic, mesotrophic, and eutrophic habitats (to misuse those terms in the way that has become widespread), and we included marine as well as freshwater forms. Test animals were exposed to stresses from contaminants under several sets of environmental conditions. These stresses were presented singly and in combinations to worms with and without sediment. The worms were exposed in both monospecific and mixed cultures because the latter, more natural state has a profound effect on ecological parameters, such as growth and respiration, in these animals. We also examined lethal and sublethal effects. The results showed that the whole group cannot be classified as pollution-tolerant and that responses to stress from contaminants such as heavy metals are very complex when parameters such as respiration rate are examined. As expected, the results are of little or no use in predicting the distribution and abundance of tubificids in the field (see also Resh and McElravy, Chapter 5).

The limitations of toxicology when extrapolated to the field are well-understood. The problem has not been resolved and, I would suggest, is irrelevant. I have recently abandoned my opposition to the toxicological approach in favor of a combined approach in which both toxicologists and field ecologists retain their individual advantages by refusing to move toward some grey or compromise area (Brinkhurst, 1985). I see good reason to manage point source emissions by simple toxicological effluent testing, backed up by carefully designed field monitoring to ensure that the permitted discharge levels are having their desired effect. Of course, we have to decide just what level of ecosystem change we are prepared to accept (if any) in order to have a standard against which to evaluate performance. I think toxicology should stay in the very best controlled laboratories we can provide and that tests be performed on culture species of known genetic lineage (or on biochemical or physiological systems rather than whole organisms if these can be provided cheaply and in large numbers). The results of such tests are immediately interpretable and demonstrable in a legal or management setting; exceeding the standard is a violation of the permit. It will take a while for standards for specific effluents, which lead to acceptable levels of ecosystem protection, to be determined by careful field work. Eventually, we may acquire enough general knowledge to set standards on an ecological basis, specifying the nature of the community that would be acceptable rather than accepting a blanket, say, 50% reduction in diversity.

The field work would have to be done rigorously, not involving unscientific indices but involving quite simple multivariate, nonparametric statistical tests that can be made available to anyone as packaged software (see Cooper and Barmuta, Chapter 11). Economies can be made by adjusting sieve size and replicate number, limiting the study to one or two microhabitats, and focusing on precision rather than accuracy (see Resh and McElravy, Chapter 5). Process-oriented ecological work calls for knowing the true abundance, biomass, and processes in an ecosystem, but the field practitioner can accept a lower level of accuracy.

Taking mobile laboratories into the field is not the answer either. In some instances, mobile laboratories have provided improved working conditions that can be contrasted with older permanent facilities, but this is irrelevant. Installing the laboratory beside the receiving water and using locally available organisms may seem to provide instant relevance. However, this is achieved at the expense of a lack of knowledge of the population genetics and toxicological background of the species being used, in contrast to focusing on specially cultured organisms that provide a well-known test bed. Impacts on the natural community should be assessed by field work. Toxicological problems should be identified using controlled experimental procedures on well-known laboratory cultures that simply represent living systems. These are surrogates for the endemic fauna.

12.4. Putting The Benefit in Cost/Benefit

Discussions about the cost/benefit of benthic surveys often are limited to a consideration of means of cost reduction without considering derived benefits in comparison with alternative methods. It might be worth calculating the cost per data point used in a cooperative study in progress in Vancouver Harbor, Burrard Inlet, British Columbia. The study involved simultaneous sediment chemistry, sediment toxicology, and benthic community analyses. The field costs were common to all three aspects of the study, so these costs were ignored. The sediment toxicology work produced 700 data points (20 stations, five replicates, seven variables) at a cost of $80,000 (Can.) paid to a consulting company. The cost, approximately $114 per data point, included analyses, salaries, overhead, profit, and production of a noninterpretive report. Another consultant was provided with triplicate, unsorted benthic samples taken from the same 20 stations. This consultant's costs included sorting, quality control analysis, shipping various taxonomic groups to appropriate experts (interestingly, it has proven cheaper in the end to use the best taxonomists directly rather than just having them verify specimens identified by relatively unskilled staff), providing the results as ASCII data files, and reporting the results. Two hundred twenty-one taxa were found, which means that 13,260 data points were obtained for a sum of $15,000, or $1.13 each. The sediment chemistry data consisted of three replicates of 31 variables (26 inorganic, five organic), or 32 if the cost of polycyclic aromatic hydrocarbon (PAH) analyses is added. These analyses were needed but excluded on the basis of cost! Hence, two estimates are available: $10.39 per data point without PAH analyses, or $11.16 with PAH analyses. At the very least, the benthic data can be seen to be cost-competitive, and they also reflect real-world conditions.

Ohio EPA (1987) did a similar cost/benefit study. It revealed not only that the *combined* cost of fish and macroinvertebrate surveys (~$1,400 to $1,600) was lower than for either physical/chemical (~$1,500 to $1,700) or bioassay work (~$3,000 to $12,000), but also that the information generated by the former was more useful than the latter in decision-making. Although Ohio EPA (1987, p. 17, 19) noted that ". . . chemical and bioassay tools are needed for adequate environmental evaluation and regulation and their value should not be diminished . . . ," the cost comparison showed ". . . that the common perception of biological field data as being 'prohibitively expensive' . . ." and, therefore, not usable on a routine basis is unfounded (Ohio EPA, 1987, p. 19).

It is worth noting that in the Vancouver Harbor study, the elapsed time between sampling and reporting was controlled by a delay in obtaining organic chemistry data rather than benthic data; this is in contrast to the frequently heard suggestion that benthic work causes such problems. The delay was caused because forensic samples took precedence over others in an overworked chemistry laboratory. Therefore, the processing time of benthic samples is not the only factor to be considered.

12.4.1. King Arthur's Index

I must confess I was a little heartened to read that toxicologists think that there has not been much fundamental change in their philosophy over 50 years (Buikema and Voshell, Chapter 10). I thought this to be true only of the field work arena, especially in terms of field methods. The search for a Holy Grail index that can be determined by the underfunded without exerting their minds persists and is as futile now as it was in King Arthur's Britain. Simplified indices, often based on an assumption that groups (such as insect families) behave in a uniform way, can only be applied to obvious, simple examples of gross disturbance (see Resh and Jackson, Chapter 6; Norris and Georges, Chapter 7).

Hynes (1960) wrote an elegant plea in defense of interpreting the basic data matrix of relative abundance of taxa (identified as far as possible) against stations (with replicate sampling) and against measures such as the Trent Index and Saprobien system. His style of work can be traced all the way back to Richardson (1929) in America (at least) and probably earlier in Europe (see Cairns and Pratt, Chapter 2). Everyone knew that increases in numbers and species could be related to mild pollution, that moderate pollution could produce changes in taxa so that diversity remained similar but species composition shifted, and that eventually species richness declined abruptly and numbers of some tolerant forms increased dramatically. Some taxonomic groups, such as stoneflies, had certain physiological characteristics (e.g., intolerance of low oxygen) typical of the majority of species, so that presence or absence of any species might be regarded as indicating certain changes in the habitat. Stoneflies that live in slower streams, or leaf-packet-dwelling species, are an exception here. Other taxa like the tubificids, thought to occupy the other end of the scale, turn out to include species that are limited to cold, oxygen-rich, unproductive habitats and other species that occupy zones immediately below sewers and septic patches of streams. Indeed, the various trophic zones (again using the term loosely) of the Laurentian Great Lakes could be recognized by their tubificid faunas in the same way that some chironomids traditionally have been used. Brinkhurst (1974) argues the pros and cons of lake typology, a philosophy closely related to biomonitoring.

Specific local habitats were known to produce unique situations where conditions were suddenly much worse, owing to geological or other features. Western Lake Erie is shallow, and weather conditions are such that it does not become thermally stratified for much more than a week at a time. The load of oxygen-demanding decomposing organic matter introduced into the system slowly increased over the years, but the bottom layer of water did not become deoxygenated during such brief periods of stratification. Eventually, in 1953, the threshold point was reached, and a period of calm weather lasted long enough so that all the oxygen was exhausted in the trapped, lower layer of cool water. A sudden dramatic change in the ecosystem resulted, with a massive mortality of mayfly larvae (Britt, 1955a,b) and an increase in worms (Carr and Hiltunen, 1965). The uninformed might have looked for an unusually high level of waste disposal into the system or a breakdown at a treatment plant. The true cause was a slow deterioration that had passed unnoticed.

In a long-term field study of the recovery of a British river from gross organic pollution, industrial discharges, and hot-water effluents, one stretch of the river showed a positive relationship between BOD and flow, whereas another showed the opposite (Brinkhurst, 1965). Slugs of organically loaded water would become trapped in stagnant side channels, only to pass downstream as a lethal plug of destruction when flushed out. The macrobenthos, tabulated on a log-scale basis of relative abundance (because the logarithmic nature of some population changes was well-understood even if few statistical tests were used), reflected the sudden improvement brought about by upgrading the sewage works. The study also revealed that the system shifted rapidly from the old state, in which the whole river contravened the sewage effluent standards in force, to a new state, in which it became an ordinary, polluted river. No further improvement followed beyond simple annual variations that appeared to be related to climate (temperature and rainfall). The engineers, expecting a slow recovery, were delighted with the higher-than-expected initial response, but were then despondent about the failure to see any additional improvement after the first year. Few of these features or the detailed stories about selected spots along the river could have been discovered with a chemical or toxicological approach.

12.4.2. Through Diversity to Clusters

It was not long before a revolution began. The intuitive processes used in reading data matrices were turned into mathematical and statistical methods. This involved recognizing the mental processes that the "experts" were using to make diagnoses and attempting to make them generally available. The demystifying process is bound to be iterative, so the early stages of development will be subject to considerable criticism. The first major step was to use diversity indices to reduce matrices to a single value per station. Such indices can be used to document the stage at which taxa are severely reduced, but they also can be masked by collecting on a sand bank instead of a riffle. In addition, two samples containing the same number of species in roughly the same proportions will appear to be similar, even if a large difference exists in the constituent species. Many government agencies appear to be satisfied that this first approximation is sufficient for their purposes and seem unwilling to exploit the data to the maximum extent.

The next step of development was to find a method that included the taxonomic information contained in the matrix, so classification and ordination methods were applied. These methods were then assailed by the suggestion that the determination of clusters was subjective and that there were too many indices to choose from. Recently, these questions were addressed.

In one solution, problems of dealing with the underlying properties of the data set itself and the spatial and temporal relationships between taxa are avoided. Dendrograms that result from the cluster analysis of benthic, sediment, or chemical data are compared statistically in order to test significant differences between the dendrograms. In another test, statistically significant clusters within a dendrogram can be identified (Nemec and Brinkhurst, 1988a,b), thus largely removing the subjective element from cluster analyses. The significant groupings of stations then can be overlaid on the original site map (Brinkhurst, 1991) to provide visual evidence of anthropogenic stress that can be understood much more readily than the results of principal component and time-series analyses, etc. (as valuable as those are to the professional scientist; see Norris and Georges, Chapter 7, and Burd et al., 1990). We could hope that, in the near future, we might persuade our managers of the utility and applicability of such measures and finally persuade them that diversity was but one step down a path.

The next stage of the process could be the application of dynamic systems theory to applied ecology. The question we ask is: "If we decide to intervene in the environment, either to harvest it or to pollute it (or as generally happens, to do both), then what is likely to happen?" (Allen 1985). Much contemporary modeling is descriptive and is based on the present with no reference to the past or the future. Changes in systems may be more like one current view of evolutionary changes, that is, a system suddenly goes down one of two alternative paths that diverge abruptly. Therefore, average conditions no longer suffice to locate the state of the system, so conventional modeling is bound to fail at some point. We need to discover why ecosystems remain stable when they do.

Future developments in modeling may significantly alter our current perspective. Chaos theory may be one of them. Chaos may be defined as a type of randomness intrinsic to a system rather than being caused by outside noise or interference (Pool, 1989). The orbit of Pluto is a much-cited example, in which small changes in initial position can result in very large differences in future location because its orbit is complex and unpredictable over the long term. Population biologists have tried to exploit this theory since 1970, but no example has been universally accepted. Simple theoretical systems with two or three species only demonstrate chaos when some parameter is pushed far from natural levels, but this may be attributed to the crude simplicity of these low-dimensional systems. Because most ecosystems contain at least the seeds of chaos within them, it is thought that human interference may be increasing pressure or stress and sending systems into chaos. Even well-intentioned restoration or treatment might have this effect because proper action is not always intuitively obvious. It may be that Club of Rome-type predictions of gloom are merely too simplistic after all and that biology suddenly will become respectable by the injection of mathematicians and physicists ready to apply their theoretical approaches to biology. Here, I part company with my own ability to follow this particular debate.

12.5. Meanwhile, Back to Rapid Assessment

We have, therefore, progressed in our ability to codify and systematize our intuitive, elitist methods of the past. The roots of modern statistical procedure lie in quantitative work, somewhat older than some authors might suggest (see Patrick 1949 on diatom communities in rivers, for instance). What, then, is to be made of those twin bogey men: cost and the need for rapid assessment?

The idea of cost/benefit already has been discussed. Too often, such ideas focus on the cost of benthic work without reference to the relative benefit. We all know of contracts our friends have lost because an agency hired a consultant willing to work for less than all others. Only later does the agency discover why the consultant was so cheap. The point here is to remember that, although we are always concerned with precision, accuracy can be adjusted to the severity of the impact being investigated.

I believe that an experienced biologist in almost any discipline can give a fairly accurate account of a small system, such as a stream or pond, by simple examination. It is the same skill demonstrated by the diagnosing family physician. The value placed on the assessment is a matter of professional standing. In my childhood in Britain, the physician was God, and his decisions were immutable. However, I do not advocate a return to reliance on authoritative decision making without a second opinion (famous doctors could be wrong!) but, unfortunately, the field biologist is not awarded even a fraction of such status.

Massive point source disturbances, such as toxic metal spills or chronic sewage discharges, can be detected easily and even documented in a qualitative manner by indicating the replacement of stoneflies and mayflies by worms, leeches, and isopods, or some other shift in the community. Where I disagree with the "quick assessment" people is another example of opposition to drifting into the grey zone of compromise. None of the indices that use fake statistics add anything to the bare information presented with expert interpretation. Movement from an elitist diagnosis to a forensic determination of health requires a properly conducted biological survey, backed up with a well-planned chemical-analysis program as indicated by the initial rapid assessment. These data will, we hope, be capable of arrangement in symmetrical matrices susceptible to statistical analysis. So-called "economic" rapid surveys sometimes involve spotty analytical procedures. Contaminants may be sought where some expectation of significant levels exists, but may be ignored at other sites. As more of us use packaged statistics on desktop computers, the need to avoid this procedure will become obvious. How often are consultants handed stacks of data, gathered at great cost over many years by a variety of agencies, only to find that the data are so lacking in consistency that they cannot be analyzed? Proponents of rapid assessment would do well to make sure that they benefit from the exhaustive debates on these issues in the older literature. Most of the ideas are being reworked, often in apparent ignorance of past syntheses (e.g., Hynes, 1960; Resh and Jackson, Chapter 6). One of the problems here is the exponential increase in literature, described in Marshall (Chapter 3) and our inability to absorb more than a fraction of it. All too much of this mass is, however, philosophically repetitive, so that truly significant work can get lost.

12.6. The Computer

Those lured into the computer age via the use of packaged computer programs will tend to realize that symmetrical matrices are required for many statistical methods. They will soon realize that missing data points create such havoc that they will be less inclined to use a shotgun approach to deciding on the parameters to be measured at each station within a survey.

On the debit side of the assessment of computers, we find those administrators who are more interested in impressive databases than they are in the statistical significance of the results of analyses performed on those data. The mere acquisition of information can become a goal in itself. This was true of the impact assessment process, in which a tendency arose to list every conceivable potential problem without any reference to risk assessment. Another powerful new tool is the geographical information system (GIS). Maps can be very powerful interpretive tools, and they appeal to those with a better visual sense than a mathematical one. A problem with simplification may exist when few data points are used to extrapolate over wide areas, leading to undue optimism about areas excluded from zones of impact simply by the absence of data.

12.7. Research and Application: Bridging the Gap

The resurrection of rapid assessment techniques provides a reminder of the need to go beyond suggesting the possible application of research efforts to actually working with applied problems in a research environment. Even more important is the presentation of those ideas in a medium appropriate to our applied colleagues. The North American Benthological Society is a major avenue for this. From time to time, the hordes of eager young stream ecologists at these meetings make one wonder where the applied people went, but the Society has grappled with this problem and will always need to do so.

We should expect strong resistance to change among administrators and environmental managers. Regulations must be subjected to political scrutiny and must be applied as evenly as possible so as not to confer a competitive edge on some at the expense of others in the same economic sector. In North America, our judicial system is essentially adversarial, and environmental need cannot justify using any means to achieve our ends; we could be wrong. Our future work should involve much more risk assessment, not risk cataloging. Environmental impact assessment should include prediction and should lead to follow-up work to test those predictions (Rosenberg et al., 1981; Hecky et al., 1984).

Schindler (1987) makes a powerful case for the sort of biomonitoring that we should expect to see developed over the next two decades. His experience is that some changes in species composition in communities indicate stress, whereas measurement of diversity and functional parameters such as respiration do not (but see Reice and Wohlenberg, Chapter 8). Mesocosms have been useful for pelagic communities, less so for benthos. Life-table effects are useful, paleolimnology will be expanded (see Walker, Chapter 9), especially in non-point source applications, and morphological anomalies need to be experimentally related to their causes (see Johnson et al., Chapter 4). Schindler's paper, and the work of the Experimental Lakes Area group of the Freshwater Institute in Winnipeg, Manitoba, provides a magnificent example of the lack of any real barrier between applied and pure research. It also demonstrates the ability of scientists to say unpopular things in print from within a government agency (although we should note that Dr. Schindler now works at the University of Alberta!).

Odum (1989) makes a plea for the formation of interdisciplinary groups capable of taking holistic approaches in terms of non-point source problems. His examples are taken mostly from landscape management, but the principles of what he says clearly are applicable to aquatic systems. Indeed, in one example, he shows how machine-intensive agriculture, which relies on fertilizers and pesticides, has resulted in the export of 25 million tons of chemical-laden soil into the Illinois River every year, material that settles into ponds and lagoons along the stream bed. This load now exceeds the amount of raw sewage that used to be discharged from the city of Chicago. Odum also calculates the value that can be associated with "undeveloped" natural ecosystems in terms of nonmarketed services and shows that it is of the same order of magnitude as the agricultural products marketed from the same area. Non-point sources must be managed at the input-to-production part of systems, not at the output (or effluent) end of systems. A good example is the recently proposed 50% reduction in packaging in Canada, to be voluntary at first, but later supposedly to be mandated by law. Odum suggests that reductionist science coupled with an output management regime is responsible for our fascination with laboratory studies in applied science instead of holistic, ecological approaches that are more relevant.

Holistic, ecological study is often undermined by administrative practice. Many government programs start out with a holistic intent, but as each sector pushes for the continuation of its own activity under the umbrella of the new program, the pie becomes divided up into pieces that are never reassembled. When the fundamental needs of the original proposal become confused, strange projects emerge. For example, the agency planning a detailed study of the lower Fraser River in British Columbia found itself trying to balance the merits of two conflicting approaches. Because of a shortage of funds to carry out the work that each self-interest group wished to see perpetuated, the agency had to decide between doing all the sector studies in one half of the zone of interest in one year and in the second half during the second year, or funding 50% of the sectors for the whole stretch of the river in year one and the other aspects in year two. Apparently, no one was willing to pare down the study to its core elements. Instead, participants tended to want to incorporate programs that were recognizable as part of each group's specific mandate. Many such examples of multidisciplinary studies that have never been integrated and tested in terms of an end result exist. However, one exception, the Southern Indian Lake, Manitoba, reservoir study (Hecky et al., 1984) should be mentioned, but I doubt that the future holds much promise in this regard. In Canada, at least, government multidisciplinary scientific groups are being disbanded, apparently to facilitate accounting procedures.

12.8. The Wider Context

Thus, let us put benthic biomonitoring into a broader perspective. None of us can predict the future, as is demonstrated by our lack of power to influence the politics of science or the stock market. At best, we can extrapolate from present technology in a linear fashion, rather than being able to foresee quantum leaps. Also, predictions offering great insights are mostly too revolutionary to gain wide acceptance, being easily rejected as misguided rambling. Take, for example, the stance of those who defend the expansion of clear cutting in forests. Those who have benefited most by past procedures and thinking have the most invested in maintaining the status quo. Thus, many industrialists who can be justifiably proud of their historical success become bewildered as society adopts new standards of behavior. They may become reactionaries who claim to be responsible and ecologically aware, but factual and unemotional in assessing environmental matters. In contrast, they categorize environmentalists as uninformed and emotional.

A need "to be balanced in supporting the needs of both industry and environment" is invoked to refute the ideas of those with serious misgivings about the wisdom of many production processes, the outcome of which cannot safely be predicted on the basis of previous experience and practice. If, for instance, the forestry industry historically has been as conscious of future needs as its advertisements claim, why does it have to expand into all the remaining old-growth forest? Furthermore, what happens once those forests have been completely "harvested" to save jobs?

What predictions about 1992 would have been made in 1960, at about the time I became involved in biomonitoring? One would not have predicted continental drift, hydrothermal vents, the greenhouse effect, the ozone layer problem, destruction of the Amazon, the fuss about asbestos, and the pervasive use of electronics in various forms that so shapes our lives in the 1990s. If we examine recent human history, the accelerating rate of change—rather than just change itself—is the norm. In contrast, most of our mental processes, and perhaps our political and religious thinking, attempt to emphasize stability; most of our thinking is linear. In his recent public appearances, Paul Ehrlich has talked about our puzzling lack of awareness of the scale and implications of changes to our world. An evolutionary explanation may lie in our sensory and intelligence system, which is very sensitive to potentially catastrophic and immediate threats, but soon damps out low-level background information such as that based on a slow but steady change in state. Changes in our ecosystem happen slowly enough so that they do not cause releases of adrenalin.

Little change has occurred in the suite of potentially catastrophic global environmental scenarios that once were identified by the Club of Rome study. This study was but one of a series of "gloom and doom" predictions that assailed us in the late 1960s and which have now reappeared in the media, along with improvements in the economy after a period of high interest rates and high unemployment in Western society. Some predictions are coming into even sharper focus. Others, such as destruction of the ozone layer, are now in general circulation. Overpopulation remains the major driving force behind the message of the pessimists. The human population is now over five billion, on its way to an original estimate of 10 billion, which now has been revised upwards to 15 billion. Some estimates put global population levels for a sustainable, reasonable living standard as low as six billion. Aside from the Four Horsemen of the Apocalypse, I am unaware of the forces that are supposed to level off the population once it reaches 10 to 15 billion (or any other number for that matter). Few nonscientists seem to have benefited from our educational system enough to understand the problem. However, two simple riddles tell the story clearly. If you place a rice grain on the first square of a chess board, two on the next, four on the next etc., how many grains go on the last square? I believe that the number is larger than the standing stock of rice grains on earth. Remember the riddle about the pond containing a mysterious one-leafed plant? This plant doubles its size every day, until on day 100 it fills the pond. When does it half fill the pond? On day 99 of course, which is why the change is imperceptible until the last few days. So what sort of world will we be biomonitoring by 2001? The lack of awareness of this problem reaches even to a senior scientist I once met who, although a field biologist, seriously suggested that no population problem existed because all the human beings in the world could be accommodated standing up on one small island off New Zealand. Surely, it is absurd to consider humans without "an ecology," or to ignore the amount of land area needed to support a self-sufficient peasant, aside from the multiplication factor required to include a North American or Japanese and his or her personal share of every road, school, hospital, factory, mine, oil well, and forest! The increase in population is coupled with an information system and a political revolution around the globe that is motivating huge numbers of people to bring their living standards up to our casually wasteful extravagance. These people seemingly are unaware that our position has been built by using the resources of the rest of the world.

12.8.1. On to the Year 2200

The upbeat optimists will advise that your future will involve traveling to work in a "smart" car containing sufficient electronic sensors and microchips to make it safer and easier to navigate than ever before. You may travel by train instead, without driver and conductor, just as in the Skytrain in Vancouver, perhaps one powered by magnetic levitation (MAGLEV). Your home and office also will be filled with sensors, programmable outlets of information that allow control of systems from any point, instead of the traditional systems of today's buildings. You may travel to your next conference in a "smart" plane with propellers on an unducted fan engine. Better still, because telephone, television, and computer have become a single system, you may simply arrange a conference call instead of traveling at all. A screen may not even exist, but rather a headset that changes the view as your head shifts. Everyone sits in assigned places at conference tables in each center, with a name tag on each place as usual. By turning your head to an empty chair with a name tag in front of it, the head set will project an image (maybe a hologram) of the individual so named, giving a near-perfect sensation of "being there" while sitting in your own conference center. Computers, modems, and the fax machine already make it possible for work to be done without all the commuting we used to take for granted. Meanwhile, robot technicians will be performing routine BOD analyses, and benthic biologists eventually may be able to determine taxonomy and relative abundance of samples by passing them, unsorted, through a DNA reader.

Your health will have been maintained through computerized diagnoses using worldwide databanks. One or two of your organs may have been removed and rebuilt in the biochemical and genetic laboratory while you are maintained temporarily on life-support systems, thereby avoiding rejection problems. Many chronic diseases will have been genetically cured *in utero* after a prebirth screening of your susceptibility.

Out in the environment, if one exists, you may whir up and down streams in your own portacopter. You will use electronic data-gathering instruments that store their information and are self-contained and self-powered. We will probably have had to come to terms with portable atomic energy sources of a microscopic scale to power these devices; battery-operated models already are available. You will do preliminary checks on many variables by tuning into land-viewing satellites, such as those currently used to direct oceanographic vessels. You will use stress probes, instruments already available that measure cheaply the production of certain proteins by stressed organisms to detect anthropogenic effects in the field. They also can be used to determine the health of human beings.

12.8.2. But Who Wants It?

Remember, this scenario depends on having an environment to go to. Have you ever noticed that in older movies about the future, people whiz about among tower-like buildings in their little personal transporters? The buildings are connected by bridges but you never see the ground. No rice fields, no rivers, no birds appear. House plants and the odd pet may exist, perhaps, and, I assume, houseflies, cockroaches, fleas, and mice, but no environment. It is the final victory of man over nature. We have truly gone forth and multiplied, and our seed has spread over all the earth. This may be the stuff of fantasy, but serious plans exist for the expansion of Tokyo over the ocean. As land there reaches \$150,000 per m^2, it is becoming economically feasible to contemplate building a city on stilts. The architects' drawings show buildings, streets, airports, and eight golf courses, perhaps made of artificial turf. Even the sea bed is uniform and tidy, with a picturesque monoculture of seaweed (undoubtedly edible) in one corner. Other Asian societies are trying to recreate Western industrial societies, often with little heed to the environmental consequences, and many still promote population increase. Major cities such as Lima, Mexico City, and Rio de Janeiro may have great opportunities for some of the population, but they also have large areas of decay that do not fit into futuristic concepts of city life.

12.9. Who Needs Progress?

The question we should be answering at this point, perhaps, is: "Do we really need all sorts of new techniques and equipment, or do we still need to educate ourselves to use the existing methods properly and to communicate the urgency for betterment to managers who decide what questions should be asked?" A great many scientists seem to be willing to produce toxicity tests in a world in which the number of chemical products to be evaluated increases faster than we are able to test them. The whole exploding field of non-point source effects of man as a geophysical force is rapidly rendering the reductionist, output side of management totally redundant. The quite terrifying population explosion, still not recognized as the fundamental force that it is, even by otherwise well-informed people, creates an urgency that will increase in the near future. Perhaps a need exists for greater political awareness among applied biologists and an activism that may put career motives at risk.

References

Allen, P.M. 1985. Ecology, thermodynamics, and self-organization: towards a new understanding of complexity. In Ecosystem theory for biological oceanography, eds. R.E. Ulanowicz and T. Platt. *Canadian Bulletin of Fisheries and Aquatic Sciences* 213:3–26.

Brinkhurst, R.O. 1965. Observations on the recovery of a British river from gross organic pollution. *Hydrobiologia* 25:9–51.

Brinkhurst, R.O. 1974. *The Benthos of Lakes*. Macmillan Press, London.

Brinkhurst, R.O. 1985. The threefold path. In Proceedings of the Tenth Annual Aquatic Toxicity Workshop, Halifax, NS, November 7–10, 1983, eds. P.G. Wells and R.F. Addison. *Canadian Technical Report of Fisheries and Aquatic Sciences* 1368:3–9.

Brinkhurst, R.O. 1991. Benthic biology of the western Canadian continental shelf. *Continental Shelf Research* 11:737–54.

Britt, N.W. 1955a. Stratification in western Lake Erie in summer of 1953: effects on the *Hexagenia* (Ephemeroptera) population. *Ecology* 36:239–44.

Britt, N.W. 1955b. *Hexagenia* (Ephemeroptera) population recovery in western Lake Erie following the 1953 catastrophe. *Ecology* 36:520–2.

Burd, B.J., A. Nemec, and R.O. Brinkhurst. 1990. The development and application of analytical methods in benthic marine infaunal studies. *Advances in Marine Biology* 26:169–247.

Carr, J.F. and J.K. Hiltunen. 1965. Changes in the bottom fauna of western Lake Erie from 1930 to 1961. *Limnology and Oceanography* 10:551–69.

Chapman, P.M. and R.O. Brinkhurst. 1984. Lethal and sublethal tolerances of aquatic oligochaetes with reference to their use as a biotic index of pollution. *Hydrobiologia* 115:139–44.

Hecky, R.E., R.W. Newbury, R.A. Bodaly, K. Patalas, and D.M. Rosenberg. 1984. Environmental impact prediction and assessment: the Southern Indian Lake experience. *Canadian Journal of Fisheries and Aquatic Sciences* 41:720–32.

Hynes, H.B.N. 1960. *The Biology of Polluted Waters*. Liverpool Univ. Press, Liverpool, England.

Nemec, A.F.L. and R.O. Brinkhurst. 1988a. Using the bootstrap to assess statistical significance in the cluster analysis of species abundance data. *Canadian Journal of Fisheries and Aquatic Sciences* 45:965–70.

Nemec, A.F.L. and R.O. Brinkhurst. 1988b. The Fowlkes-Mallows statistic and the comparison of two independently determined dendrograms. *Canadian Journal of Fisheries and Aquatic Sciences* 45:971–5.

Odum, E.P. 1989. Input management of production systems. *Science* 243:177–82.

Ohio EPA (Environmental Protection Agency). 1987. *Biological Criteria for the Protection of Aquatic Life. Vol. I. The Role of Biological Data in Water Quality Assessment.* Surface Water Section, Division of Water Quality Monitoring and Assessment, Ohio Environmental Protection Agency, Columbus, OH. (Updated February 15, 1988).

Patrick, R. 1949. A proposed biological measure of stream conditions based on a survey of the Conestoga Basin, Lancaster County, Pennsylvania. *Proceedings of the Academy of Natural Sciences of Philadelphia* 101:277–341.

Pool, R. 1989. Ecologists flirt with chaos. *Science* 243:310–3.

Richardson, R.E. 1929. The bottom fauna of the Middle Illinois River 1913–1925. Its distribution, abundance, valuation and index value in the study of stream pollution. *Bulletin of the Illinois Natural History Survey* 17:387–475.

Rosenberg, D.M., V.H. Resh, S.S. Balling, M.A. Barnby, J.N. Collins, D.V. Durbin, T.S. Flynn, D.D. Hart, G.A. Lamberti, E.P. McElravy, J.R. Wood, T.E. Blank, D.M. Schultz, D.L. Marrin, and D.G. Price. 1981. Recent trends in environmental impact assessment. *Canadian Journal of Fisheries and Aquatic Sciences* 38:591–624.

Schindler, D.W. 1987. Detecting ecosystem responses to anthropogenic stress. *Canadian Journal of Fisheries and Aquatic Sciences* 44 (Supplement 1):6–25.

Author Index

Subject Index